EUROPEAN CENTRES OF EXPERTISE

SENSORS

5th Edition

SPRINGER-SCIENCE+BUSINESS MEDIA, B.V.

This Guide has been prepared by Dr. Enrico Pigorsch, Ramon Bardolet and the Metra team.

Originally published by Metra Martech Limited in 1997
Softcover reprint of the hardcover 5th edition 1997

British Library Cataloguing in Publication Data
European Centres of Expertise SENSORS
5th Edition

ISBN 978-94-010-6518-4 ISBN 978-94-009-0065-3 (eBook)
DOI 10.1007/978-94-009-0065-3

Also in the same series:
- Advanced Materials
- Water and Waste Water Treatment

CONTENTS

ACTIVITIES

Main activities

- Support to SME's in fields such as technology transfer and industrial innovation
- Search of business partners
- Search, analysis and synthesis of scientific, technical, economic and patent-related information
- Monitoring of worldwide technological developments, and evaluation of emerging industrial trends
- Monitoring of competitors activities
- Information on Companies
- Providing promptly information from a worldwide range of published articles (patents, standards, reports, market studies, etc.).

Field of activities

- Direct access to more than 3'000 international databases
- Consulting and support to companies based on information processing in :
  - Physics, chemistry, electronics, materials, health...
  - Engineering, telecommunications, robotics, safety, manufacturing, instrumentations...
  - Markets, opportunities, competition, forecast...
  - Patents, trademarks, licences

Professional affiliation

- Member of TII (European Association for the Transfer of Technologies, Innovation and Industrial Information)
- Member of SPRINT (Strategic Programme for Innovation and Technology Transfer of the European Community) with partners in Germany, Spain, France, Italy, Austria and Belgium
- Association with TE-CH, (a group for Technology Transfer in Switzerland)
- Associated with "SwisSearch", an information service of the Swiss Bank Corporation

Miscellaneous activities

- Consulting to industry for computerized information systems
- Conception, realization and host of VIDEOTEX applications by the ARCANTEL Corp., CENTREDOC's partner

Publications

- INFOMAT, an information service on new materials and coatings
- Revue des inventions horlogères (RIH)
- SENSORS-INFORMATION
- CENTREDOC Bulletin

RECHERCHE / ANALYSE / SYNTHÈSE
D'INFORMATIONS
CENTREDOC
TRANSFERT DE TECHNOLOGIE

SENSORS-INFORMATION

CENTREDOC publishes a monthly bulletin summarizing updated, selected references on sensors related to:

- general publications and synthesis
- practical, theoretical and experimental publications
- market studies
- books, reports
- exhibitions, conferences

Annual subscription: SFr. 330.-

Adresse

Centre suisse de recherche, d'analyse
et de synthèse d'information
Rue Breguet 2

CH-2007 NEUCHATEL

Tél: +41 (038) 205 131
Fax: +41 (038) 254 873

Structure

Direction : B. Chapuis

Conseillers scientifiques et techniques : O.D. Barrelet, Ph. Girard, F. Glauser, C. Henguely, F. K. von Willisen

OSCA

Optical Sensors Collaborative Association

OSCA's objective is to encourage the application and commercial exploitation of optical sensing. It provides information to its members on advances in optical sensing technology, organises collaborative work programmes and aids the development of standards and codes of practice to guide users of the technology. Current activities include:

* Regular meetings on a broad range of optical sensing techniques and applications

* An EC-supported international project on the safe use of optical systems in potentially explosive atmospheres

* Technical literature current awareness service and reports on international conferences

Membership is open to instrumentation manufacturers, research associations and users of sensors and instrumentation. Academic institutions may become affiliate members.

For further information please contact:

OSCA Manager - Patricia West
Sira Ltd, South Hill, Chislehurst, Kent BR7 5EH, England
Tel +44 (0)181 467 2636 Fax +44 (0)181 467 6515

1. Introduction

This is the 5th edition of the Metra Martech Directory "EUROPEAN CENTRES OF EXPERTISE - SENSORS." The entries represent a survey of European sensors development. The new edition contains 425 detailed profiles of companies and research institutions in 22 countries. This is reflected in the diversity of sensors development programmes described, from sensors for physical parameters to biosensors and intelligent sensor systems.

We do not claim that all European organisations developing sensors are included, but this is a good cross section from an invited list of participants. If you see gaps or omissions, or would like your organisation to be included, please send details.

The data base invites the formation of effective joint ventures by identifying and providing access to specific areas in which organisations offer collaboration. This issue is recognised to be of great importance and most entrants include details of collaboration offered and sought. We hope the directory on Sensors will help you to find the right partners with whom you can cooperate successfully and reach new markets.

The directory on Sensors is part of Metra Martech's long term programme to participate in the development of advanced technologies in Europe. The aim of these directories in the series "European Centres of Expertise" is to promote scientific and technological cooperation and integration between European industry, research bodies and universities. Other directories in the series cover:

- Advanced Materials
- Water & Waste Water Treatment

New titles are being considered all the time and we would welcome your enquiries and suggestions.

We are most grateful for the support of a number of organisations, journals and associations whose publicity material is included here.

This directory was prepared by a team headed by Enrico Pigorsch, PhD Chemistry and Ramon Bardolet, Environmental Consultant, and included at various stages:

Charles McGinley
Andrew Clive
Paul Lewington
Jill Sukolski

Catherine Brady
Hilma Strudwick
Nicola Otake
Holger Tiepmar

CLUB MICROCAPTEURS CHIMIQUES

The goal of this French association is to promote the development of **miniaturisable chemical sensors** and their possible extension to the detection of biochemical species. The members can participate in workshops on defined topics with invited foreign speakers. A triannual information letter (in French) is distributed. Every two years the CMC2 confers a prize for research on sensors.

Present members: BRGM, CEA, LETI, CEMAGREF, DGA/CEB, C4I, CORECI, CYLERGIE, ETR, GIP Ultrasons, GDF, INERIS, LIPHA, MOTOROLA, NEREIDES, PHILIPS, PSA, Radiometer Analytical, Revenu Canada, Rhône-Poulenc, SAT, SAGEM, SAUR, SCHLUMBERGER, CERES, THOMSON.CSF, ENS (Lyon,Cachan), ENSM, EESPCI, INPG, ENSPG, ENSERG, ENSEEG, FUL, INSA, INSERM, IEMN, INRA, ECL, Ecole Polytechnique, Universités: Clermont Ferrand, Lyon 1, Bordeaux1, Genève, Rennes 1, Rouen, Chambéry, Troyes, Maine, St Etienne, Grenoble, Libre de Bruxelles, Montpellier 2, Paris 6.

Board Chairman : Prof. P. CLECHET

Head Office Address: IFoS (Ingénierie et Fonctionnalisation des Surfaces) UMR/CNRS 5621
LPCI (Laboratoire de Physicochimie des Interfaces)
Ecole Centrale de Lyon - BP 163
F 69131 ECULLY CEDEX France
Phone: 04 72 18 62 41 Fax: 04 78 33 15 77 e-mail:capteurs@cc.ec-lyon.fr

2. The European Sensors Markets

2.1. Overview

It is extremely difficult to arrive at an accurate estimate of sensor market size. As well as all the normal difficulties in market research, in this field there are particular difficulties of definition. For example, what does the sensor include? Is it the sensing head only, or does it include some or all related signal processing electronics? What do published figures include?

Market size and growth rate

On the assumption that the sensor includes all power supply and signal-conditioning electronics necessary to give a high level (digital or 0 - 4 mA analogue) output signal, built-in displays and all necessary mechanical housings, it is estimated from an examination of published figures that the European market for sensors in 1996 is of the order of $5 billion. The largest sector is for flow measurement, being of the order of 20% of the total: three other important sectors, each with around 15%, are those for pressure, level and chemical/gas sensors; the remaining 35% is taken by sensors for position, speed, temperature, radiation, etc.

Technological and organisational changes have been numerous and rapid in the recent past. The near future is expected to be similar. Annual growth rates for the next few years are expected to be about 5% in value, and almost double this in volume.

Trends in the Market

The Metra Martech questionnaire used to collect data for this directory asked the question: As we approach the year 2000, what is your main area of concern for the sensors industry in Europe? 40% of the participants answered that question. The views expressed serve to confirm trends in the market noted by industry observers. The replies are shown graphically in Figure 1 and 2. In particular, we can point out:

<u>Prices of Sensors</u>

These are continuing to fall; many of the respondents mentioned this and commented on the related cost pressures on them. This trend is expected to continue, resulting in the much larger growth forecast in market volume compared to value.

<u>Microprocessor-based Sensors</u>

There is a marked trend towards the use of intelligent sensors. In the past the main reasons for this have been increased measurement accuracy, programmability, decreased inventory costs from the larger turndown available, and decreased maintenance costs from self-diagnostics.

Developing Sensor Technology Areas

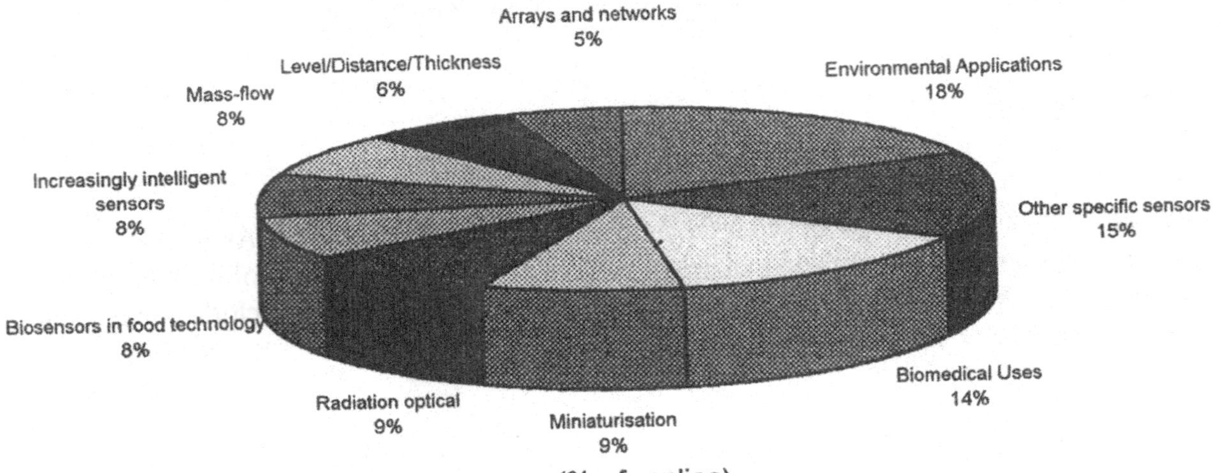

(% of replies)

Figure 1

Concerns About Sensors Business Aspects

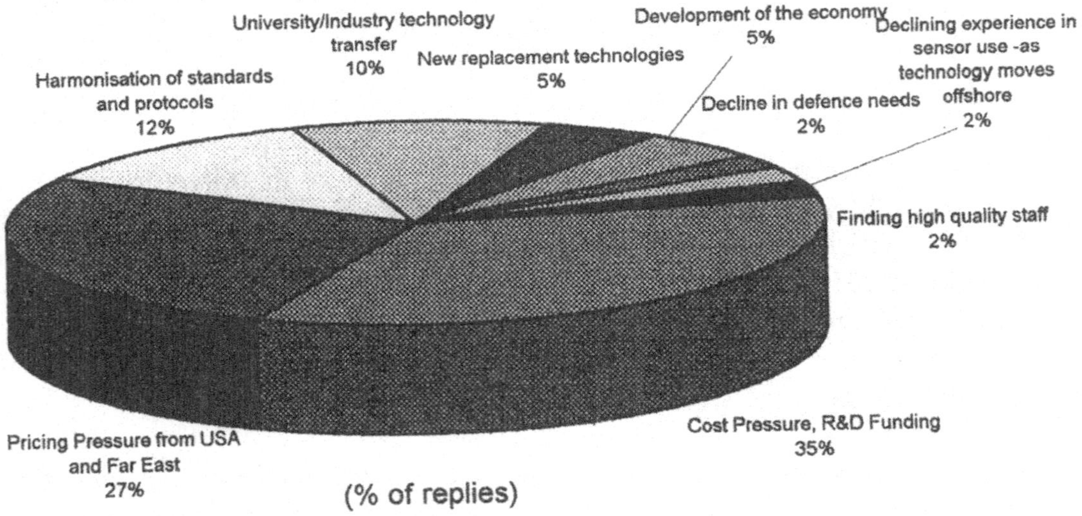

(% of replies)

Figure 2

A further reason for using microprocessor-based sensors is the possibility of two-way digital communications. Until recently, many organisations have hesitated to install sensors with these communications possibilities as they were afraid of being locked into a sensor manufacturer's proprietary standard. However, interoperable products from many manufacturers are now available following open standards (such as Profibus, WorldFIP and P-NET, the three alternatives in the new European 'standard', EN50170, announced on 1 June 1996, which must be followed by European public utilities), and the rate of installation of these products has accelerated considerably in the last year. When fully international open standards are completed this trend should accelerate further. Respondents mentioned harmonisation of communication protocol standards as one of their main area of concern.

Globalisation

Major non-European suppliers, such as Fisher-Rosemount, Fuji and Honeywell have high levels of sensor sales in Europe. Equally major European suppliers, such. as ABB Kent-Taylor, Danfoss and Endress+Hauser,are now very active in non-European markets. A related factor is the marked increase in industry concentration by mergers and acquisitions;examples are the Fisher-Rosemount merger in 1992; the purchase of Eckardt in 1993 by Siebe, the parent of Foxboro; and the purchase in 1994 of Fischer & Porter by Elsag Bailey Process Automation. There are, however, still a large number of successful smaller local suppliers in the market. Many of the respondents mentioned high levels of price competition from manufacturers from the USA and the Far East.

Ease of Use

Buyers are showing a marked preference for those types of sensors which are easier to use: this is shown, not only in the trend towards intelligent sensors, but also towards those sensors which are easier to install and maintain. Examples of this are the replacement of mechanical flowmeters by those, such as electromagnetic, vortex, Coriolis and ultrasonic, with no moving parts and therefore lower maintenance requirements; and the trend away from servogauge and capacitive level sensors towards the easier-to-install, non-invasive ultrasonic and radar types. 5% of the respondents expressed fear that new technologies, such as ultrasonics, might lead to the demise of their industries.

Automotive

The use of sensors in cars is producing a growth in the automotive sensors market; all commentators are agreed on a considerable volume growth in the future, but some fear that cost pressures in this high volume market may imply a small value growth. Part of this growth is due to environmental concerns about exhaust gas pollution; other factors which will produce greater use of sensors in cars are the availability of cheap micromachined silicon sensors for pressure, acceleration and temperature measurement; and the development of automotive sensor systems multiplex on a single bus.

Biomedical

The biomedical sensors market is forecast to grow very rapidly over the next few years; compound annual growth rates of as high as 14% have been suggested by some commentators. Measurement of blood parameters is the main application; and disposable sensor sales are expected to grow faster than sales of reusable ones.

Sensors - European Expertise
Signals sensed

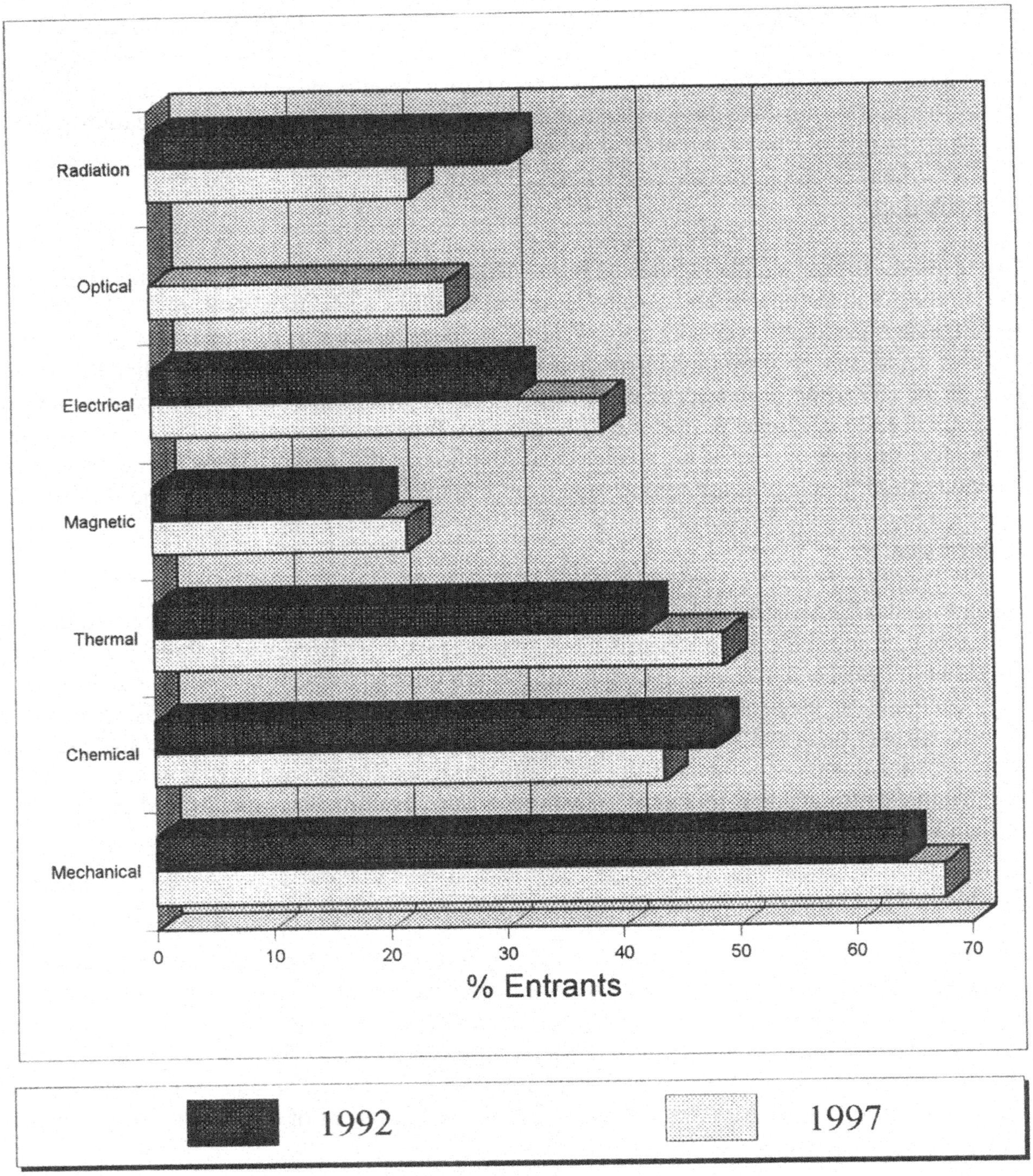

Metra Data Base Records

Figure 3

2.2 Analysis of Entrants' Activities

Signal Sensed

The Metra Martech Sensors Directory represents a major survey of the industry and an analysis of the entries gives an overview of developments in European sensor technology activities.

Entrants were asked to provide details of the types of sensors under development. A summary of their responses is shown in Figure 3. This chart also includes the comparable results from the previous edition of the directory.

This edition shows that sensors for mechanical signals still play a major role in sensor development programmes and their relative importance has increased. The trend is still towards the development of pressure, flow and level sensors.

Thermal sensors are also increasing in relative importance. Over half of the entrants develop or manufacture this type of sensors.

Chemical sensors are the third most important sensor type in this survey, but their relative importance has decreased. Most of the chemical sensors are used to detect gases.

Electrical and magnetic sensors are more developed than before. Optical sensors were included with radiation types in the previous edition. The separation of the two here shows the importance of optical sensing.

Application areas

Organisations were asked to provide details of the target applications for sensors under development. These are summarised in Figure 4 overleaf. The application of sensors ranges from process control to meteorology.

In this edition the most frequently mentioned areas are the manufacturing and machinery industries. Their relative importance has considerably increased.

Applications in telecommunications, consumer products and medicine were also considerably more often mentioned than before.

The other application areas show a decrease in importance, especially applications in the automotive industry. The initial surge of research in this sector has changed to more mature development by the few large suppliers.

Sensors - European Expertise
Application Areas

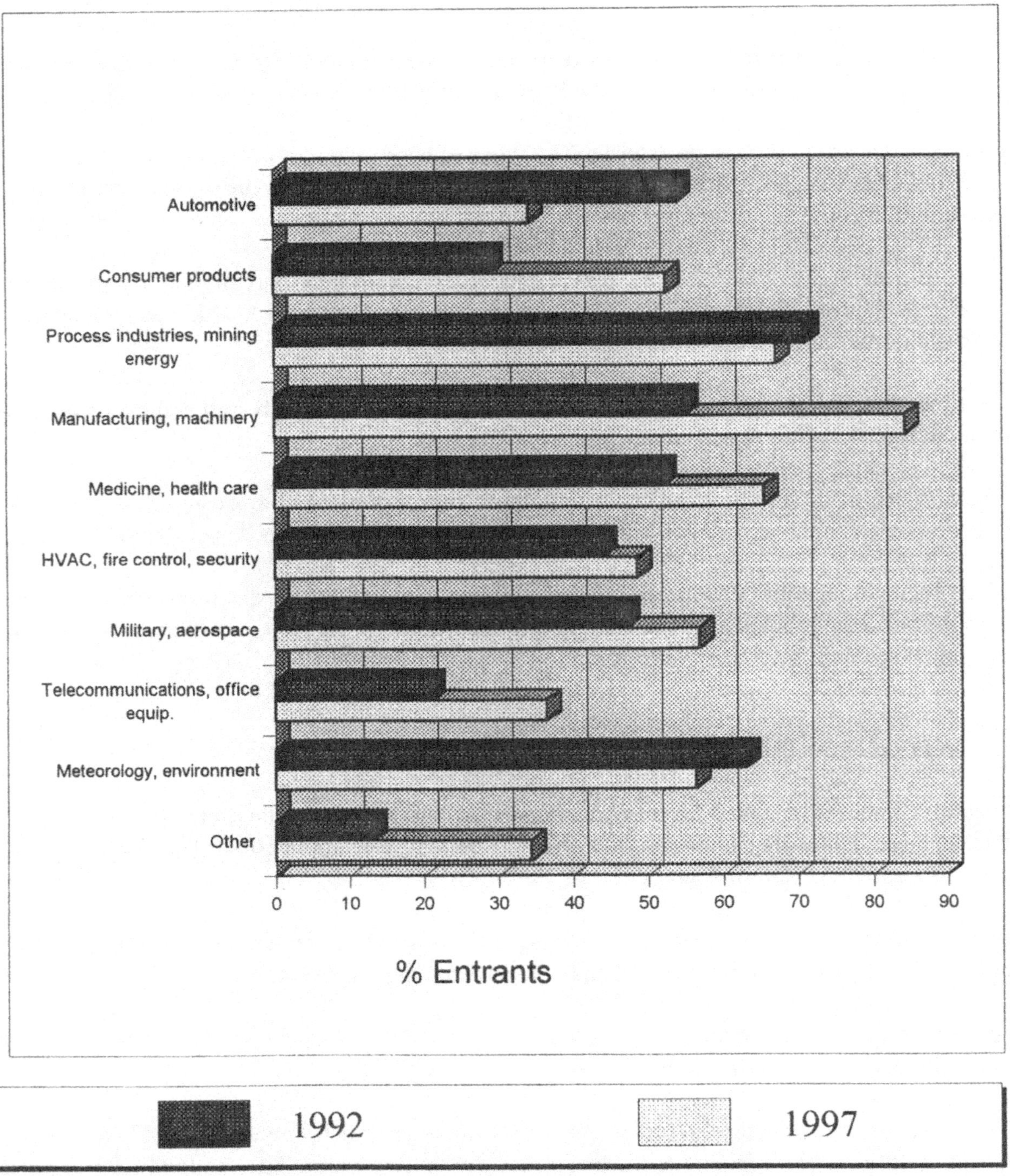

Metra Data Base Records

Figure 4

3. How to Use the Directory

The entries follow a standard format. The information is as supplied by participating organisations with format editing by Metra Martech.

The entries are grouped by country in alphabetical sequence. In general we have used the organisation title provided. If possible the full name is given in the address box. Where a company name consists of initials followed by further names, the company is indexed on the initials.

Section 4 contains the information about each company. In particular, the types of signal sensed and their areas of application are shown in grid form. The list of headings is:

Signal Sensed

- Radiation
- Optical
- Electrical
- Magnetic
- Thermal
- Chemical
- Mechanical

Application Areas

- Automotive
- Consumer products
- Process industries, mining, energy
- Manufacturing, machinery
- Medicine, health care
- HVAC, fire control, security
- Military aerospace
- Telecommunications, office equipment
- Meteorology, environment

In Section 5, the information on the application areas and signals sensed is cross referenced by the entry number of the organisation. The application areas for a particular signal sensed are dealt with in the order listed above, before moving on to the application areas for the next signal sensed. It is possible to locate the entries for each country by referring to the entry number.

Section 6 lists the organisations appearing in the guide. As before, the entries are grouped by country in alphabetical order.

4. Company Information

Austria

Entry 001

001 AVL Medical Instruments

Kleiststraße 48a
A-8070 Graz

Tel: +43 316 987 364
Fax: +43 316 987 635
E-mail:

Affiliation:
Contact: C Ritter
J Harper
D Karpf, OEM - contracts

Organisation information
No of employees 101-1000
Turnover (ECU*) 5m-50m
Sensor technology activities information
No of employees 26-100
Turnover (ECU*) 1m-5m
*1ECU=DM 1.89,FFr 6.30,£0.78,Lira 1909,Peseta 159.1

Organisation activities
Research and development on sensors, sensor materials or applications
Manufacture of finished sensors
Manufacture of sensor-based products

Application \ Signal Sensed	Radiation	Optical	Electrical	Magnetic	Thermal	Chemical	Mechanical
Automotive							
Consumer products							
Process industries,mining energy							
Manufacturing, machinery							
Medicine, health care		•				•	
HVAC, fire control, security							
Military, aerospace							
Telecommunications,office equip.							
Meteorology environment							
Other							

Main Activities

AVL Medical Instruments develops and produces sensors and instruments for critical care analysis in the medical field.

Specific Expertise

Sensors for use in biological liquids (whole blood, serum, plasma, etc.). Biosensors.

Collaboration Offered

Sensors for monitoring, clinical chemistry analysers and dialysis. Instruments for the analysis of clinical critical care parameters.

Belarus

Entry 002

002 Minsk Research Institute of Radiomaterials

Kuzhevatova 86
22-0115 Minsk

Tel: +375 278 14 00
Fax: +375 278 37 05
E-mail: khitko@irmabelpak.minsk.by

Affiliation: State Interpries
Contact: Khirko Valentin, Director
Demchenko Alexander I, Deputy

Organisation information

No of employees 101-1000
Turnover (ECU*) 1m-5m

Sensor technology activities information

No of employees 26-100
Turnover (ECU*) <0.5m

*1ECU=DM 1.89,FFr 6.30,£0.78,Lira 1909,Peseta 159.1

Organisation activities

Research and development on sensors sensor materials or applications
Manufacture of finished sensors
Manufacture of sensor-based products

Application \ Signal Sensed	Radiation	Optical	Electrical	Magnetic	Thermal	Chemical	Mechanical
Automotive			•				•
Consumer products							•
Process industries,mining energy					•		
Manufacturing, machinery			•				
Medicine, health care						•	•
HVAC, fire control, security						•	•
Military, aerospace		•				•	•
Telecommunications,office equip.		•			•	•	•
Meteorology environment							•
Other					•	•	

Main Activities

Minsk Research Institute of Radiomaterals (IRMA) was founded in 1982 and specialises in development and production of functional materials and devices on the base of these materials for radio electronics, chemical, machine building industries.

Specific Expertise

Contactless current sensors, rotation frequency and run-out amplitude sensors, pressure sensors, linear displacement sensors, rotation angle sensors.

Collaboration Offered

S&T collaboration.

Collaboration Sought

Participate in joint S&T R&O.

Belgium

Entries 003 to 011

003 Delft Sensor Systems

Westerring 21
B-9700 Oudenaarde
Belgium

Tel: +32 55 333 811
Fax: +32 55 316 895
E-mail: abo@oip.be

Affiliation: Delft Instruments
Contact: A Boumdouha, Marketing Manager

Organisation information

No of employees 101-1000
Turnover (ECU*) 5m-50m

Sensor technology activities information

No of employees 26-100
Turnover (ECU*) 0.5m-1m

*1ECU=DM 1.89,FFr 6.30,£0.78,Lira 1909,Peseta 159.1

Organisation activities

Research and development on sensors, sensor materials or applications.
Manufacture of sensor-based products.

Application \ Signal Sensed	Radiation	Optical	Electrical	Magnetic	Thermal	Chemical	Mechanical
Automotive							
Consumer products							
Process industries,mining energy							
Manufacturing, machinery		•	•		•		
Medicine, health care							
HVAC, fire control, security							
Military, aerospace		•			•		
Telecommunications,office equip.							
Meteorology, environment							
Other *							

Main Activities

Delft Sensor Systems develops and manufactures night vision equipment for applications in the defence industry.

Specific Expertise

Night vision equipment.
Holography.

004 Fondation Universitaire Luxembourgeoise

Avenue De Longwy 185
B-6700 Arlon

Tel: +32 63 23 08 57
Fax: +32 63 23 08 00
E-mail: nicolas@ful.ac.be

Affiliation:
Contact: J Nicolas, Head of Dept

Organisation information
No of employees 26-100
Turnover (ECU*)
Sensors technology activities information
No of employees 1-10
Turnover (ECU*) <0.5m
*1ECU=DM 1.89,FFr 6.30,£0.78,Lira 1909,Peseta 159.1

Organisation activities
Research and Development

Application \ Signal Sensed	Radiation	Optical	Electrical	Magnetic	Thermal	Chemical	Mechanical
Automotive							
Consumer products							
Process industries, mining, energy							
Manufacturing, machinery							
Medicine, health care							
HVAC, fire control, security							
Military, aerospace							
Telecommunications,office equip.							
Meteorology, environment	•	•	•	•	•	•	•
Other *							

Main Activities

The research in "Environmental Monitoring" at F.U.L. is based on the fact that, for management purposes, it is becoming increasingly necessary to set up efficient systems for continuous control of the indicators reflecting the state of the environment. Both field data acquisition techniques and data treatment are tackled. Particular attention is given to "intelligent measurement", bio-sensors and event forecast algorithms. The team of researchers tries to integrate the three steps of monitoring:
Measurement, treatment and decision making support.

Specific Expertise

- Smart sensor for soil quality measurement, based on a ceramic porous cell.
- Smart sensor for determination of various components of solar radiation.
- Measurement of fog extinction coefficient on a short distance by means of infrared laser.
- Electronic nose based on an array of sensors sensitive to odorous gas in the environment.

Collaboration Offered

We offer know how in environmental problems and in field instrumentation. Possibility to find an application in the environment for sensors and instruments developed in a more general way.

Collaboration Sought

Our company seeks expertise in some specific sectors of electronics:
Optoelectronics, high frequency, low signal detection..

005 IMEC

Kapeldreef 75
B-3110 Leuven

Tel: +32 16 281 211
Fax: +32 16 281 501
E-mail: hermans@imec.be

Affiliation:
Contact: L Hermans, Group Head

Organisation information

No of employees	101-1000
Turnover (ECU*)	5m-50m

Sensor technology activities information

No of employees	26-100
Turnover (ECU*)	1m-5m

*1ECU=DM 1.89,FFr 6.30,£0.78,Lira 1909,Peseta 159.1

Organisation activities
Research and development on sensors, sensor materials or applications.
Testing, consulting, design.
Manufacture of finished sensors.

Application \ Signal Sensed	Radiation	Optical	Electrical	Magnetic	Thermal	Chemical	Mechanical
Automotive		●		●	●	●	●
Consumer products		●		●	●	●	●
Process industries,mining energy		●		●	●	●	
Manufacturing, machinery		●		●	●		
Medicine, health care		●			●	●	
HVAC, fire control, security		●			●		
Military, aerospace		●			●		
Telecommunications,office equip.		●			●		
Meteorology, environment		●			●	●	
Other *							

Main Activities

R&D and micron CMOS technology; VLSI design methodologies; opto-electronics; micro machining; sensors and actuators; IC packaging technology; MMIC

Specific Expertise

CMOS based sensors for: radiation detection; optical appreciation. Chemical sensors: thick and thin film based. Biosensors and immuno sensors.

Collaboration Offered

Contract R&D

Collaboration Sought

Partners with specific experience.

006 IRM SA

Avenue du Progrès, 28
B-4432 Alleur

Tel: +32 41 39 90 10
Fax: +32 41 63 40 97
E-mail:

Affiliation:
Contact: J Lückers, MD
P Balthasart. Development
J C Uyttenhove, Strategic Mkting
D Ramelot, Process & Quality
B Mairy, Sales International

Organisation information

No of employees	26-100
Turnover (ECU*)	5m-50m

Sensor technology activities information

No of employees	26-100
Turnover (ECU*)	>5m

*1ECU=DM 1.89,FFr 6.30,£0.78,Lira 1909,Peseta 159.1

Organisation activities
Research and development on sensors sensor materials or applications
Testing, consulting, design
Manufacture of systems used in the maufacture, testing or development of sensors
Manufacture of finished sensors
Manufacture of sensor-based products

Application \ Signal Sensed	Radiation	Optical	Electrical	Magnetic	Thermal	Chemical	Mechanical
Automotive		●					
Consumer products							
Process industries,mining energy	●	●	●		●	●	●
Manufacturing, machinery							●
Medicine, health care							
HVAC, fire control, security					●	●	
Military, aerospace							
Telecommunications,office equip.							
Meteorology environment						●	
Other (specify)							

Main Activities

IRM develops, manufactures and markets measuring systems for industrial applications; IRM is specialised in metal industry.

Specific Expertise

Dimensional controls: for cold and hot products; according to application, use of optical triangulation and shadow techniques with back or front lighting, or 3D vision.
NDT: detection of internal and surface defects of strips by EMAT techniques.
Temperature: mapping of surface temperature distribution, with line or area scanning systems including application oriented data handling software.
Gas analysis: ultra-high accuracy systems including sampling; sample preparation; analysis; automatic correction and reference gases.
Non contact speed and length: laser doppler and speckle effect based scanners.

Collaboration Offered

Cooperation in the areas of research and development technology, know-how and marketing.

Collaboration Sought

Joint activities in the area of new product development and marketing of joint items.

007 Sensy SA

8 Avenue Maurice
B-1050 Brussels

Tel: +32 2 647 3800
Fax: +32 2 640 1414
E-mail:

Affiliation:
Contact: S Sendronice

Organisation information
No of employees 26-100
Turnover (ECU*) 1m-5m
Sensor technology activities information
No of employees 11-25
Turnover (ECU*) 0.5m-1m
*1ECU=DM 1.89,FFr 6.30,£0.78,Lira 1909,Peseta 159.1

Organisation activities
Research and development on sensors sensor materials or applications
Manufacture of systems used in the manufacture, testing or development of sensors
Manufacture of finished sensors
Manufacture of sensor-based products

Application \ Signal Sensed	Radiation	Optical	Electrical	Magnetic	Thermal	Chemical	Mechanical
Automotive							•
Consumer products							•
Process industries,mining energy							•
Manufacturing, machinery							•
Medicine, health care							•
HVAC, fire control, security							
Military, aerospace							•
Telecommunications,office equip.							
Meteorology environment							•
Other							

Main Activities

SENSY has obtained approvals from the Weighing Authorities of most European countries, amongst others of the PTB (Germany), the most prestigious one of all.
Furthermore, SENSY developed a data acquisition system, which, connected with load cells, gives the opportunity of dynamic weighing (traffic control and non-stop weighing), as well as using it as an intelligent load limiter.
SENSY is at the moment working on the load cell of tomorrow, using the newest technologies, using among other technologies, the laser and integrated intelligent ones, which combines high performance in matters of accuracy as well as competitive prices.

Specific Expertise

References load cells (to be used for collaboration of the other load cells); thick film technology sensors; thin film technology sensors.

Collaboration Offered

We offer technical and/or commercial ventures with companies involved in the same fields.

Collaboration Sought

Actually we seek more for commercial collaboration.

008 Solstat Operations

123 Rue P. Pastur
B-6043 Ransart

Tel: +32 71 34 12 79
Fax: +39 71 34 38 56
E-mail:

Affiliation: CERELUX S.A. Luxembourg
Contact: Fr. de Paun, Director
J. van Hulse, Managing Director
J.C. Maton, Plant Director

Organisation information

No of employees 1-25
Turnover (ECU*) <1m

Sensor technology activities information

No of employees 1-10
Turnover (ECU*) <0.5m

*1ECU=DM 1.89,FFr 6.30,£0.78,Lira 1909,Peseta 159.1

Organisation activities
Research and development.
Testing, consulting, design.
Manufacture of finished sensors.

Application \ Signal Sensed	Radiation	Optical	Electrical	Magnetic	Thermal	Chemical	Mechanical
Automotive					•		
Consumer products					•		
Process industries,mining energy							
Manufacturing, machinery							
Medicine, health care					•		
HVAC, fire control, security					•		
Military, aerospace					•		
Telecommunications,office equip.							
Meteorology, environment							
Other *					•		

* Food industry

Main Activities

Electronic ceramics.

Specific Expertise

Ceramic sensors: PTC; sizes from 0.5 to 12 mm.

009 TCS sprl

Av. Générale Michel 1E
B-6000 Charleroi

Tel: +32 71 270 381
Fax: +32 71 316 735
E-mail:

Affiliation:
Contact: Fr. Lantsoght, Director
Thierry Evrard, Business Manager

Application \ Signal Sensed	Radiation	Optical	Electrical	Magnetic	Thermal	Chemical	Mechanical
Automotive							
Consumer products							
Process industries,mining energy					•		
Manufacturing, machinery							
Medicine, health care							
HVAC, fire control, security							
Military, aerospace							
Telecommunications,office equip.					•		
Meteorology, environment							
Other *					•		

* Food industry

Organisation information
No of employees 1-25
Turnover (ECU*) <1m

Sensor technology activities information
No of employees 1-10
Turnover (ECU*) <0.5m
*1ECU=DM 1.89,FFr 6.30,£0.78,Lira 1909,Peseta 159.1

Organisation activities
Testing, consulting, design.
Manufacture of sensor-based products.

Main Activities

Manufaturing of temperature and humidity recorders and OEM products.
Consulting in telecommunications and in electronics (including ASIC's).
Development in the above areas.

Specific Expertise

High accuracy measurements using NTC.
Recorders (temperature and hunidity ~).

Collaboration Offered

TCS would like to share their previous experience. The know-how should be very useful to companies concerned with the recording of temperature and hunidity,its analysis and the quick transmission of the data recorded.

Collaboration Sought

Looking for an organisation with a sound reference in international distribution networks or other partners at European level.

010 Thermibel SA

Rue du Roeuld ZI
B-7110 Staepy-Bracquegrico

Tel: +32 64 67 54 32
Fax: +32 64 67 63 36
E-mail:

Affiliation: Thermo-est
Contact: Jean Pistis, General Manager

Organisation information

No of employees 1-25
Turnover (ECU*) 1m-5m

Sensor technology activities information

No of employees 11-25
Turnover (ECU*) 1m-5m

*1ECU=DM 1.89,FFr 6.30,£0.78,Lira 1909,Peseta 159.1

Organisation activities

Research and development on sensors, sensor materials or applications
Testing, consulting, design
Supply of raw materials or semi-finished products used to make sensors
Manufacture of systems used in the manufacture, testing or development of sensors
Manufacture of finished sensors
Manufacture of sensor-based products

Application \ Signal Sensed	Radiation	Optical	Electrical	Magnetic	Thermal	Chemical	Mechanical
Automotive					•		
Consumer products					•		
Process industries,mining energy					•		
Manufacturing, machinery					•		
Medicine, health care							
HVAC, fire control, security							
Military, aerospace							
Telecommunications,office equip.							
Meteorology environment							
Other							

Main Activities

Thermibel SA produce electronic temperature detectors.

Specific Expertise

All production industries.

011 VITO

Flemish Institute for Technological Research
Boeretang 200
B-2400 MOL

Tel: +32 14 33 55 11
Fax: +32 14 33 55 99
E-mail: vito@vito.be

Affiliation:
Contact: B Leemans, Mkting & Communication
F de Schutter, Project Leader
J Kretzschman,Scientific Director

Organisation information

No of employees	101-1000
Turnover (ECU*)	5m-50m

Sensor technology activities information

No of employees	1-10
Turnover (ECU*)	0.5m-1m

*1ECU=DM 1.89,FFr 6.30,£0.78,Lira 1909,Peseta 159.1

Organisation activities
Research and development on sensors sensor materials or applications

Application \ Signal Sensed	Radiation	Optical	Electrical	Magnetic	Thermal	Chemical	Mechanical
Automotive							
Consumer products							
Process industries,mining energy						•	
Manufacturing, machinery							
Medicine, health care							
HVAC, fire control, security							
Military, aerospace							
Telecommunications,office equip.							
Meteorology environment							
Other (specify)							

Main Activities

VITO - The Flemisch Institute for Technological Research does research in the field of chemical sensors for applications in different industries.

Specific Expertise

Areas of expertise: on-line sensor (chemical) for liquid metals (A, Cu, Zn); high temperature gas sensors.
Specific materials: ceramic proton conductors; ß-alumina ion conductos; molten salt based ion conductors.

Collaboration Offered

Expertise in sensor development on the fabrication of the necessary ceramic element (ion conductors); expertise in on-line measuring in industrial liquid metal melts; melting processes.

Czech Republic

Entries 012 to 013

012 Czech Technical University of Prague

Faculty of Electrical Engineering
Department of Measurement
Technicka 2
CZ-166 Praha 6

Tel: +42 2 24352188
Fax: +42 2 3119929
E-mail: ripka@feld.cvuvt.cz

Affiliation:
Contact: P Ripka, Industrial Consultant
K Draxer, Pressure Sensors
J Fischer, Optical Sensors

Organisation information

No of employees	101-1000
Turnover (ECU*)	<1m

Sensor technology activities information

No of employees	1-10
Turnover (ECU*)	<0.5m

*1ECU=DM 1.89,FFr 6.30,£0.78,Lira 1909,Peseta 159.1

Organisation activities
Research and development on sensors sensor materials or applications
Testing, consulting, design
Manufacture of systems used in the manufacture, testing or development of sensors

Application \ Signal Sensed	Radiation	Optical	Electrical	Magnetic	Thermal	Chemical	Mechanical
Automotive				•			
Consumer products		•					
Process industries,mining energy							
Manufacturing, machinery							
Medicine, health care							
HVAC, fire control, security							
Military, aerospace				•			•
Telecommunications,office equip.							
Meteorology environment							
Other							

Main Activities

Industrial consultations on sensor application; calibration of pressure sensors; development and calibration of magnetic sensors; optical sensors; contactless current sensors (AC and DC).

Specific Expertise

Calibration of magnetic sensors (incl. noise and temperature stability); development of low-noise fluxgate sensors and portable magnetometers; calibration of pressure sensors.

Collaboration Offered

Customised fluxgate sensors (miniature, ultra low-noise; low-cost).

013 Czech Technical University of Prague

Department of Microelectronics
Technicka 2
CZ-16627 Prague

Tel: +42 2 2435 2267
Fax: +42 2 2431 0792
E-mail: husak@feld.cvut.cz

Affiliation:
Contact: Husak Miroslav, Deputy HoD

Organisation information
No of employees 101-1000
Turnover (ECU*) 1m-5m

Sensor technology activities information
No of employees 11-25
Turnover (ECU*) <0.5m
*1ECU=DM 1.89,FFr 6.30,£0.78,Lira 1909,Peseta 159.1

Organisation activities
Research and development on sensors sensor materials or applications

Application \ Signal Sensed	Radiation	Optical	Electrical	Magnetic	Thermal	Chemical	Mechanical
Automotive							
Consumer products							
Process industries,mining energy				•	•		•
Manufacturing, machinery							
Medicine, health care				•	•		•
HVAC, fire control, security							
Military, aerospace							
Telecommunications,office equip.							
Meteorology environment							
Other (specify)							

Main Activities

General electronics, microelectronics, technological systems, radioelectronics, measurement technology, control engineering and robotics, computer science and engineering, telecommunication engineering, electric power engineering.

Specific Expertise

Research activities (projects) include: semiconductor microsystem structures; sensor signals processing; sensor signals wireless transmission; sensor control systems; biomedical sensors; flowmeters; temperature sensors; pressure sensors.

Collaboration Offered

The sensor research and development - basic principles of sensors and microsensors, basic applications in measurement, regulation and control, signal processing of sensors signal, pressure, temperature sensors and biosensors for biomedical development.

Denmark

Entries 014 to 020

014 Danfoss A/S

DK-6430 Nordborg

Tel: +45 74 882222
Fax: +45 74 490949
E-mail:

Affiliation:
Contact: BF Iwersen, Mktg

Organisation information
No of employees >100
Turnover (ECU*) >5m
Sensors technology activities information
No of employees >100
Turnover (ECU*) >5m
*1ECU=DM 1.89,FFr 6.30,£0.78,Lira 1909,Peseta 159.1

Organisation activities
Manufacture of finished sensors

Application \ Signal Sensed	Radiation	Optical	Electrical	Magnetic	Thermal	Chemical	Mechanical
Automotive							
Consumer products							
Process industries, mining, energy					•	•	•
Manufacturing, machinery					•		•
Medicine, health care							
HVAC, fire control, security					•		•
Military, aerospace							
Telecommunications,office equip.							
Meteorology, environment					•		•
Other *							

Main Activities

Supplier of a broad range of components for industrial automation. Sold via a worldwide distribution network.

Specific Expertise

Industrial Controls Department: Pressure and temperature sensors, on/off transmitters and transducers. Other parts of the company are involved in chemical sensors.

Collaboration Offered

Joint selling of products or sub-supply of sub-assemblies.

Collaboration Sought

To increase the company's product programme with a mix of bought in products, products manufactured under a licence or similiar.

015 Dantec Elektronik

Medicinsk og Videnskabeligt
Maleudstyr A/S
Tonsbakken 16-18
DK-2740 Skovlunde

Tel: +45 44923610
Fax: +45 42 846136
E-mail:

Affiliation:
Contact: E I Hayes, Sales Mgr
K Askegaard, Mktg Manager

Organisation information

No of employees >100
Turnover (ECU*) >5m

Sensors technology activities information

No of employees >100
Turnover (ECU*) >5m
*1ECU=DM 1.89,FFr 6.30,£0.78,Lira 1909,Peseta 159.1

Organisation activities

Research and Development
Manufacture of finished sensors

Application \ Signal Sensed	Radiation	Optical	Electrical	Magnetic	Thermal	Chemical	Mechanical
Automotive					●	●	●
Consumer products							●
Process industries, mining, energy							
Manufacturing, machinery							
Medicine, health care					●		
HVAC, fire control, security					●		●
Military, aerospace							
Telecommunications,office equip.							
Meteorology, environment							
Other *							

Main Activities

The Measurement Technology Division of Dantec has a very long experience in development, production and marketing of laser-based measuring equipment for industry and science. The instruments are used to study flow, turbulence, vibration, size and concentration of particles. Instrumentation dedicated to the industry measures velocity and length of moving surfaces eg. paper.

Specific Expertise

Optical, optoelectronic and fibre optic sensors applied to industry and science; fibre optic laser Doppler anemometers for non-instrusive fluid flow (velocity and turbulence) measurements; fibre optic laser vibrometer for non-contact motion studies; fibre optic particle dynamic analysers for non-instrusive particle velocity, size and concentration studies; optical sensor for measurement of velocity and length of moving surfaces; constant temperature anemometers for velocity and temperature measurements.

016 Delta Lys & Optik

Hjortekaersvej 99
DK-2800 Lyngby

Tel: +45 45 888 333
Fax: +45 45 870 810
E-mail: coatings@delta.dk

Affiliation: Delta Dansk Elektronik, Lys & Akustik
Contact: Mr F Poulsen, Head of Department
Mr H Fabricius, Research Engineer
Ms H Bay Rasmussen, Export Engineer

Organisation information
No of employees 101-1000
Turnover (ECU*) <1m
Sensor technology activities information
No of employees 11-25
Turnover (ECU*)
*1ECU=DM 1.89,FFr 6.30,£0.78,Lira 1909,Peseta 159.1

Organisation activities
Supply of raw materials or semi-finished products used to make sensors

Application \ Signal Sensed	Radiation	Optical	Electrical	Magnetic	Thermal	Chemical	Mechanical
Automotive							
Consumer products							
Process industries,mining energy							
Manufacturing, machinery							
Medicine, health care		•					
HVAC, fire control, security							
Military, aerospace							
Telecommunications,office equip.							
Meteorology environment							
Other							

Main Activities

DELTA Light & Optics is a manufacturer of optical thin film products, interference filters for fluorescence applications as well as Gradient Index Filters (smart-coatings). We also offer consultancy in connection with custom designd optical coatings for special applications.

Specific Expertise

We have developed the Gradient Index Filters (smart coatings) which are very applicable in colour recognition systems, colour detectors, equalisation of signals, adapting of sensors according to the sensitivity of the human eye.

Collaboration Offered

Consulting and advisory information when designing optical filters for special applications as well as the manufacture of prototypes and larger series.

017 DME - Danish Micro Engineering AS

Transformervej 12
DK-2730 Herlev

Tel: +45 42 849211
Fax: +45 42 849197
E-mail:

Affiliation:
Contact: C Sander, MD

Organisation information
No of employees 26-100
Turnover (ECU*) 1m-5m
Sensors technology activities information
No of employees 11-25
Turnover (ECU*)
*1ECU=DM 1.89,FFr 6.30,£0.78,Lira 1909,Peseta 159.1

Organisation activities
Research and Development
Manufacture of finished sensors

Application \ Signal Sensed	Radiation	Optical	Electrical	Magnetic	Thermal	Chemical	Mechanical
Automotive	•						
Consumer products							
Process industries, mining, energy	•						
Manufacturing, machinery	•						
Medicine, health care							
HVAC, fire control, security							
Military, aerospace							
Telecommunications,office equip.							
Meteorology, environment							
Other *							

Main Activities

Development and production of various optical sensors, STM (Scanning Tunneling Microscropes) and of software and hardware for other companies.

Specific Expertise

DME has developed 4 different optical displacement sensors. 3 of these measure the distance to an object by means of a single light spot. The fourth sensor measures the distance to 11 points on the object, making possible a calculation of the profile of the object. Sensors can be interconnected two and two for measurement of the thickness of an object. Distances as well as thicknesses from a few microns to 200 mm can be measured.

Collaboration Offered

We offer development of custom designed sensors.

018 DTH - Danmarks Tekniske Hojskole

Technical University of Denmark
Electric Power Engineering Department
Building 325
DK-2800 Lyngby

Tel: +45 45 252525
Fax: +45 45 886111
E-mail:

Affiliation:
Contact: Prof O Tynnesen, Head of Dept

Organisation information
No of employees >100
Turnover (ECU*)

Sensors technology activities information
No of employees 1-10
Turnover (ECU*)
*1ECU=DM 1.89,FFr 6.30,£0.78,Lira 1909,Peseta 159.1

Organisation activities
Research and Development

Application \ Signal Sensed	Radiation	Optical	Electrical	Magnetic	Thermal	Chemical	Mechanical
Automotive							
Consumer products							
Process industries, mining, energy			•	•			
Manufacturing, machinery							
Medicine, health care							
HVAC, fire control, security							
Military, aerospace							
Telecommunications,office equip.							
Meteorology, environment							
Other *							

Main Activities
Research and development, education of undergraduate and graduate MSc students within the field of Electrical Engineering.

Specific Expertise
Research and development on sensors for high voltage current and voltage measurement for relay, control and metering in electrical production, transmission and distribution systems.
Development of sensor materials and systems for magneto-optic and electro-optic sensors.
Development of high voltage optical instrument transformers (current and voltage transformers).

Collaboration Offered
Exchange of guest researchers.

Collaboration Sought
Exchange of information, visits, arrangement of guest researcher stays.

019 Ferroperm A/S

Piezoceramics Division
Hesreskouves 6
DK-3990 Kuistrård

Tel: +45 491 381 90
Fax: +45 491 381 88
E-mail:

Affiliation:
Contact: Wanda W Wolny, Div Manager
B Anderson, Sales
A James, Materials Development

Organisation information
No of employees 26-100
Turnover (ECU*) 1m-5m
Sensor technology activities information
No of employees 26-100
Turnover (ECU*) 1m-5m
*1ECU=DM 1.89,FFr 6.30,£0.78,Lira 1909,Peseta 159.1

Organisation activities
Research and development on sensors, sensor materials or applications.
Testing, consulting, design.
Supply of raw materials or semi-finished products used to make sensors.

Application \ Signal Sensed	Radiation	Optical	Electrical	Magnetic	Thermal	Chemical	Mechanical
Automotive							●
Consumer products							
Process industries,mining energy							●
Manufacturing, machinery							●
Medicine, health care			●				
HVAC, fire control, security							
Military, aerospace			●				●
Telecommunications,office equip.							
Meteorology, environment			●				
Other *							

Main Activities

Ferroeperm manufacture piezoceramic materials and optical interference filters for sensor applications among others full line of PZT - materials, high temperature Bismuth Titanate, operating temperature up to 400°C, highly aniotropic lead metaniobate and lead titanate.

Specific Expertise

Piezoelectric ceramics - PZT and for high temperature applications up to 500°C. Electrostrictive materials.

Collaboration Offered

Development/modification of materials, special date for samples etc during development phase, technical literature. Extended testing eg field tests of materials.

020 Reson System AS

Fabriksvangen 13
DK-3550 Slangerup

Tel: +45 47 380022
Fax: +45 47 380066
E-mail:

Affiliation:
Contact: M Raun, Mktg

Organisation information
No of employees 26-100
Turnover (ECU*) >5m
Sensors technology activities information
No of employees 26-100
Turnover (ECU*) >5m
*1ECU=DM 1.89,FFr 6.30,£0.78,Lira 1909,Peseta 159.1

Organisation activities
Research and Development
Manufacturer of sensor-based products
Testing, consulting, design
Manufacture of finished sensors

Application \ Signal Sensed	Radiation	Optical	Electrical	Magnetic	Thermal	Chemical	Mechanical
Automotive							
Consumer products							
Process industries, mining, energy							•
Manufacturing, machinery							•
Medicine, health care							
HVAC, fire control, security							•
Military, aerospace							
Telecommunications,office equip.							
Meteorology, environment							
Other *							

Main Activities
Reson System AS is a specialised engineering group which designs underwater acoustic or air based transducers and hydrophones.
Reson also designs reactive sonar products for off-shore, industrials and military applications, and acts as an innovative partner in developing new acoustic products.

Specific Expertise
Underwater acoustic and air based transducers can be provided for almost any application. Standard frequencies range from 1 kHz to 2 MHz. Based on a modular concept, Reson has the flexibility to supply competitive solutions with short lead times, as one offs and also at 10,000's of a kind. Reson also supplies omni-directed reference hydrophones which covers a frequency band from 1Hz to 500kHz.
Completely automated test systems for transducers and hydrophones can be provided.
All relevant transducer parameters, covering transmitting and receiving sensitivity, polar plots, impedance measurements and reciprocity measurements can be documented from 1kHz to 2MHz.

Collaboration Offered
Acoustic expertise and production capability.

Collaboration Sought
Complementry market input.

Finland

Entries 021 to 031

021 Bio-Orbit Oy

P O Box 36
FIN-20521 Turku

Tel: +358 24 10 11 00
Fax: +358 24 10 11 23
E-mail:

Affiliation:
Contact: M Kauneela, Sales Manager
S Wahlbeck, CEO

Organisation information
No of employees 1-25
Turnover (ECU*) 1m-5m

Sensors technology activities information
No of employees 1-10
Turnover (ECU*) 1m5m
*1ECU=DM 1.89,FFr 6.30,£0.78,Lira 1909,Peseta 159.1

Organisation activities
Manufacturer of sensor-based products

Application \ Signal Sensed	Radiation	Optical	Electrical	Magnetic	Thermal	Chemical	Mechanical
Automotive							
Consumer products							
Process industries, mining, energy		●	●			●	
Manufacturing, machinery		●	●			●	
Medicine, health care		●	●			●	
HVAC, fire control, security							
Military, aerospace							
Telecommunications,office equip.							
Meteorology, environment		●	●			●	
Other *							

Main Activities

Manufacture and sales of Intruments and Reagents on worldwide basis.

Specific Expertise

- Luminescence.
- Bioluminescence.
- Luminometers.
- ATP hygiene monitoring.
- Toxicity test systems.

Collaboration Offered

Know-how, epertise
Patents

Collaboration Sought

Financial

022 Hidex Oy

Lemminkäisenk.20B
FIN-20520 Turku

Tel: +358 2 2500057
Fax: +358 2 2410075
E-mail: jukhaa@hidex.pp.fi

Affiliation:
Contact: Jukka Haaslahti, MD

Organisation information
No of employees 1-25
Turnover (ECU*) <1m

Sensor technology activities information
No of employees 1-10
Turnover (ECU*)
*1ECU=DM 1.89,FFr 6.30,£0.78,Lira 1909,Peseta 159.1

Organisation activities
Research and development on sensors sensor materials or applications
Manufacture of sensor-based products

Application \ Signal Sensed	Radiation	Optical	Electrical	Magnetic	Thermal	Chemical	Mechanical
Automotive							
Consumer products							
Process industries,mining energy							
Manufacturing, machinery							
Medicine, health care	●	●	●				
HVAC, fire control, security							
Military, aerospace							
Telecommunications,office equip.							
Meteorology environment							
Other (specify)	●	●	●				

Main Activities

Marketing, R&D and production; development co-ordination.

Specific Expertise

Liquid scintilation counting; gamma counting; luminometers; light detection; radiation detection.

Collaboration Offered

R&D for OEM products; expert designs for the above mentioned detectors.

Collaboration Sought

Consumables for the above mentioned analysers.

023 Hollming Ltd

Electronics
Naulakatu 3
FIN-33100 Tampere

Tel: +358 3 2450111
Fax: +358 3 2130188
E-mail:

Affiliation:

Contact: P Juuti, Director
O Laaksonen, Dev Mgr

Organisation information

No of employees 26-100
Turnover (ECU*) >5m

Sensors technology activities information

No of employees 26-100
Turnover (ECU*) 1m-5m

*1ECU=DM 1.89,FFr 6.30,£0.78,Lira 1909,Peseta 159.1

Organisation activities

Research and Development
Manufacturer of sensor-based products
Testing, consulting, design
Manufacture of finished sensors

Application \ Signal Sensed	Radiation	Optical	Electrical	Magnetic	Thermal	Chemical	Mechanical
Automotive							
Consumer products							
Process industries, mining, energy							•
Manufacturing, machinery							
Medicine, health care							•
HVAC, fire control, security							
Military, aerospace							•
Telecommunications,office equip.							
Meteorology, environment							
Other *							

Main Activities

Active and passive sonars: multibeam echosounders, side scan sonars, passive and active ASW-sonars; ultrasound healthcare products; space electronics; water temperature profile measurements.

Specific Expertise

Hydroacoustic sensors for various sonar applications; US-sensors for healthcare applications.

Collaboration Offered

Sound/ultrasound transducers. Beamforming H/W and S/W. Signal processing H/W and S/W.

Collaboration Sought

New system product areas. Reference delivery possibilities. User know-how.

024 Kytöla Oy

PI5
FIN- 40951 Muurame

Tel: +358 14 631 499
Fax: +358 14 631419
E-mail: ari.riimimaki@kytola.com

Affiliation:
Contact: Ari Riimimäki, Sales Manager

Organisation information
No of employees 26-100
Turnover (ECU*) 5m-50m
Sensor technology activities information
No of employees 26-100
Turnover (ECU*) >5m
*1ECU=DM 1.89,FFr 6.30,£0.78,Lira 1909,Peseta 159.1

Organisation activities
Manufacture of finished sensors

Application \ Signal Sensed	Radiation	Optical	Electrical	Magnetic	Thermal	Chemical	Mechanical
Automotive							
Consumer products							
Process industries,mining energy						•	•
Manufacturing, machinery							
Medicine, health care							
HVAC, fire control, security							
Military, aerospace							
Telecommunications,office equip.							
Meteorology environment							
Other							

Main Activities

Flow meter manufacturing: oval gear; variable area.
Water in oil analysers.

Specific Expertise

- Mechanical sensors
- Chemical sensors

025 Labko Ab Oy

Labkotie 1
FIN-36240 Kangasala

Tel: +358 3 28 55 111
Fax: +358 3 28 55 320
E-mail: http://www.labko.fi/

Affiliation:
Contact: Ari Tuunanen, R&D Manager
Mikko Kaskela, MD

Organisation information
No of employees 26-100
Turnover (ECU*) 1m-5m
Sensor technology activities information
No of employees 26-100
Turnover (ECU*) 1m-5m
*1ECU=DM 1.89,FFr 6.30,£0.78,Lira 1909,Peseta 159.1

Organisation activities
Research and development on sensors sensor materials or applications
Testing, consulting, design
Manufacture of finished sensors
Manufacture of sensor-based products

Application \ Signal Sensed	Radiation	Optical	Electrical	Magnetic	Thermal	Chemical	Mechanical
Automotive							
Consumer products							
Process industries,mining energy			•				
Manufacturing, machinery							
Medicine, health care							
HVAC, fire control, security							
Military, aerospace							
Telecommunications,office equip.							
Meteorology environment			•				
Other							

Main Activities

The Labko Group is a manufacturer of level measurement instruments, oil separators and grease traps. Labkotec Ab Oy specialises in measuring level, volume and quality of liquids.

Specific Expertise

Areas of expertise relating to sensors technology: level measuring, monitoring and control electronics; level switcher; water/oil interface sensors; water and oil leakage alarm devices. Also our product range includes equipment for measuring pH, redox, conductivity, turbidity and dissolved oxygen.

Collaboration Offered

Expertise in solving problems related in measuring level, volume and quality of liquids. Special know-how in measuring level of interface between different liquids.

Collaboration Sought

Potential applications offering measurement challenges in our business area.

026 Tampere University of Technology

Measurement and Information Technology
PO Box 692
FIN-33101 Tampere

Tel: +358 3 365 2485
Fax: +358 3 365 2171
E-mail: halttune@ee.tut.fi

Affiliation: Automation Dept., Tampere University of Technology
Contact: Olle Aumala, Head of Laboratory
Jouko Halttunen, Assoc. Prof.
Heikki Jokinen, Lab. Engineer

Organisation information
No of employees 1-25
Turnover (ECU*) <1m
Sensor technology activities information
No of employees 1-10
Turnover (ECU*) <0.5m
*1ECU=DM 1.89,FFr 6.30,£0.78,Lira 1909,Peseta 159.1

Organisation activities
Research and development on sensors sensor materials or applications
Testing, consulting, design
Supply of raw materials or semi-finished products used to make sensors

Application \ Signal Sensed	Radiation	Optical	Electrical	Magnetic	Thermal	Chemical	Mechanical
Automotive			•		•		
Consumer products					•		
Process industries,mining energy		•	•	•	•		•
Manufacturing, machinery		•	•	•			
Medicine, health care							
HVAC, fire control, security							
Military, aerospace							
Telecommunications,office equip.			•				
Meteorology environment		•	•		•		
Other							

Main Activities

Main activities are teaching and research especially in the areas of measurement technology, measurement information processing and sensors.

Specific Expertise

Sensors for flow measurements, uncertainty and quality of sensors and measurements.
Measurement information processing and modelling.

Collaboration Offered

Joint research work, further education.

Collaboration Sought

Joint research, financing, contacts.

027 Vaisala Oy

PL 26
FIN-00421 Helsinki

Tel: +358 9 89491
Fax: +358 9 8919227
E-mail:

Affiliation:

Contact: P Detolen, President
A Sudmailailen, Sales Mgr

Organisation information

No of employees >100
Turnover (ECU*) >5m

Sensors technology activities information

No of employees 26-100
Turnover (ECU*) >5m

*1ECU=DM 1.89,FFr 6.30,£0.78,Lira 1909,Peseta 159.1

Organisation activities

Research and Development
Manufacturer of sensor-based products
Testing, consulting, design
Manufacture of finished sensors

Application \ Signal Sensed	Radiation	Optical	Electrical	Magnetic	Thermal	Chemical	Mechanical
Automotive					•	•	•
Consumer products					•	•	•
Process industries, mining, energy					•	•	•
Manufacturing, machinery					•	•	•
Medicine, health care							
HVAC, fire control, security					•	•	•
Military, aerospace					•	•	•
Telecommunications,office equip.					•	•	•
Meteorology, environment					•	•	•
Other *							

Main Activities

Vaisala develops and manufactures environmental measurement systems and instruments. applications areas include: meterology, environmental sciences, road and aviation traffic, and industrial measurements. The Company was established in 1936. The company's customers are meteorological services, research institutes, defence forces, road and airport authorities, and industry in almost 100 countries. Vaisala also develops and manufactures measurement solutions for meteorology, the environmental sciences and industry: meterological systems - including atmospheric measurement systems for weather research and forecasting; traffic safety systems - including complete road environmental measurement systems; sensor systems - including integrated measurement systems for industry, research, telecommunications and energy management.

Specific Expertise

Weather observation systems; radiosondes advanced weather warning systems (for air and road traffic); industrial humidity measurement; automotive sensors - sensor systems division: sensors; instruments for measurement of physical parameters in gases and liquids; relative humidity; temperature; dew point and pressure - special technologies: micromechanical, thin film, ceramic sensors; micromachining of silicon; digital signal processing; laser optics; laser and electronics beam welding; computer aided design, manufacture and testing.

Collaboration Offered

As applicable, resources of a globally operating company.

Collaboration Sought

Unique expertise in some speciality areas.

028 Valmet Automation Kavanni Ltd

PL 177
FIN-87101 Kajaani

Tel: +358 86 32 11
Fax: +358 86 32 12 15
E-mail: valmet.com

Affiliation: Valmet Ltd
Contact: N Farrand, Mktg Manager
P Oikarinen, Mktg Assistant

Organisation information
No of employees 101-1000
Turnover (ECU*) >50m

Sensors technology activities information
No of employees >100
Turnover (ECU*)
*1ECU=DM 1.89,FFr 6.30,£0.78,Lira 1909,Peseta 159.1

Organisation activities
Research and Development
Manufacturer of sensor-based products
Testing, consulting, design
Manufacture of finished sensors

Application \ Signal Sensed	Radiation	Optical	Electrical	Magnetic	Thermal	Chemical	Mechanical
Automotive							
Consumer products							
Process industries, mining, energy	•	•	•		•	•	•
Manufacturing, machinery							
Medicine, health care							
HVAC, fire control, security							
Military, aerospace							
Telecommunications,office equip.							
Meteorology, environment							
Other *							

Main Activities

We manufacture sensors and measurments applications for pulp and paper industry.

Specific Expertise

Valmet Automation Kavaani Ltd provides solutions for better process managememt throught the innovative application of measurement and control technology : shear/ore, optical, microwave consistency transmitters; pressure, level and temperature transmitters; digital contractors.

029 Vision Systems Limited

PO Box 353
FIN-40101 Jyvaskyla

Tel: +358 14 617 722
Fax: +358 14 617 844
E-mail:

Affiliation:
Contact: Mr K Rahkola, MD

Organisation information
No of employees 1-25
Turnover (ECU*) 1m-5m
Sensor technology activities information
No of employees 1-10
Turnover (ECU*) <0.5m
*1ECU=DM 1.89,FFr 6.30,£0.78,Lira 1909,Peseta 159.1

Organisation activities
Testing, consulting, design
Manufacture of finished sensors

Application \ Signal Sensed	Radiation	Optical	Electrical	Magnetic	Thermal	Chemical	Mechanical
Automotive							
Consumer products							
Process industries,mining energy		•					
Manufacturing, machinery		•					
Medicine, health care							
HVAC, fire control, security							
Military, aerospace							
Telecommunications,office equip.							
Meteorology environment							
Other							

Main Activities

Manufacture of machine vision systems and devices, technical consulting, domestic distribution and import and export of industrial measurement and quality control equipment.

Specific Expertise

Machine vision based sensors, measurement and quality control systems for industrial automation (paper, steel, wood, glass, plastic). CCD-camera based on-line dimension measurement and control devices.

030 VTT Electronics

P O Box 11012
FIN-02044 VTT

Tel: +358 94 561
Fax: +358 94 52 25 93
E-mail: ari.lehto@vtt.fi

Affiliation:

Contact: A Lehto, Sensor Mgr
T Wiik, IC Processing

Organisation information

No of employees 101-1000
Turnover (ECU*) 5m-50m

Sensors technology activities information

No of employees 11-25
Turnover (ECU*) 1m-5m

*1ECU=DM 1.89,FFr 6.30,£0.78,Lira 1909,Peseta 159.1

Organisation activities

Research and Development
Testing, consulting, design
Manufacture of finished sensors

Application \ Signal Sensed	Radiation	Optical	Electrical	Magnetic	Thermal	Chemical	Mechanical
Automotive		•			•		•
Consumer products							•
Process industries, mining, energy							
Manufacturing, machinery							
Medicine, health care							
HVAC, fire control, security		•			•		•
Military, aerospace		•					
Telecommunications,office equip.		•					
Meteorology, environment	•	•					
Other *							

Main Activities

Silicon sensors and microsystems. Integrated optics, packaging.

Specific Expertise

Silicon capacitive pressure sensor, elctrically tuneable Fabry-Perot interferometers, electrically modulatable IR-Sources, bolometers, thermopiles, phodiodes, integrated circuits.

Collaboration Offered

Sensor and microsystems desighn and fabrication (phototypes).

Collaboration Sought

Laboratory and field testing of the prototypes.

031 VTT - Valtion Teknillinen Tutkimuskeskus

Technical Research Centre of Finland
Medical Engineering Laboratory
Kanslerinkatu 8 Hervanta, PO Box 316
SF-33101 Tampere

Tel: +358 3 3163330
Fax: +358 3 3163319
E-mail:

Affiliation:
Contact: J Lekkala
H Joki

Organisation information
No of employees 26-100
Turnover (ECU*) >5m

Sensors technology activities information
No of employees 11-25
Turnover (ECU*) 1m-5m
*1ECU=DM 1.89,FFr 6.30,£0.78,Lira 1909,Peseta 159.1

Organisation activities
Research and Development
Testing, consulting, design

Application \ Signal Sensed	Radiation	Optical	Electrical	Magnetic	Thermal	Chemical	Mechanical
Automotive				●		●	●
Consumer products							
Process industries, mining, energy	●						●
Manufacturing, machinery							
Medicine, health care				●		●	●
HVAC, fire control, security					●		
Military, aerospace	●						
Telecommunications,office equip.							
Meteorology, environment	●						
Other *							

Main Activities

The Medical Engineering Laboratory belongs to the Division of Information Technology in VTT's organisation. The Laboratory employs a full-time staff of 65 persons of which approximately 50 have an academic degree.
The research and development and testing activities of the Laboratory are carried out in the following sections: analyser technology; clinical engineering services; medical information systems; medical instrumentation; technology for independent living.

Specific Expertise

The Medical Engineering Laboratory has been active in biosensor research for several years. The focus of the research is in rapid diagnostics and instrumentation especially in biosensor technology used in vitro in biomedicine and immunology.
However, during the recent years a more common biosensor has been approached, and also other promising application areas, like environmental and process monitoring have been probed.
At the moment biosensor research goals comprise the developments of an immunosensor based on direct measurement (without labels) and a gas sensor. The main interest is in immobilisation of biomolecules and Langmuir-Blodgett monomolecular films. The basic detection methods employed so far are optical technologies, especially Surface Plasmon Resonance (SPR) phenomenon, and utilisation of mass sensitive resonator structures (bulk acoustic wave crystals, surface acoustic wave sensors and a plate mode sensor) as microbalance.

Collaboration Offered

Know-how in optical detection methods and Langmuire-Blodgett films.

Collaboration Sought

Immobilisation of immunoproteins surface treatments and protein -28 films.

France

Entries 032 to 083

032 Auxitrol

1 rue d'Anjou
BP 241
F-92603 Asnière sur Seine

Tel: +33 1 47 906 281
Fax: +33 1 40 862 646
E-mail:

Affiliation: Esterline
Contact: Fabrice Lemarquis, Aerospace Equipment Division Manager

Organisation information
No of employees
Turnover (ECU*)
Sensor technology activities information
No of employees
Turnover (ECU*)
*1ECU=DM 1.89,FFr 6.30,£0.78,Lira 1909,Peseta 159.1

Organisation activities
Research and development.
Testing, consulting, design.
Manufacture of systems used in the manufacture, testing or development of sensors.
Manufacture of finished sensors.

Application \ Signal Sensed	Radiation	Optical	Electrical	Magnetic	Thermal	Chemical	Mechanical
Automotive					•		
Consumer products							
Process industries,mining energy	•				•		•
Manufacturing, machinery							
Medicine, health care							
HVAC, fire control, security							•
Military, aerospace	•				•		•
Telecommunications,office equip.							
Meteorology, environment							
Other *			•				

* Nuclear

Main Activities

Auxitrol is a manufacturer and supplier of a wide variety of components for the measurement, control and switching techniques applied in instrumentation and automation.

Auxitrol's Aerospace Division activities include:
Design,qualification and manufacture of all temperature sensors for space, aerospace and defence applications to individual specification - product supports, spares, A.O.G.

033 Besançon Instruments

7 Rue Denis Papin
BP 1117
F-25002 Besançon Cedex

Tel: +33 3 81 80 40 50
Fax: +33 3 81 88 88 17
E-mail:

Affiliation: Legris Industries
Contact: V Boscampomi, Managing Director

Organisation information
No of employees 11-25
Turnover (ECU*) 1-5m

Sensor technology activities information
No of employees 11-25
Turnover (ECU*) 1-5m
*1ECU=DM 1.89,FFr 6.30,£0.78,Lira 1909,Peseta 159.1

Organisation activities
Testing, consulting, design.
Manufacture of systems used in the manufacture, testing or development of sensors.

Application \ Signal Sensed	Radiation	Optical	Electrical	Magnetic	Thermal	Chemical	Mechanical
Automotive							
Consumer products							
Process industries,mining energy						•	•
Manufacturing, machinery							
Medicine, health care							
HVAC, fire control, security							
Military, aerospace						•	•
Telecommunications,office equip.							
Meteorology, environment							
Other *							

Main Activities

Besançon Instruments manufactures and sells test,control and calibration equipment.

Specific Expertise

The company manufactures and sells calibration equipment for process isntruments for checking and calibrating: pressure gauges; switches; thermometers; thermostats; transmitters; regulators; converters; positioners; recorders and sensors.

Collaboration Offered

Know-how in special electro-pneumatic calibrators on request.

Collaboration Sought

The company requires products to complete its range.

034 Biomerieux SA

ENS Lyon
UMR 103
46 Allée d'Italie
F-69384 Lyon Cedex 7

Tel: +33 4 72 72 83 59
Fax: +33 4 72 72 85 33
E-mail:

Affiliation:
Contact: B Mandrand, Director of Laboratory

Organisation information
No of employees >100
Turnover (ECU*) >5m

Sensor technology activities information
No of employees 1-10
Turnover (ECU*) -
*1ECU=DM 1.89,FFr 6.30,£0.78,Lira 1909,Peseta 159.1

Organisation activities
Research and development.

Application \ Signal Sensed	Radiation	Optical	Electrical	Magnetic	Thermal	Chemical	Mechanical
Automotive							
Consumer products							
Process industries,mining energy							
Manufacturing, machinery							
Medicine, health care						•	
HVAC, fire control, security							
Military, aerospace							
Telecommunications,office equip.							
Meteorology, environment							
Other *							

Main Activities

Reactives.

Specific Expertise

Immunosensor (at research level) - detection of antigens on semiconductors..

Collaboration Sought

Biosensors for chemicals, enzymes and antigens.

036 C4i - Centre de Compétence

et de Conception de Circuits Intégrés
French Geneva Campus
F-74166 Archamps

Tel: +33 4 50 31 57 20
Fax: +33 4 50 31 57 21
E-mail: c4i@c4imail.in2p3.fr

Affiliation: CNRS
Contact: J-M Thénard, Director
J Martin

Organisation information
No of employees 1-25
Turnover (ECU*) <1m

Sensors technology activities information
No of employees 1-10
Turnover (ECU*) <0.5m
*1ECU=DM 1.89,FFr 6.30,£0.78,Lira 1909,Peseta 159.1

Organisation activities
Research and Development
Testing, consulting, design

Application \ Signal Sensed	Radiation	Optical	Electrical	Magnetic	Thermal	Chemical	Mechanical
Automotive			●	●		●	
Consumer products			●	●		●	
Process industries, mining, energy							
Manufacturing, machinery							
Medicine, health care	●						
HVAC, fire control, security							
Military, aerospace							
Telecommunications,office equip.							
Meteorology, environment						●	
Other *							

Main Activities

Design of sensors and integrated circuits.

Specific Expertise

Hall effect and chemical devices (collaboration with Ecole Centrale de Lyon - ECL)

Collaboration Sought

R & D or mico systems devices.

037 CEA- Commissariat à l'Energie Atomique

LETI - Laboiratoire d'Electronique de Technologie et d'instrumentation
17 Rue des Martyres
F-38054 Grenoble Cedex 09

Tel: +33 4 76 88 30 78
Fax: +33 4 76 88 51 05
E-mail:

Affiliation:
Contact: G Delapierre, Mechanical, Chemical & Magnetic Sensors
C Lucas, Infrared Sensors
G Thomas, Nuclear & X-Ray Sensors
D Duret, Magnetometry

Organisation information
No of employees > 100
Turnover (ECU*) > 5m

Sensor technology activities information
No of employees > 100
Turnover (ECU*) > 5m
*1ECU=DM 1.89,FFr 6.30,£0.78,Lira 1909,Peseta 159.1

Organisation activities
Research and development.

Application \ Signal Sensed	Radiation	Optical	Electrical	Magnetic	Thermal	Chemical	Mechanical
Automotive	•			•			•
Consumer products				•			•
Process industries,mining energy	•			•		•	•
Manufacturing, machinery	•						•
Medicine, health care	•			•			
HVAC, fire control, security						•	
Military, aerospace	•			•			•
Telecommunications,office equip.				•		•	
Meteorology, environment							
Other *							

Main Activities

LETI is one of the divisions of the French Atomic Energy Commission. It is a public applied research laboratory working on electronics and microelectronic technologies. Research and development projects are defined according to internal initiatives or external demands. The transfer methods and licensing rules are adapted to specifications of each partner and of each project. The projects are generally co-financed by industry and CEA. Public authorities often participate.

Specific Expertise

Main technologies now available in LETI: thin film and micromachining for mechanical, chemical and magnetic sensors; silicon integrated electronics; silicon integrated optics; visible and infrared detection; nuclear detection; X-ray imaging; fast integrated circuits; magnetometry: total field or directional magnetometers using resonance (NMR, EPR) or superconductors.

Collaboration Offered

Mechanicalm chemical and magnetic micromachined sensors. Infrared sensors. Nuclear and X-ray sensors. Magnetometry.

038 Cemagref

Groupement de Montpellier
361 rue J-F Breton, B P 5095
F-34033 Montpellier Cedex 1

Tel: +33 5 46 76 35 795
Fax: +33 5 46 76 35 795
E-mail: veronique.bellon@cemapref.fr

Affiliation:
Contact: V Bellon-Maurel

Organisation information
No of employees 101-1000
Turnover (ECU*) 5m-50m
Sensors technology activities information
No of employees
Turnover (ECU*) 0.5m-1m
*1ECU=DM 1.89,FFr 6.30,£0.78,Lira 1909,Peseta 159.1

Organisation activities
Research and Development

Application \ Signal Sensed	Radiation	Optical	Electrical	Magnetic	Thermal	Chemical	Mechanical
Automotive							
Consumer products							
Process industries, mining, energy							
Manufacturing, machinery							
Medicine, health care							
HVAC, fire control, security							
Military, aerospace							
Telecommunications,office equip.							
Meteorology, environment	•	•	•				•
Other *							

Main Activities

Cemagref is involved in research in Agriculture, environment and food industry engineering:
Food and farm equipment, water resource management, land management, equipment for water and environment.

Specific Expertise

- Near infrared spectroscopy
- Infrared spectroscopy
- Vision systems
- Information fusion
- Texture sensors (mechanical)

039 Chauvin Arnoux

190 Rue Championnet
F-75876 Paris Cedex 18

Tel: +33 1 44 85 44 85
Fax: +33 1 46 27 73 89
E-mail:

Affiliation:
Contact: Mr. Arnoux, Managing Director

Organisation information
No of employees > 100
Turnover (ECU*) > 5m

Sensor technology activities information
No of employees 26-100
Turnover (ECU*) 1-5m
*1ECU=DM 1.89,FFr 6.30,£0.78,Lira 1909,Peseta 159.1

Organisation activities
Research and development.
Manufacture of finished sensors.
Manufacture of sensor-based products.

Application \ Signal Sensed	Radiation	Optical	Electrical	Magnetic	Thermal	Chemical	Mechanical
Automotive			•				•
Consumer products			•				
Process industries,mining energy			•		•		•
Manufacturing, machinery			•		•		•
Medicine, health care							•
HVAC, fire control, security					•		
Military, aerospace			•		•		•
Telecommunications,office equip.							
Meteorology, environment			•		•		•
Other *							

Main Activities

Chauvin Arnoux is a French manufacturer of measuremnt instruments. Its main activities include:
measurement;
control and automation;
current clamps;
multimeters;
sensors;
panel meters;
temperature controllers;
recorders; relays etc.

Sales and manufacturing are supported by five subsidiaries in USA and Europe and by more than 50 agents throughout the world.

040 CiTy sensors SA

72 Rue Bernard Iske
F-92350 Le Plessis Robinson

Tel: +33 1 46 30 37 23
Fax: +33 1 46 32 51 82
E-mail:

Affiliation:
Contact: Jean-Michel Darmon, President

Organisation information
No of employees 1-10
Turnover (ECU*) <= 0.5m
Sensor technology activities information
No of employees 1-10
Turnover (ECU*) <=0.5m
*1ECU=DM 1.89,FFr 6.30,£0.78,Lira 1909,Peseta 159.1

Organisation activities
Manufacture of finished sensors.

Application \ Signal Sensed	Radiation	Optical	Electrical	Magnetic	Thermal	Chemical	Mechanical
Automotive							
Consumer products							•
Process industries,mining energy							
Manufacturing, machinery							•
Medicine, health care							•
HVAC, fire control, security							
Military, aerospace							•
Telecommunications,office equip.							
Meteorology, environment							
Other *							•

* OEM

Main Activities

CiTy sensors designs, manufactures and sells thick film strain gauges, ceramic pressure, strain and force sensors.

Specific Expertise

Specific features of ceramic sensors are: excellent mechanical properties (no hysteresis); agressive media compatibility; long term stability.

Main applications: hydraulic systems; compressors and pumps; medical equipments; agricultural spraying systems; ink yet systems; military applications; aircraft equipment.

Collaboration Offered

Possibility to design specific sensors for mid or large volume OEM equipment.

Collaboration Sought

Co-operation in manufacture (requires typical thick film, hybrid deposition equipment).

041 COFRETH

BP 4601
F-95746 Nanterre Cedex

Tel: +33 1 41 20 10 00
Fax: +33 1 41 20 10 10
E-mail:

Affiliation: Subsidiary of Lyonnaise des Eaux
Contact: Simon Seys, R&D Director

Organisation information
No of employees 1-10
Turnover (ECU*) 1-5m
Sensor technology activities information
No of employees 1-10
Turnover (ECU*) 0.5-1m
*1ECU=DM 1.89,FFr 6.30,£0.78,Lira 1909,Peseta 159.1

Organisation activities
Research and development.

Application \ Signal Sensed	Radiation	Optical	Electrical	Magnetic	Thermal	Chemical	Mechanical
Automotive							
Consumer products							
Process industries,mining energy			•		•		•
Manufacturing, machinery							
Medicine, health care							
HVAC, fire control, security			•		•		•
Military, aerospace							
Telecommunications,office equip.							
Meteorology, environment			•			•	
Other *							

Main Activities

Scientific and technical research in the field of energy. Five themes are developed: automation; sensors and instrumentation; home automation systems; environment; artificial intelligence and expert systems.

Specific Expertise

Combustion sensor: principle involved - semiconductor; ion sensors: principle involved - ISFET (ion sensitive field effect transistor).

Collaboration Offered

All products and services.

Collaboration Sought

ISFET (ion sensitive field effect transistor).

042 Coreci

2-4 rue J Despartment
F-65008 Lyon
France

Tel: +33 4 78 77 03 00
Fax: +33 4 78 77 03 20
E-mail:

Affiliation: Groupe Coreci
Contact: S Kaplan, Maktg Director
R Valncogne, R&D Director

Organisation information
No of employees 101-1000
Turnover (ECU*) 5m-50m

Sensors technology activities information
No of employees 11-25
Turnover (ECU*) 1m-5m
*1ECU=DM 1.89,FFr 6.30,£0.78,Lira 1909,Peseta 159.1

Organisation activities
Manufacturer of sensor-based products
Manufacture of finished sensors

Application \ Signal Sensed	Radiation	Optical	Electrical	Magnetic	Thermal	Chemical	Mechanical
Automotive					•		
Consumer products							
Process industries, mining, energy					•	•	•
Manufacturing, machinery					•		
Medicine, health care							
HVAC, fire control, security						•	
Military, aerospace							
Telecommunications,office equip.							
Meteorology, environment					•		•
Other *							

Main Activities

Process control bistrament manufacturing:

- Sensors: Temperature, humidity, gas, premiere
- Transmitters:
- Temperature and process controllers.
- Monitoring and recording units.

Specific Expertise

- Process and control instruments : general applications
- HIAC, food and beverage : special applications for rlative humidity sensors (capacitive sensor).

Collaboration Sought

Gas sensoring.

043 CSO Mesure

70 rue des Martyrs
F-38000 Grenoble

Tel: +33 4 76 46 85 88
Fax: +33 4 76 46 84 56
E-mail:

Affiliation:
Contact: Lizet Jacques, Sales Manager

Organisation information
No of employees 1-25
Turnover (ECU*) <1m
Sensor technology activities information
No of employees 1-10
Turnover (ECU*)
*1ECU=DM 1.89,FFr 6.30,£0.78,Lira 1909,Peseta 159.1

Organisation activities
Research and development on sensors sensor materials or applications
Manufacture of finished sensors

Application \ Signal Sensed	Radiation	Optical	Electrical	Magnetic	Thermal	Chemical	Mechanical
Automotive							
Consumer products							
Process industries,mining energy							
Manufacturing, machinery		•					
Medicine, health care							
HVAC, fire control, security							
Military, aerospace		•					
Telecommunications,office equip.							
Meteorology environment							
Other							

Main Activities

CSO Mesure develops, manufactures and sells optical and opto-electronic measuring systems using sensor technologies of modern optics such as integrated optics on silicon (CEA/LETI Licence) for non contact displacement, vibration, thickness measurement. The company studies engineering solutions of customised optical measuring systems, develops optical measuring devices from customers' technical specifications.

Specific Expertise

Micro Laser Interferometers using integrated optics on silicon. Confocal Laser micrometers using our micro laser interferometer to measure thickness of transparent materials. Wide-hand "Lamfolameter" calibrating laser sources across a wavelength range of 0.4μm to 1.8μm. Development of stabilised laser sources.

Collaboration Sought

Commercial collaboration and distribution in the USA.

044 CYLERGIE

Chemin de la Forrestière
F-69130 Ecully

Tel: + 33 4 72 86 09 80
Fax: + 33 4 78 43 30 59
E-mail:

Affiliation: Lyonnaise des Eaux
Contact: Archenault, Research

Organisation information
No of employees 1-25
Turnover (ECU*) 1m-5m

Sensor technology activities information
No of employees 1-10
Turnover (ECU*) <0.5m
*1ECU=DM 1.89,FFr 6.30,£0.78,Lira 1909,Peseta 159.1

Organisation activities
Research and development on sensors, sensor materials or applications.
Testing, consulting, design.

Application \ Signal Sensed	Radiation	Optical	Electrical	Magnetic	Thermal	Chemical	Mechanical
Automotive							
Consumer products							
Process industries,mining energy		•	•		•	•	
Manufacturing, machinery							
Medicine, health care							
HVAC, fire control, security							
Military, aerospace							
Telecommunications,office equip.							
Meteorology environment		•	•			•	
Other							

Main Activities

Research and development on sensors and their applications to water industry.
Testing and consulting for environmental sensors (water and gas).

Collaboration Offered

Know-how in the design and final tests of sensors to environmental survey.

Collaboration Sought

Research programme.

045 Desgranges et Huot

BP 125
56 Rue des Ecoles
F-93300 Aubervilliers

Tel: +33 1 48 33 91 02
Fax: +33 1 48 33 65 90
E-mail:

Affiliation: Instuments & Mesures Industrielles, France
Contact: Luc Dargent, Marketing Manager

Organisation information

No of employees	26-100
Turnover (ECU*)	> 5m

Sensor technology activities information

No of employees	26-100
Turnover (ECU*)	1-5m

*1ECU=DM 1.89,FFr 6.30,£0.78,Lira 1909,Peseta 159.1

Organisation activities
Research and development.
Testing, consulting, design.
Manufacture of systems used in the manufacture, testing or development of sensors.
Manufacture of finished sensors.

Application \ Signal Sensed	Radiation	Optical	Electrical	Magnetic	Thermal	Chemical	Mechanical
Automotive							•
Consumer products							
Process industries,mining energy							•
Manufacturing, machinery							•
Medicine, health care							
HVAC, fire control, security							•
Military, aerospace							•
Telecommunications,office equip.							
Meteorology, environment							
Other *							

Main Activities

Manufacturers of pressure standards: dead weight testers; transfer standards; calibrators; pressure generators; pressure controllers.

Specific Expertise

Development and manufacture of high accuracy pressure measuring, calibration and control devices.

046 Ecole Centrale de Lyon

Laboratoire de Physicochimie des Interfaces
BP 163
F-69131 Ecully Cedex

Tel: + 33 4 72 18 62 61
Fax: + 33 4 78 33 15 77
E-mail: capteurs@cc.ec-Lyon.fr

Affiliation:
Contact: N. Jaffrezic, CNRS Researcher
C. Martelet, Professor
J. M. Chovelon, Conf. Master

Organisation information
No of employees 26-100
Turnover (ECU*) <1m
Sensor technology activities information
No of employees 11-25
Turnover (ECU*) <0.5m
*1ECU=DM 1.89,FFr 6.30,£0.78,Lira 1909,Peseta 159.1

Organisation activities
Research and development on sensors sensor materials or applications.

Application \ Signal Sensed	Radiation	Optical	Electrical	Magnetic	Thermal	Chemical	Mechanical
Automotive							
Consumer products							
Process industries,mining energy							
Manufacturing, machinery							
Medicine, health care						•	
HVAC, fire control, security						•	
Military, aerospace							
Telecommunications,office equip.							
Meteorology environment						•	
Other							

Main Activities

Our laboratory is mainly focussed onto physico chemical treatments allowing to obtain functional surfaces on materials with specific properties. It includes grafting processes, self assembled layers, Langman -Blodgett techniques, sol gel processes, inclusion in polymeric layers, ionic implantation.

Specific Expertise

Our area of expertise includes obtaining specific surfaces or membranes for various transducers such as ISFET's, optical fibers, micro-interdigitated circuits, carbon fibres (as amperometric microelectrodes). We develop specific membranes for detection of alcaline ions, nitrate...., pesticides, immunospecies, urea, glucose lactate.

Collaboration Offered

ISFET design. Physicochemical treatments for elaboration of specific membranes for microsensors.

Collaboration Sought

Signal treatment, system intergration.

047 Ecole des Mines de St Etienne

SPIN-Sciences des Processus Industriel et Naturel
158, cours Fauriel
F-42023 St Etienne

Tel: + 33 4 77 42 0123
Fax: + 33 4 77 42 0000
E-mail:

Affiliation:
Contact: R Lalauze, Spin Centre Director
C Dijolat, Research Manager

Organisation information

No of employees 101-1000
Turnover (ECU*) 1m-5m

Sensor technology activities information

No of employees 11-25
Turnover (ECU*) <0.5m

*1ECU=DM 1.89,FFr 6.30,£0.78,Lira 1909,Peseta 159.1

Organisation activities

Research and development on sensors, sensor materials or application.
Testing, consulting, design.

Application \ Signal Sensed	Radiation	Optical	Electrical	Magnetic	Thermal	Chemical	Mechanical
Automotive						•	
Consumer products							
Process industries,mining energy						•	
Manufacturing, machinery							
Medicine, health care		•					
HVAC, fire control, security							
Military, aerospace		•				•	
Telecommunications,office equip.							
Meteorology environment		•	•			•	
Other (specify)							

Main Activities

Engineer's school with research centres
Ecole des Mines is a graduate engineering school. The school emphasises applied industrial research as a key factor in its factor involvement with industry. The main activities are: industrial and process engineering; environmental engineering; computer science engineering; materials science engineering.

Specific Expertise

Metal oxide chemical sensors, biochemical sensors, optical sensors (spectrometers).
Development of biosensors for monitoring the quality of water/air and controlling processes.

Collaboration Offered

Research and development, prototype construction, evaluation and testing, education and training.
Research and development on sensors (biosensors and chemical sensors); CEE partner for research contract in biomedical and environmental areas; testing, consulting and design of biosensors and chemical sensors.

Collaboration Sought

Research and development in complementary areas: microelectronics data treatment, manufacturers.
Industrial applications or industrial partner for research contracts; university partner for joint research.

048 EFFA

116 avenue du Belvédère
F-93310 Le Pré-Saint-Gervais

Tel: +33 1 48 46 87 34
Fax: +33 1 48 46 55 95
E-mail:

Affiliation: Druck Limited
Contact: Touzin, Commercial Manager

Organisation information
No of employees 1-25
Turnover (ECU*) 1m-5m

Sensor technology activities information
No of employees 1-10
Turnover (ECU*) 1m-5m
*1ECU=DM 1.89,FFr 6.30,£0.78,Lira 1909,Peseta 159.1

Organisation activities
Research and development of sensors, sensor materials or applications.
Testing, consulting, design.
Manufacture of finished sensors.

Application \ Signal Sensed	Radiation	Optical	Electrical	Magnetic	Thermal	Chemical	Mechanical
Automotive							•
Consumer products							•
Process industries,mining energy							•
Manufacturing, machinery							
Medicine, health care							•
HVAC, fire control, security							•
Military, aerospace							•
Telecommunications,office equip.							
Meteorology, environment							•
Other *							•

* Nucelar - Research, etc.

Main Activities

Since 1978, EFFA campany has specialised in manufacturing standard pressure sensors based on the use of a displacement sensor using eddy current principle (inductive, non-ferromagnetic).

Specific Expertise

EFFA is specialised in very low, absolute, guage and differential pressure measurement with a very high standard of precision and stability. It also manufactures high precision atmospheric pressure transmitters.
On demand, the company can make pressure sensors able to work from 1.8° K up to 350° C, high resolution specific weight meters (based on Archimede's principle).

049 ELSAG Bailey SA

100 Rue de Paris
F-91342 Massy Cedex

Tel: +33 1 64 47 20 21
Fax: +33 1 64 47 20 16
E-mail:

Affiliation:
Contact: Mr. Lemaire, Commercial Director
Mr. Baudet, Marketing Manager

Organisation information
No of employees
Turnover (ECU*)
Sensor technology activities information
No of employees
Turnover (ECU*)
*1ECU=DM 1.89,FFr 6.30,£0.78,Lira 1909,Peseta 159.1

Organisation activities
Research and development.
Testing, consulting, design.
Manufacture of finished sensors.
Manufacture of sensor-based products.

Application \ Signal Sensed	Radiation	Optical	Electrical	Magnetic	Thermal	Chemical	Mechanical
Automotive							
Consumer products					•		
Process industries,mining energy							•
Manufacturing, machinery							•
Medicine, health care							
HVAC, fire control, security							•
Military, aerospace							
Telecommunications,office equip.							
Meteorology, environment							•
Other *					•		•

* Nuclear

Main Activities

The company has long experience in production and marketing in the field of measurement and control. It is one of the few companies in the world to design, manufacture and sell all products and automatic systems employed to run industrial processes: transmitters; systems and instruments; control valves. The Transmitter Department develops and manufactures pressure, flow and liquid level transmitters based on a pressure measurement, as well as electromagnetic flowmeter units. These instruments are supplied to numerous fields of industry: chemical industries; water distribution and processing; gas distribution; sugar industries; food industries; pulp and paper steel plants. The company also supplies some special instruments to the nuclear industry and navy.

Specific Expertise

Signal sensed:- Pressure transmitter: a delta P across an orifice plate to measure flow; a delta P as the height of a liquid column; other delta Ps; a gauge or absolute line or tank pressure measurement.
Basic technology:- Delta P: inductive cell; gauge or absolute pressure sensors: piezoresistive; electromagnetic flow meter: Faraday's law.

Collaboration Offered

Pressure transmitters for flow, pressure and liquid level measurement. Electromagnetic flow meters. Special pressure or flow transmitters.

Collaboration Sought

Latest development in terms of 'smart generation transmitters'.

050 Engelhard Pyro Contrôle

244 Avenue Franklin Roosevelt
F-69120 Vaulx en Velin

Tel: +33 472 14 15 40
Fax: +33 4 72 14 15 41
E-mail:

Affiliation:
Contact: R Gruhel, New Product Mgr

Organisation information

No of employees 26-100
Turnover (ECU*) 5m-50m

Sensor technology activities information

No of employees 26-100
Turnover (ECU*) >5m
*1ECU=DM 1.89,FFr 6.30,£0.78,Lira 1909,Peseta 159.1

Organisation activities

Research and development on sensors, sensor materials or applications.
Testing, consulting, design.
Supply of raw materials or semi-finished products used to make sensors.
Manufacture of systems used in the manufacture, testing or development of sensors.
Manufacture of finished sensors.

Application \ Signal Sensed	Radiation	Optical	Electrical	Magnetic	Thermal	Chemical	Mechanical
Automotive					●		
Consumer products							
Process industries,mining energy					●		
Manufacturing, machinery					●		
Medicine, health care					●		
HVAC, fire control, security					●		
Military, aerospace					●		
Telecommunications,office equip.							
Meteorology, environment					●		
Other *							

Main Activities

Engelhard Pyro Contrôl is active in engineering, construction, calibration amd marketing of temperature sensors

Specific Expertise

Engineering, construction, calibration of: thermocouple assemblies (precious metals and base metals); PT 100 sensors; mineral insulated metal sheathed thermocouples; prototypes.

051 ENSPM - Ecole Nationale Superieure de Physique de Marseille

Laboratoire d'Optique des Surfaces et Couches Minces
Domaine Uni. de Jerome
F-13387 Marseille Cedex 13

Tel: + 33 4 91 28 84 96
Fax: + 33 4 91 28 80 67
E-mail:

Affiliation:
Contact: Claude Amra, Director of Laboratory

Organisation information
No of employees 11-25
Turnover (ECU*) 0.5-1m

Sensor technology activities information
No of employees 1-10
Turnover (ECU*) <=0.5m
*1ECU=DM 1.89,FFr 6.30,£0.78,Lira 1909,Peseta 159.1

Organisation activities
Research and development.
Testing, consulting, design.

Application \ Signal Sensed	Radiation	Optical	Electrical	Magnetic	Thermal	Chemical	Mechanical
Automotive					•		
Consumer products					•		
Process industries,mining energy						•	•
Manufacturing, machinery							
Medicine, health care					•		
HVAC, fire control, security			•		•		
Military, aerospace							
Telecommunications,office equip.				•	•	•	
Meteorology, environment						•	
Other *							

Main Activities

The Ecole Superieure de Physique is one of the "Grandes Ecoles" in France and the main objective is providing education. A part of the Optical Laboratory is devoted to basic and industrial research on optical fibre sensors.

Specific Expertise

The main area of work is in optical fibre sensors. Specific sensors: on/off temperature sensors for security or medicine; chemical sensors CH_4, CO_2, etc. Sensor networks: coherence multiplexing, miniature several mode sensor; bi-refringent sensors.

Collaboration Offered

Consultancy. Research on new sensors. Development of new sensors.

Collaboration Sought

All optical fibre sensors.

052 Entran Sarl

26 Rue des Dames
F-78340 Les Claye-Sous Bois

Tel: +33 1 30 79 33 00
Fax: +33 1 34 81 03 59
E-mail:

Affiliation:
Contact: Jean Claude Provost, General Manager
Denis Delemazure, Export Sales Manager

Organisation information

No of employees 26-100
Turnover (ECU*) 1-5m

Sensor technology activities information

No of employees -
Turnover (ECU*) -
*1ECU=DM 1.89,FFr 6.30,£0.78,Lira 1909,Peseta 159.1

Organisation activities

Research and development.
Testing, consulting, design.
Manufacture of systens used in the manufacture, testing or development of sensors.
Manufacture of finished sensors.

Application \ Signal Sensed	Radiation	Optical	Electrical	Magnetic	Thermal	Chemical	Mechanical
Automotive			●		●		●
Consumer products							
Process industries,mining energy							
Manufacturing, machinery			●		●		●
Medicine, health care							●
HVAC, fire control, security							●
Military, aerospace			●				●
Telecommunications,office equip.							
Meteorology, environment							●
Other *							

Main Activities

Founded in 1970 by a team of engineers, Entran specialised in the static and dynamic measurement of physical parameters through the use of advanced high technology. Focusing on miniature and compact instrumentation, Entran has spent more than a decade in optimising microminiaturisation techniques and applying the latest semiconductor and hybrid circuit technologies to the measurement of acceleration, force and pressure in a multitude of environments. From sophisticated transducers the size of a pinhead to instruments which can survive a 50.000g pyrotechnic shock, Entran's instrumentation has taken an integral part in the development of high technology programmes around the world.

Specific Expertise

Missiles and Weaponry: design, development and control testing the Polaris, Poseidonm Trident, Cruise and MX missile systems. Aerospace: from model to flight test on 727, 747SP, 757, 767, F111, F15, F16, F18 and B-1 programmes. Design, model, full scale testing and flight of Space Shuttle, Intelsat V, Space lab and deep space probes and the Ariane. Automotive and Farm Machinery: crash testing, dummy instrumentation, track and road monitoring, scales model design and hydraulic system analysis. Biomedical: hospical, laboratory and field studies of animal and human response dynamics, patient monitoring for pressure, tremble and trauma, in sity blood pressure measurements, medical implants and depth/motion studies of fish and sea mammals. Nuclear: fast breeder reactor fuel-pin and control rod flow induced vibration and pressure studies. Nuclear reactor vibration monitoring. Reactor model studies and high energy release experiments. Pressure measurements in blast test programmes.

Collaboration Offered

New technologies.

053 ERI

Laboratoire d'Energetique et Réactivté aux Interfaces
Boîte 39 4 place Jussieu
F-75262 Paris Cedex

Tel: +33 3 44 27 36 76
Fax: +33 3 44 27 30 35
E-mail:

Affiliation: Université Paris VI
Contact: Prof P Letellier
Dr M Turmine

Organisation information
No of employees 1-25
Turnover (ECU*) <1m
Sensor technology activities information
No of employees 1-10
Turnover (ECU*) <0.5m
*1ECU=DM 1.89,FFr 6.30,£0.78,Lira 1909,Peseta 159.1

Organisation activities
Research and development on sensors, sensor materials or applications
Testing, consulting, design

Application \ Signal Sensed	Radiation	Optical	Electrical	Magnetic	Thermal	Chemical	Mechanical
Automotive							
Consumer products							
Process industries,mining energy							
Manufacturing, machinery							
Medicine, health care						•	
HVAC, fire control, security							
Military, aerospace							
Telecommunications,office equip.							
Meteorology environment						•	
Other							

Main Activities

Laboratoire d'Energétique et Réactivité aux Interfaces is involved in the perfecting of solid ISE working in water and hydro-organic media, and the miniaturisation of these solid ISE's.

Specific Expertise

- potentiometric titrations of anionic, cationic and non-ionic amphiphiles
- preparation of miniature electrodes

Collaboration Offered

We offer our technology and the knowledge of surfactant solutions in water and hydro-organic mixtures, concentrated salt solutions.

Collaboration Sought

Applications to industrial, medicinal problems.

054 ESPCI - CNRS

VBA 429
10 rue Vauquelin
F-75005 Paris

Tel: +33 1 40 79 46 30
Fax: +33 1 40 79 44 25
E-mail: thami@cis.espci.fr

Affiliation:
Contact: Thami Thierry

Organisation information
No of employees 1-25
Turnover (ECU*) <1m
Sensor technology activities information
No of employees 1-10
Turnover (ECU*) <0.5m
*1ECU=DM 1.89,FFr 6.30,£0.78,Lira 1909,Peseta 159.1

Organisation activities
Research and development on sensors sensor materials or publications

Application \ Signal Sensed	Radiation	Optical	Electrical	Magnetic	Thermal	Chemical	Mechanical
Automotive							
Consumer products							
Process industries,mining energy							
Manufacturing, machinery							
Medicine, health care							
HVAC, fire control, security							
Military, aerospace							
Telecommunications,office equip.							
Meteorology environment						•	
Other							

Main Activities

The activities of our laboratory concern the synthesis of molecular materials and the study of their corresponding condensed phases (monocrystals, thin films, liquid crystals). The physical properties are studied in collaboration in the field of electronics, opto-electronics and ion electronics (sensors).

Specific Expertise

We are developing specific molecular materials for ISFET devices. Our expertise concerns the field of the chemical sensors.

055 FGP Instrumentation

24 rue des Dames
F-78340 Les Clayes sous Bois

Tel: +33 1 30 79 65 40
Fax: +33 1 30 54 01 43
E-mail:

Affiliation:
Contact: Sean Gough, Export Manager

Organisation information
No of employees 26-100
Turnover (ECU*) 1m-5m
Sensor technology activities information
No of employees 26-100
Turnover (ECU*) 0.5m-1m
*1ECU=DM 1.89,FFr 6.30,£0.78,Lira 1909,Peseta 159.1

Organisation activities
Manufacture of finished sensors
Manufacture of sensor-based products

Application \ Signal Sensed	Radiation	Optical	Electrical	Magnetic	Thermal	Chemical	Mechanical
Automotive			•				•
Consumer products							
Process industries,mining energy							
Manufacturing, machinery			•				•
Medicine, health care							
HVAC, fire control, security							
Military, aerospace			•				•
Telecommunications,office equip.							
Meteorology environment							
Other (specify)							

Main Activities

Manufacturer of sensors for force torque pressure and acceleration measurements. FGP Instrumentation also supplies systems and measurement chains. We manufacture our own electronics for the sensors.

Specific Expertise

Sensor design (strain gauge based); custom transducers (30% of our business); automotive sensors; aerospace sensors.

Collaboration Offered

Well established French transducer company (20 years); efficient technical aftersales service for companies wishing to have an agent in France/Europe; 20 years' experience in design of sensors.

Collaboration Sought

Commercial cooperation in European countries; technical cooperation for joint development of new products.

056 Fogale Nanotech

190 Parc G Besse
F-30 035 Nimes Cedex

Tel: +33 466 38 14 85
Fax: +33 466 38 14 59
E-mail:

Affiliation:
Contact: J Leteurtre, Manager
D Roziere, Commercial

Organisation information

No of employees 1-25
Turnover (ECU*) 1m-5m

Sensor technology activities information

No of employees 11-25
Turnover (ECU*) 0.5m-1m

*1ECU=DM 1.89,FFr 6.30,£0.78,Lira 1909,Peseta 159.1

Organisation activities

Research and development on sensors sensor materials or applications
Manufacture of systems used in the manufacture, testing or development of sensors
Manufacture of finished sensors
Manufacture of sensor-based products

Application \ Signal Sensed	Radiation	Optical	Electrical	Magnetic	Thermal	Chemical	Mechanical
Automotive		•	•				
Consumer products							
Process industries,mining energy		•	•				
Manufacturing, machinery							
Medicine, health care							
HVAC, fire control, security							
Military, aerospace		•	•				
Telecommunications,office equip.		•	•				
Meteorology environment							
Other							

Main Activities

Fogale Nanotech, a company created in 1983 by researchers, is now an authority in the field of fast and precise measurement.
Industrial and scientific mastering of contactless dimensional measurement systems.

Specific Expertise

Our capability is now proven in the most demanding fields of industry and laboratory work (automobile, aeronautics, armaments, nuclear and heavy industry, synchrotron, etc.) in France, Europe as well as in the USA and Japan.

057 GIP Ultrasons

Faculté de Mèdecine LBM
2 Bis Boulevard Tonnellé
Bâtiment Vialle BP 3223
F-37032 Tours Cedex

Tel: +33 2 47 36 62 21
Fax: +33 2 47 36 61 20
E-mail: tessier@univ.tours.fr

Affiliation:
Contact: L Tessier
G Feuillard

Organisation information
No of employees 1-25
Turnover (ECU*) 1m-5m

Sensors technology activities information
No of employees 1-10
Turnover (ECU*) <0.5m
*1ECU=DM 1.89,FFr 6.30,£0.78,Lira 1909,Peseta 159.1

Organisation activities
Research and Development

Application \ Signal Sensed	Radiation	Optical	Electrical	Magnetic	Thermal	Chemical	Mechanical
Automotive							
Consumer products							
Process industries, mining, energy							
Manufacturing, machinery							
Medicine, health care							
HVAC, fire control, security							
Military, aerospace							
Telecommunications,office equip.							
Meteorology, environment							
Other *							

Main Activities

Our group gathers private and public interest for the development of ULTRASONIC INSTRUMENTATION. Our main interests are focussed on:

- Ultrasonic transducers technology
- High frequency imaging (dermotology, NDT.....)
- Flow-metering (Doppler and trausct-time)
- Specific acoustic sensors (chemistry, biochemistry)

Specific Expertise

- Piezoelectric materials : Ceramics, composites Characterization and simulation.
- Integrated probes for ultrasonic application - Radiation Field and electracoustic response calculation. Prototypes construction.
- Functionalized sensors for chemistry and biochemistry (Quartz crystal microbalance, SAW sensors)

Collaboration Offered

- Contractual research (study and/or prototypes developted)
- Partnership in industrial development
- Participation in European projects (already 3 running)

Collaboration Sought

- Know-how in microtechnologies
- Mastering in application field and market.

058 Groupe ESIEE

Cité Descartes
BP 99
2BD Blaise-Pascal
F-93162 Noisy-le-Grand Cedex

Tel: +33 1 45 92 65 00
Fax: +33 1 45 92 66 99
E-mail: demnardf@esiee.fr

Affiliation: CCIP
Contact: F Demnard,
P Bildstein, Lab LSOL Manager
F Baillieu, C2M2 Manager
D Perrin, Research Manager

Organisation information
No of employees 101-1000
Turnover (ECU*)
Sensor technology activities information
No of employees 11-25
Turnover (ECU*)
*1ECU=DM 1.89,FFr 6.30,£0.78,Lira 1909,Peseta 159.1

Organisation activities
Research and development on sensors, sensor materials or applications.
Testing, consulting, design.

Application \ Signal Sensed	Radiation	Optical	Electrical	Magnetic	Thermal	Chemical	Mechanical
Automotive			•				
Consumer products			•				
Process industries,mining energy			•				
Manufacturing, machinery			•				
Medicine, health care			•				
HVAC, fire control, security			•				
Military, aerospace			•				
Telecommunications,office equip.			•				
Meteorology, environment							
Other *							

Main Activities

Teaching and research in electronic engineering.

Specific Expertise

Microelectronics: silicon on silicon bonding, sensors based on silicon microstructures (accelerometers, magnetic field sensors, microphone type sensors, acoustic gyrometers...), micromotors and micromotion systems, integrated circuit design.

059 HEITO SA

5 Rue Gramme
F-75-15 Paris

Tel: +33 1 48 28 37 02
Fax: +33 1 48 28 34 69
E-mail:

Affiliation:
Contact: Jean-Pierre Heitzmann, Managing Director

Organisation information
No of employees -
Turnover (ECU*) -
Sensor technology activities information
No of employees -
Turnover (ECU*) -
*1ECU=DM 1.89,FFr 6.30,£0.78,Lira 1909,Peseta 159.1

Organisation activities
Manufacture of finished sensors.

Application \ Signal Sensed	Radiation	Optical	Electrical	Magnetic	Thermal	Chemical	Mechanical
Automotive					●		
Consumer products							
Process industries,mining energy					●	●	
Manufacturing, machinery					●		
Medicine, health care						●	
HVAC, fire control, security					●		
Military, aerospace					●	●	
Telecommunications,office equip.							
Meteorology, environment						●	
Other *							

Main Activities

Ever since HEITO produced the first direct reading pH meter, in France in 1948, it has continually enlarged the range of physico-chemical instruments and laboratory material. The pH and millivolt meters are complemented with conductivity meters, oxymeters and ionometers.

Specific Expertise

The HEITO workshops specialised in this work can provide all types of electrodes likely to be required. Listed below are just the most commonly used: specific electrodes; conductivity cells; dissolved oxygen sondes; combined pH electrodes; redox electrodes; temperature probes.

060 Heraeus Sensor

BP 50
F-77610 Fontenay-Tresigny

Tel: +33 1 64 42 52 52
Fax: +33 1 64 42 63 17
E-mail:

Affiliation:
Contact: B Drouilly, Genaeral Manager
S Deprez, Commercial Manager

Organisation information

No of employees	26-100
Turnover (ECU*)	1m-5m

Sensor technology activities information

No of employees	26-100
Turnover (ECU*)	1m-5m

*1ECU=DM 1.89,FFr 6.30,£0.78,Lira 1909,Peseta 159.1

Organisation activities
Research and development on sensors sensor materials or applications
Testing, consulting, design
Manufacture of systems used in the manufacture, testing or development of sensors
Manufacture of finished sensors
Manufacture of sensor-based products

Application \ Signal Sensed	Radiation	Optical	Electrical	Magnetic	Thermal	Chemical	Mechanical
Automotive			•				
Consumer products			•				
Process industries,mining energy			•				
Manufacturing, machinery			•				
Medicine, health care			•				
HVAC, fire control, security			•				
Military, aerospace			•				
Telecommunications,office equip.			•				
Meteorology environment			•				
Other			•				

*special applications

Main Activities

Research, development and production of sensing devices for temperature measurements. Calibration services.

Specific Expertise

Satellite and aerospace temperature sensors standard and thermocouples and resistance sensors (Pt100, Pt1,000). Nowadays we are developing new glass Pt100 elements and a calibration laboratory.

Collaboration Offered

Know-how and technology.

Collaboration Sought

New customers and markets.

061 IEMN - Institut d'Electronique

et de Microelectronique du Nord
Bd Poincare
Cité Scientifique
F-59650 Villeneuve d'Ascq

Tel: +33 3 20 19 78 02
Fax: +33 3 20 19 78 80
E-mail: constant@iemn.univ-lille1.fr

Affiliation:
Contact: E Constant,Head of IEMN
P Godts, Scientist

Organisation information
No of employees 101-1000
Turnover (ECU*) 5m-50m
Sensor technology activities information
No of employees 26-100
Turnover (ECU*) 1m-5m
*1ECU=DM 1.89,FFr 6.30,£0.78,Lira 1909,Peseta 159.1

Organisation activities
Research and development on sensors sensor materials or applications
Testing, consulting, design

Application \ Signal Sensed	Radiation	Optical	Electrical	Magnetic	Thermal	Chemical	Mechanical
Automotive							
Consumer products	●				●		
Process industries,mining energy	●				●		
Manufacturing, machinery	●				●		
Medicine, health care	●				●		
HVAC, fire control, security	●				●		
Military, aerospace	●						
Telecommunications,office equip.	●						
Meteorology environment	●				●		
Other (specify)							

Main Activities

Microwave correlation radiometry; short range telemetry sensors, microwave thermal imagery; bio-medical applications for diagnostic and treatment; doppler sensors; anti collision radar; IR thermal microsensors.

Specific Expertise

Technology transfer for IR sensors; radiometric sensors up to 110 GHz.
Materials: by thin film technologies; III-V semiconducting (GeAs); Si; metals; ceramics; polymer (superconducting).

Collaboration Offered

IEMN transfer; industrial relations; SME spin-off IEMN (Captec, etc.).

Collaboration Sought

Industrial contacts for development of innovative sensors.

062 Ifremer

Centre de Brest
DITI/GO
BP 70
F-29280 Plouzané

Tel: +33 2 98 22 40 95
Fax: +33 2 98 22 41 35
E-mail:

Affiliation: Ifremer, Issy Les Moulineaux, France
Contact: Jacques Legrand, Head of the Laboratory
Dominique Birot, Physico-chemical Sensors
Michel Lehaitre, Optical Sensors
Sylvie Le Bras, Optical & Laser Instruments
Jean-Marie Vercelli, Testing & Sensors Qualifications

Organisation information
No of employees >100
Turnover (ECU*) -
Sensor technology activities information
No of employees 11-25
Turnover (ECU*) -
*1ECU=DM 1.89,FFr 6.30,£0.78,Lira 1909,Peseta 159.1

Organisation activities
Research and development.
Testing, consulting, design.

Application \ Signal Sensed	Radiation	Optical	Electrical	Magnetic	Thermal	Chemical	Mechanical
Automotive							•
Consumer products							
Process industries,mining energy							
Manufacturing, machinery							
Medicine, health care							
HVAC, fire control, security							
Military, aerospace							
Telecommunications,office equip.							
Meteorology, environment						•	
Other *	•	•	•	•	•	•	•

* Oceanography/Coastal Monitoring

Main Activities

Ifremer is the French National Organisation for Reseach and Development in the domain of ocean related activites. The Oceanographic Instrument Laboratory is in charge of the development of new instruments and transfer to industry in the domain of physical oceanography and coastal environment.

Specific Expertise

The organisation is involved in the development of physico-chemical sensors for long term monitoring of sea water qualities near shore. It has developed a size and load laser sensor which gives, in situ, the histogram of particles in suspension in sea water in classes of size; nitrate sensor using flow injection analysis technique; and is involved in using optical techniques (laser, fibre optic) for application to physico-chemical sensors.

Collaboration Offered

Competence in electronics. Electro-optic, optics, sea water chemistry applied to sensors.
Qualification - test in real environment of physico-chemical sensors.

Collaboration Sought

New techniques of physico-chemical sensor and biomedical.

063 INERIS

Parc Technologique
Alata
BP 2
F-60550 Verneuil en Halatte

Tel: +33 1 44 55 66 77
Fax: +33 1 44 55 66 99
E-mail:

Affiliation:
Contact: P Villeneuve de Janti, Prod Mgr

Organisation information

No of employees	101-1000
Turnover (ECU*)	5m-50m

Sensor technology activities information

No of employees	11-25
Turnover (ECU*)	1m-5m

*1ECU=DM 1.89,FFr 6.30,£0.78,Lira 1909,Peseta 159.1

Organisation activities
Research and development.
Testing, consulting, design.
Manufacture of systems used in the manufacture, testing or development of sensors.

Application \ Signal Sensed	Radiation	Optical	Electrical	Magnetic	Thermal	Chemical	Mechanical
Automotive			●			●	
Consumer products							
Process industries,mining energy		●	●		●	●	
Manufacturing, machinery							
Medicine, health care							
HVAC, fire control, security		●				●	
Military, aerospace							
Telecommunications,office equip.							
Meteorology, environment			●			●	
Other *							

Main Activities

INERIS is a public research body of an industrial and commercial nature, founded in 1990 on the base of two institutes: The former Research and Development Centre of Charbonnages de France (CERCHAR) and the Applied Chemistry Research Institute (IRCHA).

INERIS' objective, under the auspices of the French Ministry of the Environment, is to carry out research and development, technical support and consultancy to prevent and minimise the impacts and hazards of industrial activity on man and the environment.

Specific Expertise

Sensors for toxic and flammable gases (investigating, development, evaluation) for industrial safety and domestic applications. Development of thin films microtechnologies for gases detection and selective identificaiton performances. Development of measurement techniques and signal processing. Transfer to industry through patents or know-how licences.

Collaboration Offered

Long experience in catalytic and electrochemical technologies. Good experience in transfer to industry. Laboratory facilities for evaluation of conformity to international standards or projects of standards.

Collaboration Sought

Research laboratories interested in bridging the gap between research and application products.

064 INPG

Enserg
23 Rue des Martyrs, BP 257
F-38016 Grenoble Cedex 1

Tel: +33 1 76 856 028
Fax: +33 1 76 856 080
E-mail: benech@enserg.fr

Affiliation:
Contact: P Benech, Head of Sensor Group

Organisation information
No of employees 1-25
Turnover (ECU*) <1m
Sensors technology activities information
No of employees 1-10
Turnover (ECU*) <0.5m
*1ECU=DM 1.89,FFr 6.30,£0.78,Lira 1909,Peseta 159.1

Organisation activities
Research and Development

Application \ Signal Sensed	Radiation	Optical	Electrical	Magnetic	Thermal	Chemical	Mechanical
Automotive							
Consumer products							
Process industries, mining, energy		•					
Manufacturing, machinery							
Medicine, health care							
HVAC, fire control, security							
Military, aerospace							
Telecommunications,office equip.		•					
Meteorology, environment		•					
Other *							

Main Activities

Laboratoire d'Electromagnetisme Micro-ondes et Optoelectronique works in the areas of: electromagnetism, electromagnetic compatibility, non linear optics, integrated optics for telecommunication and sensors application.

Specific Expertise

- Displacement sensors
- Chemical and biological sensors
- Doppler senors
- Vibrator Sensor
- Telecommunication passive and active components

Collaboration Offered

- Special study of integrated optical components
- Sensors application

065 INRA

Laboratoire de Génie&Microbiologie des Procèdés Alimentaires
INRA INAPG
F-78850 Thiverval Grignon

Tel: +33 1 30 81 54 87
Fax: +33 1 30 81 55 97
E-mail:

Affiliation:
Contact: G Corrieu, Director

Organisation information
No of employees 26-100
Turnover (ECU*) <1m
Sensor technology activities information
No of employees 1-10
Turnover (ECU*) <0.5m
*1ECU=DM 1.89,FFr 6.30,£0.78,Lira 1909,Peseta 159.1

Organisation activities
Research and development on sensors sensor materials or applications

Application \ Signal Sensed	Radiation	Optical	Electrical	Magnetic	Thermal	Chemical	Mechanical
Automotive							
Consumer products							
Process industries,mining energy		•			•	•	
Manufacturing, machinery							
Medicine, health care							
HVAC, fire control, security							
Military, aerospace							
Telecommunications,office equip.							
Meteorology environment							
Other (specify)							

Main Activities

Development and testing of sensors for food industry and biochemical processes (alcoholic and lactic fermentations, cheese making, ...)

Specific Expertise

Thermal and optical sensors to control milk gelatin and to predict cutting time in cheese making.
Sensors and softwares to follow and control wine making and beer making.
Sensors and softwares to follow and control lactic acid starters production, fresh cheese and yoghurt manufacture.
MID infrared to determine products composition.

Collaboration Offered

Research work, sensors tests.
Sales of devices and sensors.

Collaboration Sought

New methods and principles for sensors development.
Trials in industrial context to validate sensors.

066 INSERM CNRS Lyon

Départ. Médecine Expérimentale
8 avenue Rockefeller
F-69373 Lyon

Tel: +33 4 78 77 71 26
Fax: +33 4 78 77 71 72
E-mail: cespugli@cismsum.univ-lyon1.fr

Affiliation:
Contact: Mr Cespuglio, Director

Organisation information
No of employees 26-100
Turnover (ECU*) 5m-50m
Sensor technology activities information
No of employees 1-10
Turnover (ECU*) <0.5m
*1ECU=DM 1.89,FFr 6.30,£0.78,Lira 1909,Peseta 159.1

Organisation activities
Research and development on sensors sensor materials or applications

Application \ Signal Sensed	Radiation	Optical	Electrical	Magnetic	Thermal	Chemical	Mechanical
Automotive							
Consumer products							
Process industries,mining energy							
Manufacturing, machinery							
Medicine, health care			•			•	
HVAC, fire control, security							
Military, aerospace							
Telecommunications,office equip.							
Meteorology environment							
Other (specify)							

Main Activities

Sensors for brain and blood transmitters detection; application to the understanding and the regulation of the sleep-waking cycle; ready now - sensors for aims, NO, glucose.

Specific Expertise

Sensor for detection of: NO, glucose - to do with brain, and blood.

067 Laboratoire de Chimie Physique

d' Environment
405 Rue De Vandoeuvre
F-59500 Villers les Nancy

Tel: +33 3 83 91 63 43
Fax: +33 3 83 27 54 44
E-mail: walcariv@;cpe.cbrs-nancy-fr

Affiliation:

Contact: Prof J Bessiere, Head of Unit
A Walcarius

Organisation information

No of employees 26-100
Turnover (ECU*) <1m

Sensors technology activities information

No of employees 1-10
Turnover (ECU*) <0.5m

*1ECU=DM 1.89,FFr 6.30,£0.78,Lira 1909,Peseta 159.1

Organisation activities

Research and Development

Application \ Signal Sensed	Radiation	Optical	Electrical	Magnetic	Thermal	Chemical	Mechanical
Automotive							
Consumer products							
Process industries, mining, energy						•	
Manufacturing, machinery							
Medicine, health care							
HVAC, fire control, security							
Military, aerospace							
Telecommunications,office equip.							
Meteorology, environment			•			•	
Other *							

Main Activities

Our organisation is mainly concerned with both fundamental and applied research topics in Chemistry for the Environment, with special emphasis to the liquid-solid interfaces. In this context, some of us are developing new electrochemical sensors based on chemically modified electrodes (modifiers: zeolites, silicas, clays.....) as well as dielectric sonsores based on silica and modified silica.

Specific Expertise

Main areas:

- Electrochemical analysis of aqueous solutions (heavy metal ions, perticides).
- Analysis of mass transport reactions at the divided solids - liquid interfaces.
- Specific materials
- Zeolites, silica, and modified silica,clays.
- Electrodes modified with ion exchanger solids.

Collaboration Offered

Development of new electrochemical sensors.

Collaboration Sought

Testing in natural media.

068 LEPMI-ENSEEG

Laboratory of Electrochemistry and Physical Chemistry
of Materials and Interfaces
1130 Rue de la Pisane, BP 75
F-38402 St Martin d'Heis

Tel: +33 76 82 65 73
Fax: +33 76 82 66 70
E-mail: lepmi@ccomm.grenet.fr

Affiliation: CNRS -INPG
Contact: E Siebert, Sensor Group Mgr

Organisation information
No of employees 101-1000
Turnover (ECU*) <1m
Sensor technology activities information
No of employees 11-25
Turnover (ECU*)
*1ECU=DM 1.89,FFr 6.30,£0.78,Lira 1909,Peseta 159.1

Organisation activities
Research and Development

Application \ Signal Sensed	Radiation	Optical	Electrical	Magnetic	Thermal	Chemical	Mechanical
Automotive							
Consumer products							
Process industries, mining, energy							
Manufacturing, machinery							
Medicine, health care			•			•	
HVAC, fire control, security							
Military, aerospace							
Telecommunications,office equip.						•	
Meteorology, environment							
Other *							

Main Activities

- Solid State electrochemistry.
- Ionic conduction in solid state materials.
- Gramics.
- Electrochemical sensors for environment and biomedical applications.

Specific Expertise

- Electrochemical gas sensors (H_2, O_2, Cl_2, HCl, CO_2).
- Ion Selective electrodes (Na^+, Li^+,pH).
- Bioelectric signal measurements with surface elctrodes
- Materials
- Ceramics : ZrO_2, Nasicon, $SrCl_2$-KCl glasses.

069 Metalscan

57 Rue Mortillet
F-38000 Grenoble

Tel: +33 4 76 51 74 73
Fax: +33 4 76 54 41 67
E-mail:

Affiliation:
Contact: J C Lecomte, R&D, Ultrasonic Probes

Organisation information
No of employees 11-25
Turnover (ECU*) -
Sensor technology activities information
No of employees 11-25
Turnover (ECU*) -
*1ECU=DM 1.89,FFr 6.30,£0.78,Lira 1909,Peseta 159.1

Organisation activities
Research and development.
Testing, consulting, design.
Manufacture of systems used in the manufacture, testing or development of sensors.
Manufacture of finished sensors.

Application \ Signal Sensed	Radiation	Optical	Electrical	Magnetic	Thermal	Chemical	Mechanical
Automotive							•
Consumer products							
Process industries,mining energy							•
Manufacturing, machinery							•
Medicine, health care							
HVAC, fire control, security							
Military, aerospace							•
Telecommunications,office equip.							
Meteorology, environment							
Other *							

Main Activities

Non destructive ultrasonic testing: manufacture of standard and special ultrasonic probes for manual, semi-automatic and automatic testing; systems for ultrasonic testing (manual scanner etc.); immersion tank testing - standard and special ultrasonic testing systems; additional products for ultrasonic probes (miniature preamplifier); studies and systems (establishment and validation of inspection methods).

Specific Expertise

Sensors using the piezoelectric effect for ultrasonic signals. Application areas: iron and steel manufacture; aerospace; nuclear testing; chemical testing; laboratories; companies specialised in testing.

Collaboration Sought

Ultrasonic probes. Research and Development.

070 Motorola Semiconducteurs SA

Sensor Products Division
Avenue de General Eisenhower
Le Mirail B P 1029
F-31023 Toulouse Cedex

Tel: +33 5 61 19 91 02
Fax: +33 5 61 19 94 69
E-mail:

Affiliation:

Contact: J Trice, Operataion Mgr
L Grant, Marketing

Organisation information

No of employees 101-1000
Turnover (ECU*) >50m

Sensors technology activities information

No of employees >100
Turnover (ECU*) >5m

*1ECU=DM 1.89,FFr 6.30,£0.78,Lira 1909,Peseta 159.1

Organisation activities

Research and Development
Testing, consulting, design
Manufacture of finished sensors

Application \ Signal Sensed	Radiation	Optical	Electrical	Magnetic	Thermal	Chemical	Mechanical
Automotive						•	•
Consumer products						•	•
Process industries, mining, energy							•
Manufacturing, machinery							•
Medicine, health care							•
HVAC, fire control, security						•	•
Military, aerospace							•
Telecommunications,office equip.							•
Meteorology, environment						•	•
Other *							

Main Activities

Motorola design, manufacture and sale of semiconductor sensor components for the measurement of pressure, acceleration and chemicals. Major market segments include automotive, industrial, medical, communications and consumer.

Specific Expertise

- Medical pressure sensors - invasive and non-invasive
- Altimeter pressure sensors - automotive map/Bap and instruments.
- Airbag aceleration sensors - automotive sing and multi-axis.
- Acceleration sensors - Home security intrusion detectors.
- Fluid level pressure sensors - Automatove, industrial, Consumer.
- Chemical Gas Sensors - CO,CH_4, ALCOHOL for Consumer, Industrial fire detector systems.
- Intelligent sensors - MCU + Sensor.

Collaboration Offered

Special arrangements to develop sensor components with system partners. Technical collaboration for Sensor R&D.

Collaboration Sought

Additional system partners for commercilizing smart pressure, acceleration, and chemical sensor components.

071 Oldham

BP 962 ZI Est
F-62033 Arras Cedex

Tel: +33 321 60 25 25
Fax: +33 321 60 26 20
E-mail:

Affiliation: B. T. R. (UK)
Contact: Mr C Prat, Director Division Detection
Mr L Witrant, Manager of R&D
Dr M Taylor, Head of Electrochemical sensor development

Organisation information
No of employees >1000
Turnover (ECU*) >50m

Sensor technology activities information
No of employees >100
Turnover (ECU*) >5m
*1ECU=DM 1.89,FFr 6.30,£0.78,Lira 1909,Peseta 159.1

Organisation activities
Manufacture of finished sensors
Manufacture of sensor-based products

Application \ Signal Sensed	Radiation	Optical	Electrical	Magnetic	Thermal	Chemical	Mechanical
Automotive							
Consumer products						•	
Process industries,mining energy						•	
Manufacturing, machinery						•	
Medicine, health care							
HVAC, fire control, security		•				•	
Military, aerospace							
Telecommunications,office equip.							
Meteorology environment						•	
Other (specify)							

Main Activities

Manufacture of equipment for gas detection (toxic gases, explosive gases and oxygen). Oldham has a wide range of detectors - portable and fixed to detect over 50 gases.

Specific Expertise

Manufacturer of electrochemical and thermocatalytic sensors; developing low power thermocatalytic sensors and long-life electrochemical sensors.

Collaboration Offered

Good testing facilities in laboratory and in the field. Strong sensor electronics capabilities.

Collaboration Sought

Technical research capability.

072 Optilas Systems

4 Allée du Cantal
21 la Petite Montagne Sud
F-91019 Evry Cedex

Tel: +33 160 79 59 00
Fax: +33 160 86 96 33
E-mail:

Affiliation: Optilas International
Contact: G Arie, DG
A Touron, Communication Mgr

Organisation information
No of employees 26-100
Turnover (ECU*) 5m-50m
Sensor technology activities information
No of employees 1-10
Turnover (ECU*)
*1ECU=DM 1.89,FFr 6.30,£0.78,Lira 1909,Peseta 159.1

Organisation activities
Distribution of finished sensors.

Application \ Signal Sensed	Radiation	Optical	Electrical	Magnetic	Thermal	Chemical	Mechanical
Automotive					●		
Consumer products					●		
Process industries,mining energy					●		
Manufacturing, machinery					●		
Medicine, health care					●		
HVAC, fire control, security					●		
Military, aerospace							
Telecommunications,office equip.							
Meteorology, environment							
Other *							

Main Activities

Optilas International is a leading distributor of high tech products in Europe, such as laser, opto electronic imaging, optic components, etc.

Specific Expertise

Thermal sensors luxtron/accufibre.

073 Photonetics

52 Avenue de l'Europe
BP 39
F-78160 Marly le Roi

Tel: +33 1 39 16 33 77
Fax: +33 1 39 16 56 06
E-mail:

Affiliation:
Contact: Arnaud Quelquejay, Export Manager

Organisation information
No of employees 26-100
Turnover (ECU*) >5m
Sensor technology activities information
No of employees 11-25
Turnover (ECU*) 0.5-1m
*1ECU=DM 1.89,FFr 6.30,£0.78,Lira 1909,Peseta 159.1

Organisation activities
Research and development.
Manufacture of systems used in the manufacture, testing or development of sensors.
Manufacture of finished sensors.

Application \ Signal Sensed	Radiation	Optical	Electrical	Magnetic	Thermal	Chemical	Mechanical
Automotive							•
Consumer products							
Process industries,mining energy				•	•	•	•
Manufacturing, machinery							
Medicine, health care							
HVAC, fire control, security							
Military, aerospace				•	•		•
Telecommunications,office equip.							
Meteorology, environment							
Other *							

Main Activities

PHOTONETICS is a leading developer, manufacturer and supplier in the field of lasers, fibre optic instrumentation and sensors, and electro-optic test instruments. It has teh capability of assembling full engineering turn-key systems according to teh customer's specific requirements.

Specific Expertise

PHOTONETICS is devoted to the design, development and manufacture of photonic technologies such as fibre optic sensors. A versatile team of 20 reseach and development engineers has accumulated more than 15 years experience in the field of fibre optics and fibre optic sensors. They are active in organising international fibre optic conferences of high scientific level and in chairing international standardisation activities on fibre optic sensors. Today, PHOTONETICS has a range of fibre optic sensors, which are commercially available such as: optoflow (void fraction measurement device); spectrofip: spectrophotometry with optical fibres and same engineering prototypes such as:- FOG (fibre optic gyroscope); optonet: all optical on/off switches network; optolog: interferometric sensor system for the measurement of basic physical parameters; optovib: distributed vibration sensor.

074 Precilec

Immeuble Perisud
4 Rue François Ory
F-92128 Montrouge Cedex

Tel: +33 1 46 12 90 90
Fax: +33 1 46 56 74 21
E-mail:

Affiliation:
Contact: R Pizzamiglio, Gemeral Manager
C Fenninger, General Manager
J P Bossuto, Displacement Transducers, Sub-Assemblies

Organisation information
No of employees > 100
Turnover (ECU*) >5m

Sensor technology activities information
No of employees -
Turnover (ECU*) -
*1ECU=DM 1.89,FFr 6.30,£0.78,Lira 1909,Peseta 159.1

Organisation activities
Research and development.
Testing, consulting, design.
Manufacture of finished sensors.
Manufacture of sensor-based products.

Application \ Signal Sensed	Radiation	Optical	Electrical	Magnetic	Thermal	Chemical	Mechanical
Automotive							•
Consumer products							
Process industries,mining energy							
Manufacturing, machinery							
Medicine, health care							
HVAC, fire control, security							
Military, aerospace							•
Telecommunications,office equip.							
Meteorology, environment							
Other *							

Main Activities

Precilec was founded in 1946 to design, manufacture and market synchros and servo-mechanisms for naval and aerospace applications. In 1960 the company developed and commercialised electronic navigation and plotting equipment for the French and other navies of the world. During 1979 the Research and Development Division of Precilic employed a new team of computer microprocessor specialists to undertake the development of entirely new product lines. These complemented and completely revitalised the existing range of equipment. In 1985, as a result of a management buyout Precilec's future was secured and its strong independence maintained. Precilec manufacture 4 main product lines: displacement transducers; torque motors; complete sub-assemblies and display systems.

Specific Expertise

Displacement transducers: angular displacement transducers include high precision synchros, multi-speed resolvers and high performance RVDT and LVDT (rotary adn linear variable differential transformers). The precision attained is of the order of 10 seconds of arc.

Collaboration Offered

Displacement transducers. Sub-assemblies.

075 Protais SA

Route de Savigny
F-41800 Montoire sur la Loire

Tel: +33 2 54 85 02 67
Fax: +33 2 54 72 61 20
E-mail:

Affiliation:
Contact: P Crosnier, Pressure Sensors
B Crevenat, Optoelectronics

Organisation information
No of employees -
Turnover (ECU*) -

Sensor technology activities information
No of employees -
Turnover (ECU*) -
*1ECU=DM 1.89,FFr 6.30,£0.78,Lira 1909,Peseta 159.1

Organisation activities
Manufacture of finished sensors.

Application \ Signal Sensed	Radiation	Optical	Electrical	Magnetic	Thermal	Chemical	Mechanical
Automotive							
Consumer products							
Process industries,mining energy							•
Manufacturing, machinery							•
Medicine, health care							•
HVAC, fire control, security							
Military, aerospace							
Telecommunications,office equip.							
Meteorology, environment							
Other *							

Specific Expertise

Manufacturing pressure and level sensors. These can be applied in all industrial application areas. The principles employed are strain gauges and optoelectronics.

Collaboration Offered

Pressure sensors. Level sensors.

Collaboration Sought

Optoelectronics.

076 Scaime SA

ZI de juvigny
BP 501
F-74105 Annemasse Cedex

Tel: +33 4 50 87 78 64
Fax: +33 4 50 87 78 46
E-mail:

Affiliation:
Contact: Philippe Cartier, Sales and Marketing Director

Organisation information
No of employees >100
Turnover (ECU*) >5m

Sensor technology activities information
No of employees >100
Turnover (ECU*) >5m
*1ECU=DM 1.89,FFr 6.30,£0.78,Lira 1909,Peseta 159.1

Organisation activities
Research and development.
Manufacture of finished sensors.
Manufacture of sensor-based products.

Application \ Signal Sensed	Radiation	Optical	Electrical	Magnetic	Thermal	Chemical	Mechanical
Automotive							•
Consumer products							
Process industries,mining energy							•
Manufacturing, machinery							•
Medicine, health care							•
HVAC, fire control, security							
Military, aerospace							•
Telecommunications,office equip.							
Meteorology, environment							
Other *							

Main Activities

SCAIME is a leading company inthe field of force and load cells. Using strain gauge Wheastone Bridge technology, SCAIME is one of the only companies in teh world to design and manufacture its own strain gauge programme. SCAIME manufactures over 500.000 sensors a year, mainly sold in France, EC and USA.

Specific Expertise

Load and force sensors from 1 kg to 1000 tonnes; position sensors from 0-50 mm to 0-30 m; pressire sensors from 0-100 mbar to 0-700 bar.

077 SFERE SA

Société Française d'Etudes et de Réalisation Electroniques
3 rue Jacques Monod
BP 69
F 69684 Chassieu Cedex

Tel: +33 4 78 90 21 00
Fax: +33 4 78 90 83 90
E-mail:

Affiliation:
Contact: Mr J Mechain, Chairman
Mr Cadiere, Mkting Manager
Mr Saint-Jean, R&D Dept Manager

Organisation information

No of employees	26-100
Turnover (ECU*)	5m-50m

Sensor technology activities information

No of employees	11-25
Turnover (ECU*)	0.5m-1m

*1ECU=DM 1.89,FFr 6.30,£0.78,Lira 1909,Peseta 159.1

Organisation activities
Research and development on sensors sensor materials or applications
Supply of raw materials or semi-finished products used to make sensors
Manufacture of sensor-based products

Application \ Signal Sensed	Radiation	Optical	Electrical	Magnetic	Thermal	Chemical	Mechanical
Automotive							
Consumer products							
Process industries,mining energy			•		•		
Manufacturing, machinery							
Medicine, health care							
HVAC, fire control, security							
Military, aerospace							
Telecommunications,office equip.							
Meteorology environment			•		•		
Other (specify)							

Main Activities

Specialise in the design and manufacture of measurement transmitters and programmable digital indicators for front panels (more than 30,000 items per year, 15% of which are exported)

Specific Expertise

We make the programmable electronics work with low signals from sensors. We are increasingly selling miniaturised items (cases are smaller and smaller, rack mounted winds are no more than 4TE).

Collaboration Offered

We are commercially independent all over France. We are quite well-connected in some foreign countries (particularly Europe). Up to date, we are focusing on Asian Countries.

Collaboration Sought

We would like to find a company selling sensors which does not yet sell electronic devices. We are also seeking companies selling automation and survey functions.

078 Université de Clermont-Ferrand

LASMEA-Laboratoire des Sciences et Matériaux pour l'Electronique et d'Automatique
24, avenue des Landais
F-63177 Aubiere Cedex

Tel: +33 4 73 40 72 50
Fax: +33 4 73 40 72 62
E-mail:

Affiliation:

Contact: M Richtin, Director
JP Germain, Head Gas Sensor

Organisation information

No of employees

Turnover (ECU*)

Sensors technology activities information

No of employees

Turnover (ECU*)

*1ECU=DM 1.89,FFr 6.30,£0.78,Lira 1909,Peseta 159.1

Organisation activities

Research and Development

Application \ Signal Sensed	Radiation	Optical	Electrical	Magnetic	Thermal	Chemical	Mechanical
Automotive						•	
Consumer products							
Process industries, mining, energy						•	
Manufacturing, machinery							
Medicine, health care							
HVAC, fire control, security						•	
Military, aerospace							
Telecommunications,office equip.							
Meteorology, environment							
Other *							

Main Activities

The team "Gas Sensors" is a "scientific operation" of the Laboratory LASMEA. LASMEA means "Laboratoire des Sciences et Materiaux pour l'Electronique et d'Automatique. This is a Laboratory of the BLAISE PASCAL University. Four permenant searchers work in the team "Gas sensors."

Specific Expertise

- Research of new sensing materials for semi conductor gas sensors : organic and inorgancic thin films. (For CO, NO_3, NO_2, SO_2...... detection in air).
- Modelisation of gas/material interactions.
- Research on a selective gas sensor obtained by multisensors associated with signal processing (multi-layers integrable gas sensor.)

Collaboration Offered

- ability to interprete gas/material phenomena : the retical modelisation.
- Studies of a multisensor based on various semi-conductor materials: test and recognition methods.

Collaboration Sought

Scientific and partnership for financial support.

079 Université de Nantes

Groupe Electrochimie
2 rue de la Houssinòro
F-44072 Nantes Cedex 03

Tel: +33 2 40 74 13 10
Fax: +33 2 40 74 13 10
E-mail: el-murr@eesinto.univ-nantes.fr

Affiliation:
Contact: Nabil El Murr, Research Director

Organisation information
No of employees 101-1000
Turnover (ECU*)

Sensor technology activities information
No of employees 11-25
Turnover (ECU*) <0.5m
*1ECU=DM 1.89,FFr 6.30,£0.78,Lira 1909,Peseta 159.1

Organisation activities
Research and development on sensors sensor materials or applications
Testing, consulting, design

Application \ Signal Sensed	Radiation	Optical	Electrical	Magnetic	Thermal	Chemical	Mechanical
Automotive							
Consumer products						•	
Process industries,mining energy							
Manufacturing, machinery							
Medicine, health care						•	
HVAC, fire control, security							
Military, aerospace							
Telecommunications,office equip.							
Meteorology environment						•	
Other (specify)						•	

Main Activities

Research and development of biosensors. Fundamental research as well as development in order to bring biosensors into the market. Amperometric or potentiometric biosensors. Electgron transfer reactions involving bioproducts. Chemical and electrochemical behaviour of bioproducts used for sensing.

Specific Expertise

Biosensors; amperometric and potentiometric sensors; modified electrodes, enzymatic electrodes; renewable biosensors; food analysis using biosensors; FIA - flow injection analysis using biosensors; biosensors for routine assays which meet industrial needs (food processing).

Collaboration Offered

Expertise for the development of biosensors usable in food industry; patents on renewable biosensors; know-how for the production of renewable biosensors at industrial level.

Collaboration Sought

Help in the final phase of development of enzymatic biosensors; manufacturing biosensors for food analysis.

080 Université de Savoie - ESIGEC

Laboratoire CME
Campus Scientifique
F-73 376 Le Bourget du Lac

Tel: +33 4 79 75 88 30
Fax: +33 4 79 75 86 74
E-mail:

Affiliation:
Contact: Mr Naffrechoux, Researcher

Organisation information
No of employees 101-1000
Turnover (ECU*) <1m

Sensor technology activities information
No of employees 1-10
Turnover (ECU*) <0.5m
*1ECU=DM 1.89,FFr 6.30,£0.78,Lira 1909,Peseta 159.1

Organisation activities
Research and development on sensors sensor materials or applications

Application \ Signal Sensed	Radiation	Optical	Electrical	Magnetic	Thermal	Chemical	Mechanical
Automotive							
Consumer products							
Process industries,mining energy							
Manufacturing, machinery							
Medicine, health care							
HVAC, fire control, security							
Military, aerospace							
Telecommunications,office equip.							
Meteorology environment		•				•	
Other							

Main Activities

Research of new methods for water pollution examination.
Key words: UV absorptiometry; diode array detector; water pollution.

081 Université de St Etienne

Laboratoire du Traitment du Signal et Instrumentation
23 rue du D P Michelon
F-42023 St Etienne Cedex 2

Tel: +33 4 77 42 15 60
Fax: +33 4 77 42 18 20
E-mail: gagnaire@univ-st-etienne.fr

Affiliation:
Contact: H Gagnaire, Professor
D A Mure-Ravau, Professor

Organisation information

No of employees 26-100
Turnover (ECU*) <1m

Sensor technology activities information

No of employees 1-10
Turnover (ECU*) <0.5m

*1ECU=DM 1.89,FFr 6.30,£0.78,Lira 1909,Peseta 159.1

Organisation activities
Research and development on sensors, sensor materials or applications.
Testing, consulting, design.

Application \ Signal Sensed	Radiation	Optical	Electrical	Magnetic	Thermal	Chemical	Mechanical
Automotive							
Consumer products							
Process industries,mining energy		•					
Manufacturing, machinery							
Medicine, health care							
HVAC, fire control, security							
Military, aerospace							
Telecommunications,office equip.							
Meteorology, environment		•					
Other *							

Main Activities

Our organisation is a laboratory of the University of Saint-Etienne. Four teams are working on: optics; imagery; laser; electronics.
Optical sensors. Optical fibre sensors for chemical species.

Specific Expertise

The work is devoted to chemical sensing with fibre optic sensors.

Collaboration Offered

Research and development.

Collaboration Sought

Application of our sensors.

082 Université de Troyes

IBT
11 Rue Marie Curie
F-10000 Troyes

Tel: +33 3 25 71 46 00
Fax: +33 3 25 71 46 10
E-mail:

Affiliation:
Contact: G Dine, Executive Mgr

Organisation information

No of employees	1-25
Turnover (ECU*)	<1m

Sensors technology activities information

No of employees	1-10
Turnover (ECU*)	<0.5m

*1ECU=DM 1.89,FFr 6.30,£0.78,Lira 1909,Peseta 159.1

Organisation activities
Research and Development
Testing, consulting, design

Application \ Signal Sensed	Radiation	Optical	Electrical	Magnetic	Thermal	Chemical	Mechanical
Automotive							
Consumer products		•	•		•	•	•
Process industries, mining, energy		•	•			•	•
Manufacturing, machinery			•			•	•
Medicine, health care		•	•	•	•	•	•
HVAC, fire control, security						•	
Military, aerospace							•
Telecommunications,office equip.							
Meteorology, environment		•	•	•	•	•	
Other *							

Main Activities

- Research and development of Sensors.
- Testing, consulting and design on sensors.
- Research and development on sensor materials.
- Development on new applications of sensors.

Specific Expertise

- Consumer products, Process industries, Health Care, Pollution and bio-medical engineering.
- Human physiology, veterinary, human biology, industrial analysis laboratories.

Collaboration Offered

- New prototypes for developing and testing.
- New products for using and selling.

Collaboration Sought

- Bio-sensors in particular in the field of health and pollution.

083 Whessoe Varec SA

Rue de Bitche
F-62100 Calais

Tel: +33 3 21 96 4993
Fax: +33 3 21 34 3612
E-mail:

Affiliation: Whessoe Plc
Contact: P Versluÿs, Bus Dev Manager

Organisation information
No of employees 101-1000
Turnover (ECU*) >50m
Sensor technology activities information
No of employees >100
Turnover (ECU*) >5m
*1ECU=DM 1.89,FFr 6.30,£0.78,Lira 1909,Peseta 159.1

Organisation activities
Manufacture of finished sensors.

Application \ Signal Sensed	Radiation	Optical	Electrical	Magnetic	Thermal	Chemical	Mechanical
Automotive							
Consumer products							
Process industries,mining energy							•
Manufacturing, machinery							
Medicine, health care							
HVAC, fire control, security							
Military, aerospace							
Telecommunications,office equip.							
Meteorology, environment							•
Other *							

Main Activities

Process liquids and vapour control instrumentation.

Specific Expertise

Level sensing instrumentation; density sensing instrumentation; temperature sensing instrumentation; safety shut down systems; vapour recovery; data handling software.

Additional Short Form Entries

CEA Laboratoire des Systèmes Moléculaires Organisés
DRECAM - SCM - Bât.125
Centre de Saclay
F-91191 Gif-sur-Yvette Cedex

Contact	Annie Ruaudel
Tel:	+33 1 69 08 54 55
Fax:	+33 1 69 08 66 40
email:	

Centre d'Etudes du Bouchet
DGA/CEB
BP3
F-91710 Vert-le-Petit

Contact	Colette Tiret
Tel:	+33 1 69 90 83 45
Fax:	+33 1 64 93 52 66
email:	

CNRS
Laboratoire d'Analyse et d'Architecture des Systèmes (LAAS)
7 avenue du Colonel Roche
F-31077 Toulouse Cedex

Contact	Anne Marie Melle Gue
Tel:	+33 5 61 33 62 00
Fac:	+33 5 61 55 35 77
email:	laas-contact@:laas.fr

CNRS
Laboratoire d'Electrochimie et Chimie Analytique
URA 216
ENSCP
11 rue Pierre & Marie Curie
F-75231 Paris Cedex 05

Contact	Jacques Devynck
Tel:	+33 1 43 54 53 84
Fax:	+33 1 36 27 67 50
email:	devynck@ext.jussieu.fr

CNRS
Laboratoire des Matériaux Organiques à Propriétés Spécifiques
LMOPS - CNRS
BP 24
F-69390 Vernaison

Contact	Yves Chevalier
Tel:	+33 4 78 02 22 71
Fax:	+33 4 78 02 71 87
email:	

CNRS
Laboratoire Physique et Métrologie des Oscillateurs
LPMO
32 avenue de l'Observatoire
F-25044 Besancon Cedex

Contact	E Bigler
Tel:	+33 3 81 66 69 99
Fax:	+33 3 81 66 69 98
email:	bigler@lpmo.univ.fcomte.ft

CNRS
Service Central d'Analyses
chemin du Canal
BP22
F-69390 Vernaison

Contact	Philippe Linet
Tel:	+33 4 78 02 22 22
Fax:	+33 4 78 02 71 87
email:	

ENS Cachan
PPSM
Départment de Chimie ENS Cachan
61 avenue du Président Wilson
F-94235 Cachan Cedex

Contact	Claude Bied-Charreton
Tel:	+33 1 47 40 27 05
Fax:	+33 47 40 24 54
email:	claude@lppm2.ens.cachan.fr

Gaz de France
Direction de la Recherche
361 avenue du Président Wilson
F-93210 La Plaine St Denis

Contact	Pascale Massok
Tel:	+33 1 49 22 50 00
Fax:	+33 `1 49 22 53 75
email:	prénom.nom@gdt.edfgdf.fr

INSA - Lyon
Laboratoire de Physique de la Matière
Microcapteurs Miscosystèmes
Biomédicaux
INSA Lyon
20 avenue A. Einstein
F-69621

Contact	André Dittmar
Tek:	+33 4 78 01 33 31
Fax:	+33 4 78 77 71 75
email:	dittmar@cism.univ.lyon.fr

LIPHA SA
34 rue Saint Romain
S-69008 Lyon

Contact	Georges Perrot
Tel:	+33 4 78 77 18 18
Fax:	+33 4 78 75 39 05
email:	gperrot@lep.ens.lyon.fr

NEREIDES
4 avenue des Indes
ZA de Courtaboeuf
F-91969 Les Ulis

Contact	Frédéric Griffe
Tel:	+33 1 69 07 20 48
Fax:	+33 1 69 07 19 14
email:	100700.3566@compuserve.com

PHILIPS Laboratoire d'Electronique
Systèmes de Détection et de Contrôle Adaptatifs
BP 15
F-94453 Limeil-Brevannes Cedex

Contact	S Gourrier
Tel:	+33 1 45 10 67 00
Fax:	+33 1 33 10 67 43
email:	

PSA Peugeot-Citroën
DET MXT - CED
18 allée Faivelles
F-92250 Lagarenne-Colombes

Contact	Willy Jones
Tel:	+33 1 47 69 22 36
Fax:	+33 1 47 69 87 70
email:	

Radiometer Analytical
72 rue d'Alsace
F-69627 Villeurbanne Cedex

Contact	Jean-Jacques Fombon
Tel:	+33 4 78 03 38 38
Fax:	+33 4 78 68 88 12
email:	

Rhône Poulenc Industrialisation
24 avenue Jean Jaurès
F-69153 Decines Charpieu Cedex

Contact	Jean-Louis Linossier
Tel:	+33 4 72 05 20 00
Fax:	+33 4 72 05 25 00
email:	

SAT
Route de Quevert
BP 130
F-222103 Dinan Cedex

Contact	Pierre Morillon
Tel:	+33 2 96 85 61 86
Fax:	+33 2 96 85 61 86
email:	

SAGEM
Senseurs Optiques & Physique Appliquée - UDR 17
Centre d'Argenteuil
72-74 rue de la Tour Billy
BP 72
F-95101 Argenteuil Cedex

Contact	Xavier Boutin
Tel:	+33 1 34 26 38 00
Fax:	+33 1 34 26 37 22

SAUR
2 rue de la Bresle
F-78310 Maureopas

Contact	Lucien Hahn
Tel:	+33 1 30 13 38 17
Fax:	+33 1 30 15 28 79
email:	

Schlumberger
Gas Engineering Montrouge
(GEM)
50 avenue Jean Jaurès
BP 620-13
F-92542 Montrouge Cedex

Contact	Véronique Nouaze
Tel:	+33 1 47 46n 65 29
Fax:	+33 1 47 46 72 12
email:	

SERES
BP 87000
F-13793 Aix-en-Provence Cedex 3

Contact:	Jean-Claude Obert
Tel:	+33 4 42 16 76 76
Fax:	+33 4 42 24 30 97
email:	

Thomson - CSF
Laboratoire Central de Recherches
F-91404 Orsay Cedex

Contact	Jean-François Lipskier
Tel:	+33 1 69 33 07 72
Fax;	+33 1 69 33 07 55
email:	lipskier@lcr.thomson.fr

Université de Rennes 1
Laboratoire Radiocommunications
CNRS
Structures Rayonnantes
Université de Rennes 1
Campus de Beaulieu - Bâtiment 22
F-35042 Rennes Cedex

Contact	Jean-Loï Meury
Tel:	+33 2 99 28 63 81
Fax:	+33 2 99 28 16 59
email:	

Université de Rouen
Laboratoire des Polymères et Biopolymères
F-76821 Mt St Aignan Cedex

Contact	Jean-Marc Valleton
Tel:	+33 2 35 14 67 09
Fax:	+33 2 35 14 67 04
email:	valleton@membrane.univ.rouen.fr

Université du Maine
ENSIM
Avenue O. Messiaen
F-72017 Le mans Cedex

Contact	Michel Rousseau
Tel:	+33 2 43 83 35 93
Fax:	+33 2 43 83 37n 94
email:	mrousso@aviion.univ.lemans.fr

Germany

Entries 084 to 163

084 ABB - Asea Brown Boveri AG

Corporate Research Centre
CRH/C "Funktionskeramik/Functional Ceramics"
Postfach 10 13 32
D-69003 Heidelberg

Tel: +49 6221 596000
Fax: +49 6221 596403
E-mail:

Affiliation:
Contact: G Bauer, Head R&D

Organisation information
No of employees >100
Turnover (ECU*) >5m
Sensors technology activities information
No of employees 1-10
Turnover (ECU*) 1-5
*1ECU=DM 1.89,FFr 6.30,£0.78,Lira 1909,Peseta 159.1

Organisation activities
Research and Development
Testing, consulting, design

Application \ Signal Sensed	Radiation	Optical	Electrical	Magnetic	Thermal	Chemical	Mechanical
Automotive						•	
Consumer products							
Process industries, mining, energy						•	
Manufacturing, machinery							
Medicine, health care							
HVAC, fire control, security						•	
Military, aerospace							
Telecommunications,office equip.							
Meteorology, environment							
Other *						•	

Main Activities

Energy, electricity and environmental products.

Specific Expertise

Solid state electrochemical sensors for combustion control. Sensor systems for gas and water analysis.

Collaboration Offered

Research and consultancy for control systems using their sensors and sensor know-how.

Collaboration Sought

Manufacturers and distributors of sensors.

085 AB Elektronik GmbH

Bauteile der Elektronik
Klöcknerstrasse 4
D-59368 Werne

Tel: +49 2389 7880
Fax: +49 2389 788190
E-mail:

Affiliation: AB Electronic Products Group
Contact: HJ Zegula, MD
Ch Suff, Sales Mgr

Organisation information
No of employees >100
Turnover (ECU*) >5

Sensors technology activities information
No of employees >100
Turnover (ECU*) >5
*1ECU=DM 1.89,FFr 6.30,£0.78,Lira 1909,Peseta 159.1

Organisation activities
Research and Development
Manufacturer of sensor-based products
Testing, consulting, design
Manufacture of finished sensors

Application \ Signal Sensed	Radiation	Optical	Electrical	Magnetic	Thermal	Chemical	Mechanical
Automotive			•	•	•		•
Consumer products							
Process industries, mining, energy							
Manufacturing, machinery							
Medicine, health care				•			
HVAC, fire control, security							
Military, aerospace							
Telecommunications,office equip.							
Meteorology, environment				•			
Other *							

Main Activities

AB Elektronik GmbH is a manufacturer of: potentiometers; trimmers; rotary switches; attenuators; rotational speed sensors; ABS sensors; potentiometric sensors; level sensors; temperature sensors; current sensors; connectors and couplings.

Specific Expertise

AB Elektronik develops and manufactures sensors to customer specification. Sensors are developed to monitor steering systems: ABS and ASR systems measure both inside and outside temperatures of the vehicle, for road speed and engine management. The company also carries out its own research into new sensor concepts and continually develops a programme to improve existing sensors.

Collaboration Sought

All proposals are judged individually on their merit.

086 AST Angewandte System Technik GmbH

Masschnerstrasse 26
D-01307 Dresden

Tel: +49 351 4455-30
Fax: +49 351 459 4071
E-mail:

Affiliation:
Contact: S Arudt, Internal Sales Manager
K Wolf, External Sales Manager

Organisation information
No of employees 101-1000
Turnover (ECU*) 5m-50m
Sensor technology activities information
No of employees 11-25
Turnover (ECU*) 0.5m-1m
*1ECU=DM 1.89,FFr 6.30,£0.78,Lira 1909,Peseta 159.1

Organisation activities
Research and development of sensors, sensor materials or applications.
Manufacture of finished sensors.

Application \ Signal Sensed	Radiation	Optical	Electrical	Magnetic	Thermal	Chemical	Mechanical
Automotive							•
Consumer products							
Process industries,mining energy							
Manufacturing, machinery							•
Medicine, health care							
HVAC, fire control, security							
Military, aerospace							
Telecommunications,office equip.			•				
Meteorology, environment							
Other *							

Main Activities

Systems and components for: measurement of force, temperature, sound and vibration; weighing systems and power supplies, driving railway coach preheating systems, warming up and keeping warm systems for diesel locomotives; heating and sanitary systems.
Services: production in the electronic and mechanical field DKD calibration laboratory; EMC measurement laboratory.

Specific Expertise

Force measuring: force transducers/load cells; display and evaluation appliances; authorised calibration of force transducers; project drafting and assembling of complete measuring equipment.
Sound and vibration technique: sound and vibration measuring equipment; sound and vibration analysis.

Collaboration Offered

As above.

Collaboration Sought

Components as amplifiers, displays, slot cards, transducers for acceleration, analysers.

087 ASTECH GmbH

Nottulner Landweg 90
D-48161 Münster

Tel: +49 2534 800180
Fax: +49 2534 800187
E-mail: faoast@uni-muenster.de

Affiliation:
Contact: Dr T Weiss, MD
Mr F Hupe, MD

Organisation information
No of employees 1-25
Turnover (ECU*) <1m
Sensor technology activities information
No of employees 1-10
Turnover (ECU*) <0.5m
*1ECU=DM 1.89,FFr 6.30,£0.78,Lira 1909,Peseta 159.1

Organisation activities
Research and development on sensors sensor materials or applications
Testing, consulting, design
Manufacture of finished sensors
Manufacture of sensor-based products

Application \ Signal Sensed	Radiation	Optical	Electrical	Magnetic	Thermal	Chemical	Mechanical
Automotive							
Consumer products							
Process industries,mining energy							
Manufacturing, machinery							
Medicine, health care						•	
HVAC, fire control, security							
Military, aerospace							
Telecommunications,office equip.							
Meteorology environment						•	
Other (specify)							

Main Activities

Development and production of chemical sensors and biosensors.

Specific Expertise

Electrochemical sensors and biosensors; disposable chemical sensors based on silicon and paper technology.

Collaboration Offered

Customer specific development and design of chemical sensor based products; OEM production capacity.

088 Beck GmbH & Co KG

Druckkontrolltechnik
Maybachstrasse 11
D-70771 Leinfelden-Echterdingen

Tel: +49 711 790830
Fax: +49 711 7908383
E-mail:

Affiliation:
Contact: HP Funk, Mktg Mgr

Organisation information
No of employees 26-100
Turnover (ECU*) 1m-5m
Sensors technology activities information
No of employees 26-100
Turnover (ECU*) 1m-5m
*1ECU=DM 1.89,FFr 6.30,£0.78,Lira 1909,Peseta 159.1

Organisation activities
Research and Development
Manufacturer of sensor-based products
Manufacture of finished sensors

Application \ Signal Sensed	Radiation	Optical	Electrical	Magnetic	Thermal	Chemical	Mechanical
Automotive							•
Consumer products							•
Process industries, mining, energy							•
Manufacturing, machinery							•
Medicine, health care							•
HVAC, fire control, security							•
Military, aerospace							
Telecommunications,office equip.							•
Meteorology, environment							•
Other *							

Main Activities

Development of sensors for pressure and flow of air, gases and non aggressive liquids. Production of these sensors and distribution via representatives in Germany, Europe, United States and Far East countries. Development and manufacture of control units for sensors.

Specific Expertise

Sensing of pressures below 10 bars. Mechanical and electronic sensors for pressure in air, gases and non-aggressive liquids. Main areas are fan control in gas heated devices, liquid level sensing via pressure in household appliances like coffee machines and washing machines. Development of special solutions for customers in the medical field, office equipment or automotive applications. Development and production of electronic sensors. Sensing principle is the capacitive pressure sensing method. Sensors are temperature compensated. Development and production of low cost flow sensors for heating devices.

089 Bernstein Senso Plus GmbH

Rocker Strasse 16
D-31675 Buckerburg

Tel: +49 5722 2080
Fax: +49 5722 20855
E-mail:

Affiliation: Bernstein GmbH
Contact: R Wecke, Eng Dir

Organisation information

No of employees	>100
Turnover (ECU*)	1m-5m

Sensors technology activities information

No of employees	26-100
Turnover (ECU*)	<0.5m

*1ECU=DM 1.89,FFr 6.30,£0.78,Lira 1909,Peseta 159.1

Organisation activities
Research and Development
Manufacturer of sensor-based products
Testing, consulting, design
Manufacture of finished sensors

Application \ Signal Sensed	Radiation	Optical	Electrical	Magnetic	Thermal	Chemical	Mechanical
Automotive							•
Consumer products							
Process industries, mining, energy				•			•
Manufacturing, machinery				•			•
Medicine, health care							•
HVAC, fire control, security							
Military, aerospace							
Telecommunications,office equip.							
Meteorology, environment				•			•
Other *							

Main Activities

Bernstein GmbH manufactures a wide range of limit switches, foot switches and safety switches from the 'Classic' division, whilst 'Senso-Plus' manufactures one of the most comprehensive ranges of proximity sensors. The 'Compact' division produce one of Europes leading ranges of '1965' enclosures in ABS polycarbonate polyester and aluminium. The 'Ex' market is catered for by the Bernstein Secomex company which produces all the above products fully certified. Bernstein Coliprox Ltd imports and markets all these items in the UK.

Specific Expertise

Bernstein Senso Plus develops and manufactures a wide range of inductive and capacitive proximity switches and is developing a range of photoelectronics. Particularly significant developments have been 2-wire DC sensors and analogue output sensors.

090 BLH SR-4 Sensoren GmbH

Boellinger Str. 75
D-74078 Heilbronn

Tel: +49 7131 21511
Fax: +49 7131 21514
E-mail:

Affiliation:
Contact: H Nellmann, Sales Manager

Organisation information
No of employees 101-1000
Turnover (ECU*)

Sensor technology activities information
No of employees >100
Turnover (ECU*) >5m
*1ECU=DM 1.89,FFr 6.30,£0.78,Lira 1909,Peseta 159.1

Organisation activities
Manufacture of finished sensors
Manufacture of sensor-based products

Application \ Signal Sensed	Radiation	Optical	Electrical	Magnetic	Thermal	Chemical	Mechanical
Automotive							•
Consumer products							•
Process industries,mining energy							•
Manufacturing, machinery							•
Medicine, health care							•
HVAC, fire control, security							
Military, aerospace							•
Telecommunications,office equip.							
Meteorology environment							
Other (specify)							

Main Activities

Strain gauges for transducer manufacturing and stress analysis; special gauges for force and torque transducers; high temperature gauges for material testing up to 1000°C; software to calculate strain gauge based transducers; force transducers and load cells for consumer, health care, and manufacturing machinery.

091 Burster Präzisionsmeßtechnik GmbH

Talstraße 1-7
D-76593 Gernsbach

Tel: +49 7224 6450
Fax: +49 7224 64588
E-mail:

Affiliation:
Contact: Matthias Burster, General Manager
Christina Strohm, Marketing Services

Organisation information

No of employees	101-1000
Turnover (ECU*)	5m-50m

Sensor technology activities information

No of employees	26-100
Turnover (ECU*)	1m-5m

*1ECU=DM 1.89,FFr 6.30,£0.78,Lira 1909,Peseta 159.1

Organisation activities
Research and development on sensors sensor materials or applications
Manufacture of finished sensors

Application \ Signal Sensed	Radiation	Optical	Electrical	Magnetic	Thermal	Chemical	Mechanical
Automotive					●		●
Consumer products					●		●
Process industries,mining energy					●		●
Manufacturing, machinery					●		●
Medicine, health care							
HVAC, fire control, security							
Military, aerospace					●		●
Telecommunications,office equip.							
Meteorology environment							
Other (specify)							

Main Activities

Sensors for pressure, load, torque and displacement: pressure transducers and pressure transmitters, differential pressure transmitters, miniature load cells for tension and compression measurements, load cells in various designs for tension and compression measurements, torque transducers for static and rotational applications, displacement transducers, potentiometric.
Processing and acquisition instruments for sensor signals: panel meters for sensors, module amplifier for strain gauge and potentiometric sensors, strain gauge measuring instrument DIGIFORCE, strain gauge simulators, modular systems for the electrical data acquisition of sensor signals SEMMEG, strain gauge signal conditioners and process controllers.
Calibration instruments for temperature, voltage and current: precision temperature panel meters, temperature sensors, precision temperature instruments, portable calibrators, simulators for Pt100 and thermocouples, precision calibrators for current and voltage, DC-standards, precision calibration sources with computer interface.
Resistor and decade boxes: precision resistors, calibration resistors, precision resistance decade boxes.
Digital ohmmeters: micro-ohmmeters, test systems for the production of cable. Quality assurance for coils, wire holding devices.

Specific Expertise

Load cell development in stainless steel.

092 Dantec Measurement Technology GmbH

Am Weichselgarten 21
D-91083 Erlangen

Tel: +49 9131 777 630
Fax: +49 9131 777 644
E-mail:

Affiliation: Dantec Measurement Technology ApS
Contact: Dr E Wieczorek, MD
B Solinbert, Production Manager

Organisation information
No of employees 1-25
Turnover (ECU*) 1m-5m

Sensor technology activities information
No of employees 1-10
Turnover (ECU*) 0.5m-1m
*1ECU=DM 1.89,FFr 6.30,£0.78,Lira 1909,Peseta 159.1

Organisation activities
Manufacture of finished sensors.
Manufacture of sensor-based products.

Application \ Signal Sensed	Radiation	Optical	Electrical	Magnetic	Thermal	Chemical	Mechanical
Automotive							
Consumer products							
Process industries,mining energy							
Manufacturing, machinery							
Medicine, health care		•	•		•		
HVAC, fire control, security		•	•		•		
Military, aerospace							
Telecommunications,office equip.							
Meteorology, environment							
Other *							

Main Activities
Research and development of fluid mechanics measurement equipment and sales /marketing; research and development of flow control systems, manufacturing/sales/marketing; research and development of particle characterisation equipment/sales marketing.

Specific Expertise
Small flow rates of gases; application areas - laboratories, HVAC; hot wire anemometers.

Collaboration Offered
Flow monitoring and control systems for fume hoods; room pressure control systems.

Collaboration Sought
Distribution capability.

093 Data Instruments Next Sensors GmbH

Im Kirchenfelde 8
D-31157 Sarstedt

Tel: +49 5066 7038-0
Fax: +49 5066 65675
E-mail:

Affiliation:
Contact: A Hohgräbe, Technical Support

Organisation information
No of employees 1-25
Turnover (ECU*) 1m-5m
Sensor technology activities information
No of employees
Turnover (ECU*)
*1ECU=DM 1.89,FFr 6.30,£0.78,Lira 1909,Peseta 159.1

Organisation activities
Manufacture of finished sensors.

Application \ Signal Sensed	Radiation	Optical	Electrical	Magnetic	Thermal	Chemical	Mechanical
Automotive							
Consumer products							
Process industries,mining energy							
Manufacturing, machinery							•
Medicine, health care							•
HVAC, fire control, security							
Military, aerospace							
Telecommunications,office equip.							
Meteorology, environment							•
Other *							

Main Activities

Sales and distribution of: semiconductor pressure sensors (micromachined systems); stainless steel pressure transducers; high precision inductive position transducers. Customer support for design. Special solutions.

Collaboration Offered

Customer support for design. Special solutions.

094 Dr Staiger Mohilo & Co GmbH

Baumwasenstr 5
D-73614 Schorndorf

Tel: +49 7181 70080
Fax: +49 7181 700849
E-mail:

Affiliation:
Contact: F Winter, MD
E Mayer, Dir R&D

Organisation information
No of employees 26-100
Turnover (ECU*) >5m

Sensors technology activities information
No of employees 26-100
Turnover (ECU*) >5m
*1ECU=DM 1.89,FFr 6.30,£0.78,Lira 1909,Peseta 159.1

Organisation activities
Research and Development
Manufacturer of sensor-based products
Manufacture of finished sensors

Application \ Signal Sensed	Radiation	Optical	Electrical	Magnetic	Thermal	Chemical	Mechanical
Automotive							•
Consumer products							
Process industries, mining, energy	•						•
Manufacturing, machinery							•
Medicine, health care							
HVAC, fire control, security							
Military, aerospace							
Telecommunications,office equip.							
Meteorology, environment	•						
Other *							

Main Activities

Production of standard and OEM sensors for torque, force, pressure and turbidity. Amplifiers and signal processing as well as dedicated quality control equipment.

Specific Expertise

Sensors for extremely high or low range for torque or force. Adapted dimensions. Sensors for high pressures at elevated temperatures up to 500°C. IR sensors for process control.

Collaboration Offered

Sensors of a.m. domain, adapted to special tasks in research and development applications.

Collaboration Sought

New technologies for production of a.m. sensors.

095 Eltrotec Elektro GmbH

Seestr 9
D-73099 Adelberg

Tel: +49 7166 50300
Fax: +49 7166 50333
E-mail:

Affiliation:
Contact: Claus Hofmann, General Manager
Joclen Rausch, Sales Manager

Organisation information
No of employees 1-25
Turnover (ECU*) 5m-50m
Sensor technology activities information
No of employees 11-25
Turnover (ECU*)
*1ECU=DM 1.89,FFr 6.30,£0.78,Lira 1909,Peseta 159.1

Organisation activities
Research and development on sensors, sensor materials or applications
Manufacture of systems used in the manufacture, testing or development of sensors
Manufacture of finished sensors
Manufacture of sensor-based products

Application \ Signal Sensed	Radiation	Optical	Electrical	Magnetic	Thermal	Chemical	Mechanical
Automotive		•					
Consumer products		•					
Process industries,mining energy		•					
Manufacturing, machinery		•					
Medicine, health care							
HVAC, fire control, security							
Military, aerospace		•					
Telecommunications,office equip.							
Meteorology environment		•					
Other (specify)							

Main Activities

A production and selling company in Germany (distribution in 25 countries) of: colour recognition sensors (1-99 colours) ; fibre optics for sensors; analog light barriers; laser distance sensors; laser light barriers; frame and fork sensors; technical endoscopes; enclosures and housing for electric induststries.

Specific Expertise

Solutions for problems in optical sensors for all types of industries like: colour, dimensions, frame fork sensors, analogue sensors with fibre optics as specialities.

Collaboration Offered

Technology for sensors, application know-how, selling know-how for sensors.

Collaboration Sought

New sort of sensors for the German market. CCD - low cost sneors without software for end users.

096 FRABA Management GmbH

Bremerhavener Street 35
D-50735 Köln

Tel: +49 22 17 15 230
Fax: +49 22 17 12 44 36
E-mail: fuchs@fraba.com

Affiliation:
Contact: P Fuchs, Sales Manager

Organisation information
No of employees 26-100
Turnover (ECU*) 1m-5m
Sensors technology activities information
No of employees 11-25
Turnover (ECU*) 0.5m-1m
*1ECU=DM 1.89,FFr 6.30,£0.78,Lira 1909,Peseta 159.1

Organisation activities
Manufacturer of sensor-based products

Application \ Signal Sensed	Radiation	Optical	Electrical	Magnetic	Thermal	Chemical	Mechanical
Automotive		•					
Consumer products		•					
Process industries, mining, energy							
Manufacturing, machinery		•					
Medicine, health care							
HVAC, fire control, security							
Military, aerospace		•					
Telecommunications,office equip.							
Meteorology, environment							
Other *							

Main Activities

FRABA Management GMBH leads three independent companies: FRABA Sicherheitssysteme GMBH, FRABA Sensorsysteme GMBH, FRABA Systemtechnik GMBH.

Specific Expertise

FRABA Sensorsysteme GMBH is one of the German leaders for:-
Producing optical absolute angular encoders. It supplies sensor systems for monitoring angular and linear movement. The main product range are absolute angular encoders and linear displacement transducers.
FRABA Sicherheissysteme GMBH:-
Producing safety systems for monitoring gates and doors, ie. Pneumatic opto-electric safety switches.

Collaboration Sought

FRABA Sensorsysteme GMBH seeks:
New innovations of measuring systems for angles and distances

097 Fraunhofer-Institut für Informations - und

Datenverarbeitung (IITB)
Fraunhoferstr. 1
D-76131 Karlsruhe

Tel: +49 721 60910
Fax: +49 721 6091560
E-mail:

Affiliation:
Contact: Dr Paul, Dept Mgr
Prof Kroschel, Dept Mgr

Organisation information
No of employees >100
Turnover (ECU*) >5m

Sensors technology activities information
No of employees 11-25
Turnover (ECU*) 0.5m-1m
*1ECU=DM 1.89,FFr 6.30,£0.78,Lira 1909,Peseta 159.1

Organisation activities
Research and Development
Manufacturer of sensor-based products

Application \ Signal Sensed	Radiation	Optical	Electrical	Magnetic	Thermal	Chemical	Mechanical
Automotive							●
Consumer products							
Process industries, mining, energy	●						●
Manufacturing, machinery	●						●
Medicine, health care							
HVAC, fire control, security							
Military, aerospace							●
Telecommunications,office equip.							
Meteorology, environment							
Other *							

Main Activities

Fraunhofer-Gesellschaft is a non-profit organisation for applied research consisting of 42 independent institutes with different fields of activity. Fraunhofer-Institut für Informations - und Datenverarbeitung (IITB) deals with: pattern recognition and automatic image analysis; vision systems; digital signal processing and diagnosis; industrial automation; electronic systems realisation; human-machine communication; computer architectures and communication; system engineering environments; technical expert systems; process control.

Collaboration Offered

Vision systems. Signal processing for acoustic pressure, force, speed sensors etc. Dedicated measuring equipment for process control applications, consisting of sensors and digital signal processing. Measuring devices using models and observers.

098 Fraunhofer-Institut für Integrierte

Schaltungen
Am Weichselgarten 3
D-91058 Erlangen

Tel: +49 9131 776 471
Fax: +49 9131 776 499
E-mail: web@iis.fhg.de

Affiliation: Fraunhofer-Gesellschaft
Contact: Dr N Weber, Group Manager

Organisation information
No of employees 101-1000
Turnover (ECU*) 5m-50m

Sensor technology activities information
No of employees 1-10
Turnover (ECU*) 0.5m-1m
*1ECU=DM 1.89,FFr 6.30,£0.78,Lira 1909,Peseta 159.1

Organisation activities
Research and development on sensors, sensor materials or applications.
Testing, consulting, design.

Application \ Signal Sensed	Radiation	Optical	Electrical	Magnetic	Thermal	Chemical	Mechanical
Automotive							
Consumer products				•			
Process industries,mining energy							
Manufacturing, machinery					•		•
Medicine, health care		•			•		
HVAC, fire control, security							
Military, aerospace							
Telecommunications,office equip.	•						
Meteorology, environment							
Other *							

Main Activities
ASIC and system design and development; telecommunication; quality control; integrated sensors; audio and multimedia systems.

Specific Expertise
Development of integrated sensors in standard CMOS-technology. Types of sensors: MAGFET; current sensors; photodiodes; temperature coil on chip.

Collaboration Offered
Know how in integrated sensors and ASIC design.

Collaboration Sought
Industrial applications.

099 Fraunhofer-Institut für Mikroelektronische

Schaltungen und Systeme
nstitute for Microelectronic Circuits and Systems
Finkenstr. 61
D-47057 Duisburg

Tel: +49 203 37830
Fax: +49 203 3783266
E-mail:

Affiliation:

Contact: W Mokwa, Group Leader
Dr Hoech, Public Relation

Organisation information
No of employees >100
Turnover (ECU*) >5m

Sensors technology activities information
No of employees 11-25
Turnover (ECU*) 1-5
*1ECU=DM 1.89,FFr 6.30,£0.78,Lira 1909,Peseta 159.1

Organisation activities
Research and Development
Testing, consulting, design

Application \ Signal Sensed	Radiation	Optical	Electrical	Magnetic	Thermal	Chemical	Mechanical
Automotive				•			•
Consumer products	•			•			
Process industries, mining, energy							
Manufacturing, machinery				•			•
Medicine, health care			•		•	•	•
HVAC, fire control, security						•	
Military, aerospace							
Telecommunications,office equip.							
Meteorology, environment							
Other *						•	

Main Activities

The Fraunhofer-Institute for Microelectronic Circuits and Systems exploits the practical possibilities of microelectronics. The Institute's efforts are devoted to the design, fabrication and application of integrated circuits that go beyond standard concepts. Very large scale integration: dielectric isolation of devices (silicon on insulator); digital signal processing; CAD for system design. Industrial Electronics: planar CMOS-processes for medium voltages; smart actuators; active protection in hostile environment. Analogue Circuits: knowledge-based design system; transistor and interconnection modelling; analogue signal processing. Solid state sensors: chemical sensors (CO, H_2, NH_3, H_2S); physical sensors (temperature, light, magnetic fields, pressure); sensor readout electronics.

Specific Expertise

Physical and chemical microsensors based on silicon; smart sensors, having sensing element and corresponding circuits on one chip; specific sensors: temperature sensors; pressure sensors; magnetic field sensors; flow sensors.

Collaboration Offered

Consultancy. Research for new sensors based on silicon technology. Development of prototype sensors or sensor systems.

100 Fraunhofer-Institut für Physikalische

Messtechnik (IPM)
Heidenhofstrasse 8
D-79110 Freiburg

Tel: +49 761 88570
Fax: +49 761 8857224
E-mail:

Affiliation:
Contact: H Höfler, Fibre Optic Sensors
A Dumbs, Integrated Optics

Organisation information
No of employees 26-100
Turnover (ECU*) >5m

Sensors technology activities information
No of employees 26-100
Turnover (ECU*) >5m
*1ECU=DM 1.89,FFr 6.30,£0.78,Lira 1909,Peseta 159.1

Organisation activities
Research and Development
Testing, consulting, design
Manufacture of finished sensors

Application \ Signal Sensed	Radiation	Optical	Electrical	Magnetic	Thermal	Chemical	Mechanical
Automotive	•	•	•	•	•	•	•
Consumer products							
Process industries, mining, energy	•	•	•	•	•	•	•
Manufacturing, machinery	•	•	•	•	•	•	•
Medicine, health care	•	•			•		
HVAC, fire control, security	•				•		
Military, aerospace							
Telecommunications,office equip.							
Meteorology, environment							
Other *							

Main Activities

Development of measuring techniques with optical, fibre optic, integrated optic sensors. Application to industrial, automotive, biomedical, chemical areas.

Specific Expertise

Sensors developed in the area of fibre and integrated optics: pressure; displacement and position; liquid level; refractive index; temperature; mobility; distributed sensors; sensor networks. Design and fabrication of integrated optical components such as switches, modulators, polarisers Bragg cells, interferometers.

Collaboration Offered

Development of fibre optics sensors. Development and fabrication of integrated optic components. Consultancy.

Collaboration Sought

Fibre optic sensors. Integrated optics, new products in this area. Spectroscopy. Detectors.

101 Fraunhofer-Institut für Produktionstechnik

und Automatisierung
Nobelstraße12
D-70569 Stuttgart

Tel: +49 711/970-1800
Fax: +49 711/970-1004
E-mail: kam@qt.ipa.fhg.de

Affiliation:
Contact: Dr K Melchior, Tech. Prod. Manager

Organisation information
No of employees 101-1000
Turnover (ECU*) 5m-50m
Sensor technology activities information
No of employees 26-100
Turnover (ECU*) 1m-5m
*1ECU=DM 1.89,FFr 6.30,£0.78,Lira 1909,Peseta 159.1

Organisation activities
Testing, consulting, design.

Application \ Signal Sensed	Radiation	Optical	Electrical	Magnetic	Thermal	Chemical	Mechanical
Automotive		•	•				•
Consumer products							
Process industries,mining energy						•	
Manufacturing, machinery		•				•	
Medicine, health care							
HVAC, fire control, security							
Military, aerospace							
Telecommunications,office equip.							
Meteorology, environment							
Other *							

Main Activities

The Institute's research and development projects are focused on organizational and technological problems associated with production in industrial enterprises. The research and development projects aim at maintaining or improving the company's competitiveness and employment levels with better, cheaper and ecologically compatible production methods by finding and exhausting the entire automation and rationalization potential within the company.

Specific Expertise

Corporate mangement.
Automation.
Manufaturing technics.

102 Fraunhofer-Institut für Siliziumtechnologie

Fraunhoferstraße 1
D-25524 Itzehoe

Tel: +49 4821 17 4214
Fax: +49 4821 17 4251
E-mail: wacker@isit.fhg.de

Affiliation:
Contact: CLaus Wacker, Public Relations
Dr Ralf Dudde, Marketing

Organisation information
No of employees 101-1000
Turnover (ECU*) 5m-50m

Sensor technology activities information
No of employees 26-100
Turnover (ECU*) 1m-5m
*1ECU=DM 1.89,FFr 6.30,£0.78,Lira 1909,Peseta 159.1

Organisation activities
Research and development on sensors sensor materials or applications
Testing, consulting, design
Supply of raw materials or semi-finished products used to make sensors
Manufacture of systems used in the manufacture, testing or development of sensors
Manufacture of finished sensors

Application \ Signal Sensed	Radiation	Optical	Electrical	Magnetic	Thermal	Chemical	Mechanical
Automotive			•		•	•	•
Consumer products		•			•		
Process industries,mining energy							
Manufacturing, machinery		•	•				
Medicine, health care					•	•	•
HVAC, fire control, security							
Military, aerospace		•					•
Telecommunications,office equip.		•			•		
Meteorology environment						•	•
Other (specify)							

Main Activities

The Fraunhofer-Institute for Silicon Technology is running one of the most modern 6 inch, industry like pilot lines for research, development and production of microelectronics, sensors, actuators and microsystems based on silicon.
ISiT supplies technology for the manufacturing of sensors, discrete and integrated circuits and their combination into monolithic or hybrid integrated microsystems.
Additional working fields are the development of packaging and mounting techniques for sensor elements in application specific houses and the development of high reliability/low cost packages.

Specific Expertise

Areas of Expertise: development and design of silicon based sensors, actuators and microsystems and fabrication processes for various applications; small series production of developed devices with modern equipment in cleanrooms; development of packaging and mounting techniques for sensor elements in application specific houses; development of high reliability/low cost packages.
Specific materials: only micromachined silicon sensors.

Collaboration Offered

ISiT belong to Fraunhofer-Society, the largest organisation in Europe for applied research and development. We offer contract based R&D in the sensors and microsystems business including consulting and services for project funding. Small series production for developed devices is also possible.

103 Fraunhofer Institute for Solid State Tech.

Hansastrasse 27d
D-80686 Muenchen

Tel: +49 89 54759 240
Fax: +49 89 54759-100
E-mail: hillerich ift.fhg.de

Affiliation: Fraunhofer Gesellschaft
Contact: Dr B Hillerich, Micro Technologies & Sytems
Dr G Roeska, ASICs Development
Dr S Drost, Chemical & Bio Sensors
Mr F Kozlowski, Mechanical Sensors

Organisation information
No of employees 101-1000
Turnover (ECU*) 5m-50m

Sensor technology activities information
No of employees 26-100
Turnover (ECU*) 1m-5m
*1ECU=DM 1.89,FFr 6.30,£0.78,Lira 1909,Peseta 159.1

Organisation activities
Research and development on sensors sensor materials or applications
Testing, consulting, design
Supply of raw materials or semi-finished products used to make sensors
Manufacture of finished sensors
Manufacture of sensor-based products

Application \ Signal Sensed	Radiation	Optical	Electrical	Magnetic	Thermal	Chemical	Mechanical
Automotive							•
Consumer products							•
Process industries,mining energy						•	
Manufacturing, machinery							•
Medicine, health care						•	
HVAC, fire control, security						•	
Military, aerospace							
Telecommunications,office equip.							
Meteorology environment						•	
Other							

Main Activities

The sensor related activities of the Fraunhofer Institute for Solid State Technology Contract are: contractual research and development for sensor elements; sensor systems and MEMS as well as individual development steps (e.g. simulation or special technological steps); prototype manufacture and (small-scale) series production. Furthermore development and production of ASICs for sensors.

Specific Expertise

Chemical and biosensors: ionsensitive field effect transistor, quartz crystal microbalance (antibody, viruses, bacteria), SAW oscillators (conductance, viscosity), complete flow injection analysis systems;Gas sensors for combustible and toxic gases, for alcohol, smelling substances,and air quality based on the principles - metal oxide semiconductor and pellistor in Si micromechanics, SAW, dielectric with organic indicator; signal processing for sensor arrays.
Mechanical sensors: sensors for physical values (pressure, acceleration, force) using several Si micromachining technologies, also with integrated microelectronics (CMOS, SOI), software tools for design optimisation.
ASICs for Sensors: Sensor circuits for resistive bridges and capacitive sensors, macro cells, and high-temp circuits produced in qualified CMOS foundry.

Collaboration Offered

Fully confidential R&D work on above fields. Development and production of sensor cells, ASICs and fluidmeasuring systems.

Collaboration Sought

Sensitive layers for chemical, biochemical, and gas sensors. Technology for micro machined sensors and other MEMS devices low cost production technologies.

104 Fraunhofer Institut für Zerstörungsfreie

Prüfverfahren
Universität Gebäude 37
D-66123 Saarbrücken

Tel: +49 68 13 02 38 04
Fax: +49 68 13 72 63 6
E-mail: dobmann@izfp.fhg.de

Affiliation: Fraunhofer Society
Contact: Dr G Dobmann, Dep. Director
Dr W Gebharott, Project Mgr

Organisation information

No of employees 101-1000
Turnover (ECU*) 5m-50m

Sensor technology activities information

No of employees 11-25
Turnover (ECU*) 1m-5m
*1ECU=DM 1.89,FFr 6.30,£0.78,Lira 1909,Peseta 159.1

Organisation activities

Research and development on sensors, sensor materials or applications
Testing, consulting, design
Supply of raw materials or semi-finished products used to make sensors
Manufacture of sensor-based products

Application \ Signal Sensed	Radiation	Optical	Electrical	Magnetic	Thermal	Chemical	Mechanical
Automotive			●	●	●		●
Consumer products							
Process industries,mining energy	●	●	●	●	●		●
Manufacturing, machinery		●	●	●	●	●	●
Medicine, health care	●						●
HVAC, fire control, security							
Military, aerospace							
Telecommunications,office equip.							
Meteorology environment			●	●			●
Other (specify) *Plant monitoring	●	●	●	●	●		●

Main Activities

Research and development of sensors as a part of transducers for non-destructive testing devices in the field of: radiation (mainly for tomography inspection); ultrasonic testing (all kinds of applications); electromagnetic testing (eddy current, magnetic leakage field, dynamic micromagnetic techniques); thermal inspection (mainly for ceramics and composites); acoustic emission; medical technique. Development of NDT inspection systems (transducers, equipment, hard and software for interpretation).

Specific Expertise

- Piezoelectric Composites for ultrasonic sensors
- Eddy Current and magnetic Inductive sensors
- Hard and software intelligence for sensors.

Collaboration Offered

Monitoring and process control

Collaboration Sought

Modelling and simulation of processors.

105 Hamos Elektronik

Ruhe am Bach 5
PO Box 1243
D-82377 Penzberg

Tel: +49 8856 2011
Fax: +49 8856 1375
E-mail:

Affiliation:

Contact: Dr Koehnlechner, Gen Mgr
Ms Meyer, Mktg

Organisation information

No of employees 1-10
Turnover (ECU*) 1m-5m

Sensors technology activities information

No of employees 1-10
Turnover (ECU*) 1m-5m

*1ECU=DM 1.89,FFr 6.30,£0.78,Lira 1909,Peseta 159.1

Organisation activities

Research and Development
Manufacturer of sensor-based products
Testing, consulting, design
Manufacture of finished sensors

Application \ Signal Sensed	Radiation	Optical	Electrical	Magnetic	Thermal	Chemical	Mechanical
Automotive							
Consumer products							
Process industries, mining, energy				•			
Manufacturing, machinery				•			
Medicine, health care							
HVAC, fire control, security							
Military, aerospace							
Telecommunications,office equip.							
Meteorology, environment							
Other *							

Main Activities

Research, development, production and sales of all electronic metal detectors and separators and electronic colour sorting machines.

Specific Expertise

Metallic particles (ferrous and non-ferrous) are detected in bulk material by inductive coils (inductive ring sensor). Sensitivity is very high where particles from 0.3 mm diameter can be detected. In connection with a rejection mechanic all metallic particles can be rejected completely automatically. Sensor systems similar to these all-metal separators, are used in the plastic, food or chemical industry to protect the machines and the product.

Collaboration Offered

Development of metal detectors for any application.

106 Hartmann & Braun AG

Sensortechnik
Postfach 900507
D-60488 Frankfurt/M90

Tel: +49 69 7993350
Fax: +49 69 7992570
E-mail:

Affiliation:
Contact: GR Wagner, Sensors
B Horn, Dev&Manf

Organisation information
No of employees >100
Turnover (ECU*) >5m

Sensors technology activities information
No of employees 11-25
Turnover (ECU*) 0.5m-1m
*1ECU=DM 1.89,FFr 6.30,£0.78,Lira 1909,Peseta 159.1

Organisation activities
Research and Development
Manufacturer of sensor-based products
Manufacture of finished sensors

Application \ Signal Sensed	Radiation	Optical	Electrical	Magnetic	Thermal	Chemical	Mechanical
Automotive							
Consumer products							
Process industries, mining, energy	•				•	•	•
Manufacturing, machinery							
Medicine, health care	•				•	•	
HVAC, fire control, security	•				•	•	
Military, aerospace							
Telecommunications,office equip.							
Meteorology, environment	•				•	•	
Other *							

Main Activities

Hartmann & Braun AG is one of the major companies operating in the field of instrumentation and process automation. It was founded in 1879 and has today approximately 8,000 employees and 11 subsidiary companies in Europe, the US, Brazil, Africa and East Asia. The Sensortechnik Department was formed in 1986 to develop and market sensor technology based on the capabilities of Hartmann & Braun AG in micromechanics. It delivers to both in-house and OEM customers.

Specific Expertise

Silicon micromachined sensors; solid state chemical sensors; optical sensors.
Specific sensors: thermal conductivity sensors for determining contents of two and three phase gas mixtures; vacuum pressure sensors; bolometric sensors (eg. for pyrometry); wavelength selective radiation sensors for monitoring chemical concentrations.

Collaboration Offered

TCS series thermal conductivity sensors. VPS series vacuum pressure sensors. BRS series bolometric radiation sensors. Development and manufacturing of sensors listed above.

Collaboration Sought

OEM users for sensors series listed.

107 Heimann Optoelectronics GmbH

Weher Köppel 3
D-65197 Wiesbaden

Tel: +49 611 492 303
Fax: +49 611 492 228
E-mail:

Affiliation: EG&G USA
Contact: Dr J Schieferdecker, R&D Manager
W Schmidt, Manager Mkting& Sales
Dr M Gauer, Manager Mkting electr.

Organisation information
No of employees >1000
Turnover (ECU*) >50m
Sensor technology activities information
No of employees >100
Turnover (ECU*) 1m-5m
*1ECU=DM 1.89,FFr 6.30,£0.78,Lira 1909,Peseta 159.1

Organisation activities
Research and development on sensors sensor materials or applications
Manufacture of finished sensors

Application \ Signal Sensed	Radiation	Optical	Electrical	Magnetic	Thermal	Chemical	Mechanical
Automotive	●	●			●		
Consumer products	●	●			●		
Process industries,mining energy	●	●			●		
Manufacturing, machinery	●						
Medicine, health care	●	●			●		
HVAC, fire control, security	●	●			●		
Military, aerospace							
Telecommunications,office equip.							
Meteorology environment		●					
Other							

Main Activities

Sensors for infrared radiation; photocells for light detection; large area amorphos silicon detector arrays for medical applications; accelerometer sensors for automotive applications; vacuum sensors (micro piranis).

Specific Expertise
Infrared detectors and sensor arrays.

Collaboration Offered
Collaboration with sensor systems manufacturer. Heimann is a specialist sensor component manufacturer.

Collaboration Sought
Incorporate our product into a complete sensor.

108 Hengstler GmbH

Postfach 1151
D-78554 Aldingen

Tel: +49 7424 890
Fax: +49 7424 89370
E-mail:

Affiliation:
Contact: G Haller,Prod Mgr

Organisation information
No of employees >100
Turnover (ECU*) >5m
Sensors technology activities information
No of employees
Turnover (ECU*) 1m-5m
*1ECU=DM 1.89,FFr 6.30,£0.78,Lira 1909,Peseta 159.1

Organisation activities
Manufacture of finished sensors

Application \ Signal Sensed	Radiation	Optical	Electrical	Magnetic	Thermal	Chemical	Mechanical
Automotive							
Consumer products							
Process industries, mining, energy							•
Manufacturing, machinery							•
Medicine, health care							•
HVAC, fire control, security							
Military, aerospace							•
Telecommunications,office equip.							•
Meteorology, environment							•
Other *							

Main Activities

Manufacture of: mechanical, electro-mechanical and electronic counters and controls; time and acces control; shaft encoders, incremental/absolute; opto sensors; proximity switches.

Specific Expertise

Incremental encoders: from 1-5000 pulses; industrial standard types; miniature type; heavy duty type. Absolute encoders: single turn 9-14 bit; multi-turn 12+12 bit; high speed version; parallel or serial data transmission. Opto sensors: through beam; retro reflex; scanner; max. 20 m; polarisation filter. Proximity switches: inductive dia. 4 mm - 30 mm; capacitive; magnet/ reed switches.

109 Heraeus Sensor GmbH

Reinhard-Heraeus-Ring 23
D-63801 Kleinostheim

Tel: +49 6027 503-0
Fax: +49 6027 503-101
E-mail:

Affiliation:
Contact: R Seitz, Mktg Dir
U Wittemann, Sales Mgr

Organisation information
No of employees >100
Turnover (ECU*) >5m

Sensors technology activities information
No of employees >100
Turnover (ECU*) >5m
*1ECU=DM 1.89,FFr 6.30,£0.78,Lira 1909,Peseta 159.1

Organisation activities
Research and Development
Manufacture of finished sensors

Application \ Signal Sensed	Radiation	Optical	Electrical	Magnetic	Thermal	Chemical	Mechanical
Automotive					•		
Consumer products					•		
Process industries, mining, energy					•		
Manufacturing, machinery					•		
Medicine, health care							
HVAC, fire control, security					•		
Military, aerospace							
Telecommunications,office equip.					•		
Meteorology, environment							
Other *							

Main Activities

Manufacture and distribution of temperature sensors such as: resistance thermometer detectors; complete resistance thermometers; thermocouples; thermowires; components for temperature sensors. QuaT-temperature measurement system; calibration systems for temperature sensors; DKD-calibration service; temperature transmitters.

Specific Expertise

Temperate sensing, resistance thermometers, thermocouples, quartz oscillator temperature measuring system.

110 Hottinger Baldwin Messtechnik GmbH

Im Tiefen See 45
PO Box 4235
D-64283 Darmstadt

Tel: +49 6151 8030
Fax: +49 6151 895677
E-mail:

Affiliation:
Contact: D Franke, Mktg Dir

Organisation information
No of employees
Turnover (ECU*) >5m
Sensors technology activities information
No of employees
Turnover (ECU*) >5m
*1ECU=DM 1.89,FFr 6.30,£0.78,Lira 1909,Peseta 159.1

Organisation activities
Research and Development
Manufacturer of sensor-based products
Testing, consulting, design
Manufacture of finished sensors

Application \ Signal Sensed	Radiation	Optical	Electrical	Magnetic	Thermal	Chemical	Mechanical
Automotive							•
Consumer products							
Process industries, mining, energy							•
Manufacturing, machinery							•
Medicine, health care							
HVAC, fire control, security							
Military, aerospace							
Telecommunications,office equip.							
Meteorology, environment							•
Other *							

Main Activities

HBM develop, produce and distribute/sell products for the electrical measurement and mechanical quantities. Products means sensors, transducers, amplifiers, transmitters, scanners, indicators, interfaces, precision instruments etc. for measuring force, mass, torque (including speed and power), pressure, strain, displacement and vibration.
With these components and several accessories like mounting aids for the transducers or funtional units for the amplifiers, one can build up a complete measuring chain to solve the measurement task.

Specific Expertise

For the transducers the company uses metallic strain gauges (foil and thin film techniques) as well as the differential reluctance principle. As amplifiers the company offers DC-types as carrier frequency types.

Collaboration Offered

HBM offers load cells, force, transducers, pressure transducers, strain transducers, vibration transducers and instruments, sensors, amplifiers, function units, indicators, digital precision instruments, multipoint measuring instruments, output equipment, measured value transmission equipment. With these products complete measuring chains can be built up for solving the measurement tasks.

Collaboration Sought

Information which deals with the above mentioned products and their application fields

111 Institut Dr Friedrich Förster GmbH

In Laisen 70
D-72766 Reutlingen

Tel: +49 7121 140 215
Fax: +49 7121 140 357
E-mail:

Affiliation:
Contact: M Förster, GM, Sensor Products

Organisation information

No of employees 101-1000
Turnover (ECU*) >50m

Sensor technology activities information

No of employees 11-25
Turnover (ECU*) 0.5m-1m

*1ECU=DM 1.89,FFr 6.30,£0.78,Lira 1909,Peseta 159.1

Organisation activities

Research and development on sensors, sensor materials or applications.
Testing, consulting, design.
Manufacture of finished sensors.
Manufacture of sensor-based products.

Application \ Signal Sensed	Radiation	Optical	Electrical	Magnetic	Thermal	Chemical	Mechanical
Automotive							
Consumer products							
Process industries,mining energy							
Manufacturing, machinery							
Medicine, health care							
HVAC, fire control, security							
Military, aerospace			●	●			
Telecommunications,office equip.							
Meteorology, environment							
Other *							

Main Activities

Förster GmbH develops electrical sensors for the military and aerospace industries.

112 Institut für Physikalische Hochtechnologie

Postfach 100 239
D-07702 Jena

Tel: +49 3641 302900/657702
Fax: +49 3641 302905/657700
E-mail: hoenig@main.ipht.jena.de

Affiliation:
Contact: Prof H E Hoenig, Director
J Müller, Thermal Sensors
Dr R Mattheis, Magnetic Sensors
Dr R Riesenberg, Infrared Optics

Organisation information
No of employees 101-1000
Turnover (ECU*) 5m-50m

Sensor technology activities information
No of employees 26-100
Turnover (ECU*) 1m-5m
*1ECU=DM 1.89,FFr 6.30,£0.78,Lira 1909,Peseta 159.1

Organisation activities
Research and development on sensors sensor materials or applications
Manufacture of finished sensors

Application \ Signal Sensed	Radiation	Optical	Electrical	Magnetic	Thermal	Chemical	Mechanical
Automotive				•			
Consumer products							
Process industries,mining energy		•		•			
Manufacturing, machinery							
Medicine, health care		•		•		•	
HVAC, fire control, security					•		
Military, aerospace					•		
Telecommunications,office equip.							
Meteorology environment		•			•		
Other (specify)							

Main Activities

Thermoelectric mico sensors; infrared analysis of harmful substances; microsystems for chemo- and biotechnology, e.g. evolution machines; magnetoresistive mulitlayers and sensors; superconducting Quantum Interferometers (SQUIDs) robust to magnetic disturbances; a 10GHz voltage standard; high frequency oscillators; epitaxial HTS films - YBCO and TBCCO devices (circuits) for microwave and sensor applications; interferometric and spectral encoded fibreoptic sensors and measuring principles; distributed sensor systems; fibreoptic modules for communication and signal processing; micro mounting and packaging techniques.

Specific Expertise

Cryoelectronics; thin film microsystems; optical fibres; micro optics; optical micro systems; laser diagnostics.

Collaboration Sought

We would like to invite you to join us as a partner in research assignment and in common R&D works in the frame of national and international research programmes. We offer our know-how, service and development of demonstrator devices.

113 IMPAC Electronic GmbH

Kriftelerstr 32
D-60326 Frankfurt/M

Tel: +49 69 97373-0
Fax: +49 69 97373-167
E-mail:

Affiliation:
Contact: H Painly, Product Manager

Organisation information
No of employees 26-100
Turnover (ECU*)

Sensor technology activities information
No of employees 11-25
Turnover (ECU*)
*1ECU=DM 1.89,FFr 6.30,£0.78,Lira 1909,Peseta 159.1

Organisation activities
Research and development on sensors, sensor materials or applications.
Testing, consulting, design.
Manufacture of sensor-based products.

Application \ Signal Sensed	Radiation	Optical	Electrical	Magnetic	Thermal	Chemical	Mechanical
Automotive	•				•		
Consumer products	•				•		
Process industries,mining energy	•				•		
Manufacturing, machinery	•				•		
Medicine, health care	•				•		
HVAC, fire control, security							
Military, aerospace	•				•		
Telecommunications,office equip.							
Meteorology, environment	•				•		
Other *							

Main Activities

Measuring and control systems for non-contact and contact measurements of temperatures and for the measurement of air humidity.

114 IMT - Industrie Messtechnik Frankfurt

An der Sandelmühle 13
D-60439 Frankfurt

Tel: +49 69 5806-0
Fax: +49 69 5806-297
E-mail:

Affiliation:
Contact: F J Vogel, Assist. GM
Peter Ottens, Prod Mgr

Organisation information

No of employees 101-1000
Turnover (ECU*) 5m-50m

Sensor technology activities information

No of employees 11-25
Turnover (ECU*) 1m-5m

*1ECU=DM 1.89,FFr 6.30,£0.78,Lira 1909,Peseta 159.1

Organisation activities
Testing, consulting, design.
Manufacture of finished sensors.

Application \ Signal Sensed	Radiation	Optical	Electrical	Magnetic	Thermal	Chemical	Mechanical
Automotive							●
Consumer products							
Process industries,mining energy					●		●
Manufacturing, machinery					●		●
Medicine, health care					●		●
HVAC, fire control, security					●		●
Military, aerospace							●
Telecommunications,office equip.							●
Meteorology, environment					●		●
Other *							

Main Activities

Complete range of instruments for measuring pressure and temperature from mechanical electronical. Business relations all over Europe and 60 other countries.

Specific Expertise

Pressure transmitters based on either piezo or thinfilm technology. Temperature transmitters and PT 100.

115 Jenasensoric eV

Am Planetarium 5
D-07743 Jena

Tel: +49 3641 463022
Fax: +49 3641 463022
E-mail:

Affiliation:
Contact: Horst Ahlers, Chairman

Organisation information
No of employees 1-25
Turnover (ECU*) <1m
Sensor technology activities information
No of employees 11-25
Turnover (ECU*) 0.5m-1m
*1ECU=DM 1.89,FFr 6.30,£0.78,Lira 1909,Peseta 159.1

Organisation activities
Research and development

Application \ Signal Sensed	Radiation	Optical	Electrical	Magnetic	Thermal	Chemical	Mechanical
Automotive		●			●	●	
Consumer products		●			●	●	
Process industries,mining energy		●			●	●	
Manufacturing, machinery							
Medicine, health care		●			●	●	
HVAC, fire control, security		●			●	●	
Military, aerospace		●			●	●	
Telecommunications,office equip.							
Meteorology, environment		●			●	●	
Other *							

Main Activities

The main activities of the organisation is to: look for efficient co-operation between scientists at the University and industry; search for new research fields using the specific knowledge of the Jena/Thuringia area; transfer the results to industry. Furthermore, Jenasensoric deals with teaching and qualification of scientific workers in the sensor field (desvelopment, measurement & application).

Specific Expertise

Competence for intelligent multisensors with special sleeving and with special materials; front end sensors.

Collaboration Offered

Common projects in research and development

116 Kayser Threde GmbH

Perchtingerstrasse 3
D-81306 München

Tel: +49 89 7872065
Fax: +49 89 7872098
E-mail:

Affiliation:
Contact: H-V Heyer, Head Dev Dept

Organisation information
No of employees >100
Turnover (ECU*) >5m
Sensors technology activities information
No of employees 1-10
Turnover (ECU*) 1-5
*1ECU=DM 1.89,FFr 6.30,£0.78,Lira 1909,Peseta 159.1

Organisation activities
Manufacturer of sensor-based products

Application \ Signal Sensed	Radiation	Optical	Electrical	Magnetic	Thermal	Chemical	Mechanical
Automotive					•		•
Consumer products							
Process industries, mining, energy					•		•
Manufacturing, machinery					•	•	•
Medicine, health care							
HVAC, fire control, security							
Military, aerospace					•		•
Telecommunications,office equip.							
Meteorology, environment						•	
Other *							

Main Activities

The organisation is involved in aerospace projects (80%). Some example projects are: SOHO (SUMER, LASCO); ORFEUS - ASTROSPAS (Space Shuttle); MEDEA - SPACELAB. Other activities: data acquisition systems (mobile and stationary), including: car crash test systems, road load test systems, turbine test facilities etc.

Specific Expertise

Kayser Threde GmbH manufactures modules for sensors: transducers identification circuit; intelligent front end with DSP. Sensor/Transducer indentification: this circuit gives the sensor a unique number or a complete data set (calibration data). Works together with the ORACLE Data base. The Intelligent Front End: possibility of pre-processing sensor data; filtering, classification, linearisation, integration, FFT, etc.

Collaboration Sought

Distribution of products in USA.

117 Klaschka GmbH & Co

Elektronik & Automation
Postfach 1120
D-75229 Tiefenbronn

Tel: +49 7234 790
Fax: +49 7234 79112
E-mail:

Affiliation:
Contact: Günter Schulz, Marketing
Uwe Scobel-Freimüller, Sales

Organisation information
No of employees 101-1000
Turnover (ECU*) 5m-50m
Sensor technology activities information
No of employees 26-100
Turnover (ECU*) 1m-5m
*1ECU=DM 1.89,FFr 6.30,£0.78,Lira 1909,Peseta 159.1

Organisation activities
Research and development on sensors sensor materials or applications
Testing, consulting, design
Supply of raw materials or semi-finished products used to make sensors
Manufacture of systems used in the manufacture, testing or development of sensors
Manufacture of finished sensors
Manufacture of sensor-based products

Application \ Signal Sensed	Radiation	Optical	Electrical	Magnetic	Thermal	Chemical	Mechanical
Automotive			•	•	•		
Consumer products			•				
Process industries,mining energy			•	•	•		
Manufacturing, machinery		•	•	•	•		
Medicine, health care							
HVAC, fire control, security					•		
Military, aerospace			•				
Telecommunications,office equip.							
Meteorology environment							
Other			•				

Main Activities

The "Klaschka Elektronik & Automation" company was founded in 1964. Its development from a two-man company to a medium-sized enterprise with a staff of 250 is linked closely with the development of industrial automation. The product range embraces the entire spectrum from industrial sensors through control and monitoring systems to industrial network spanning several cells, programmable logic controllers in all performance classes as well as industrial computers.
As a partner of the large automobile manufacturers for many years, we have also made a name for ourselves outside this branch, for example in the construction of textile, printing, and special machines. Besides these, projects in the fields of pollution control, energy and water supply, and in sewage plants are increasingly gaining in significance.
Furthermore we have specialised on the the development of customer specific products, especially sensors, and controllers.

Specific Expertise

Rotation, speed and direction controlling, magnetic field plates and halle element sensors, metal thickness controlling, contactless with contact (ferrous and non ferrous metals), high temperature sensors (800°C),metal and non metal materials.

Collaboration Offered

Fabrication and development of distance sensors on an inductive on capacitive base.

Collaboration Sought

Development and fabrication of sensor elements on a micromechanical base.

118 Klaus Fischer Meß- und Regeltechnik

Bielefelder Str 37a
D-32107 Bad Salzuflen

Tel: +49 5222 9740
Fax: +49 5222 7170
E-mail: klausfischer@t-online.de

Affiliation:
Contact: C-D Pilch, MD
Mr Kerker, Mark & QC Mgr

Organisation information

No of employees	101-1000
Turnover (ECU*)	5m-50m

Sensor technology activities information

No of employees	26-100
Turnover (ECU*)	1m-5m

*1ECU=DM 1.89,FFr 6.30,£0.78,Lira 1909,Peseta 159.1

Organisation activities
Research and development on sensors, sensor materials or applications.
Testing, consulting, design.
Supply of raw materials or semi-finished products used to make sensors.
Manufacture of finished sensors.
Manufacture of sensor-based products

Application \ Signal Sensed	Radiation	Optical	Electrical	Magnetic	Thermal	Chemical	Mechanical
Automotive							
Consumer products			•		•		
Process industries,mining energy			•				•
Manufacturing, machinery		•	•		•		•
Medicine, health care							
HVAC, fire control, security							
Military, aerospace							
Telecommunications,office equip.							
Meteorology, environment			•		•		•
Other *							

Main Activities
Manufacturer, designer and trading organisation for pressure, tempeature and level sensors and transmitters.

Specific Expertise
Sensor thick film technology with "ASIC". Sensor thin film technology with "ASIC". We are using "stainless steel" materials for measuring abrasive medias like water, wastewater, gas (fluid components).

Collaboration Offered
Pre developed pressure sensor; an international development cooperation.

Collaboration Sought
Experience with ASIC design.

119 Laaser & Co GmbH

Feurigstraße 59
D-10827 Berlin

Tel: +49 30 78 41 081
Fax: +49 30 78 20 508
E-mail:

Affiliation:
Contact: Dieter Platow, Manager

Organisation information

No of employees 1-25
Turnover (ECU*) 1m-5m

Sensor technology activities information

No of employees 11-25
Turnover (ECU*) 1m-5m

*1ECU=DM 1.89,FFr 6.30,£0.78,Lira 1909,Peseta 159.1

Organisation activities
Manufacture of finished sensors

Application \ Signal Sensed	Radiation	Optical	Electrical	Magnetic	Thermal	Chemical	Mechanical
Automotive			•				
Consumer products							
Process industries,mining energy			•	•	•		•
Manufacturing, machinery			•	•			•
Medicine, health care			•				
HVAC, fire control, security							
Military, aerospace							
Telecommunications,office equip.							
Meteorology environment							
Other							

Main Activities

Manufacturer of sensors for flow, pressure, differential pressure and level.

Specific Expertise

Flow control and flow measurement; flow switches and flow meters.

Collaboration Offered

Knowledge of how to detect and meter small flow quantities, leakages, pipe breakages.

Collaboration Sought

Laser-doppler-anemometry.

120 LOT GmbH

Im Tiefen See 58
D-64283 Darmstadt

Tel: +49 6151 88060
Fax: +49 6151 896667
E-mail:

Affiliation:
Contact: Mr Hilger, MD
KE Aauifa, Mktg Director

Organisation information
No of employees
Turnover (ECU*)

Sensors technology activities information
No of employees
Turnover (ECU*)
*1ECU=DM 1.89,FFr 6.30,£0.78,Lira 1909,Peseta 159.1

Organisation activities
Manufacturer of sensor-based products
Manufacture of finished sensors

Application \ Signal Sensed	Radiation	Optical	Electrical	Magnetic	Thermal	Chemical	Mechanical
Automotive							•
Consumer products							
Process industries, mining, energy							•
Manufacturing, machinery							•
Medicine, health care							
HVAC, fire control, security							
Military, aerospace							
Telecommunications,office equip.							
Meteorology, environment							
Other *							

Main Activities

Distribution and engineering of optical and laser-optical measuring systems.

Specific Expertise

LTS: laserscanning system for diameter roundness etc. measurements. LAS: laser distance measuring systems. LPA: laser positioning and alignment system. AXION: laser interferometer high precision distance and position measuring system.

Collaboration Offered

Laser-optical dimension measurements. All products and services.

Collaboration Sought

Laser-optical and optical measuring systems.

121 Lumatec GmbH

Steinerstrasse 15
D-81396 Munchen

Tel: +49 89 7237021
Fax: +49 89 7235064
E-mail:

Affiliation:
Contact: J Patterson, Ultrafine Techn
Mr Jaeger, Gen Mgr

Organisation information
No of employees
Turnover (ECU*)

Sensors technology activities information
No of employees
Turnover (ECU*)
*1ECU=DM 1.89,FFr 6.30,£0.78,Lira 1909,Peseta 159.1

Organisation activities
Manufacturer of sensor-based products

Application \ Signal Sensed	Radiation	Optical	Electrical	Magnetic	Thermal	Chemical	Mechanical
Automotive							
Consumer products							
Process industries, mining, energy	•						•
Manufacturing, machinery	•						•
Medicine, health care							
HVAC, fire control, security							
Military, aerospace							
Telecommunications,office equip.							
Meteorology, environment							
Other *	•						•

Main Activities

Lumatec specialises in the development and manufacture of liquid light guides. It has recently invented a self-illuminating flexible liquid light guide, which it is believed will have various interesting sensing applications. A data sheet is in preparation and will be ready shortly. This entry is submitted by Ultrafine Technology Ltd, distributor of Lumatec products in the UK: 16 Foster Road, Chiswick, London Telephone: +44 181 9952303.

Collaboration Offered

Self-illuminating flexible liquid light guides.

122 Micro-Epsilon Messtechnik GmbH & Co KG

Königbacherstr 15
D-94496 Ortenburg

Tel: +49 8542 168-0
Fax: +49 8542 168 90
E-mail:

Affiliation:
Contact: Johann Salzberger, Mkting Manager

Organisation information
No of employees 101-1000
Turnover (ECU*) 5m-50m
Sensor technology activities information
No of employees 26-100
Turnover (ECU*) 1m-5m
*1ECU=DM 1.89,FFr 6.30,£0.78,Lira 1909,Peseta 159.1

Organisation activities
Research and development on sensors sensor materials or applications
Manufacture of finished sensors
Manufacture of sensor-based products

Application \ Signal Sensed	Radiation	Optical	Electrical	Magnetic	Thermal	Chemical	Mechanical
Automotive							•
Consumer products							
Process industries,mining energy							•
Manufacturing, machinery							•
Medicine, health care							
HVAC, fire control, security							
Military, aerospace							
Telecommunications,office equip.							
Meteorology environment							
Other (specify)							

Main Activities

Development, manufacturer and marketing of displacement sensors, displacement measuring systems and thickness measuring systems.

Specific Expertise

Displacement sensors using eddy currents, inductive, capacitive and optical principles.

Collaboration Offered

All products and services.

123 M/A-Com Etac

Karl Goetz Str 5
D-97424 Schweinfurt

Tel: +49 9721 79 75 70
Fax: +49 9721 79 75 99
E-mail: blaxilla@corp.macom.com

Affiliation:
Contact: Dr W Weidmann, GM
A Blaxill, Mkting Manager

Organisation information

No of employees	>1000
Turnover (ECU*)	>50m

Sensor technology activities information

No of employees	11-25
Turnover (ECU*)	1m-5m

*1ECU=DM 1.89,FFr 6.30,£0.78,Lira 1909,Peseta 159.1

Organisation activities
Research and development on sensors sensor materials or applications
Supply of raw materials or semi-finished products used to make sensors
Manufacture of finished sensors
Manufacture of sensor-based products

Application \ Signal Sensed	Radiation	Optical	Electrical	Magnetic	Thermal	Chemical	Mechanical
Automotive	●						
Consumer products							
Process industries,mining energy	●						
Manufacturing, machinery	●						
Medicine, health care							
HVAC, fire control, security	●						
Military, aerospace							
Telecommunications,office equip.							
Meteorology environment	●						
Other (specify)							

Main Activities

Design and manufacture of SMART microwave sensors at X band and K band.

Specific Expertise

Doppler radar; FMCW radar; high resolution, carrier based pulsed radar (HRR); carrier free, pulsed time domain radar (TDR).

Collaboration Offered

Microwave part of system only - no DSP.

Collaboration Sought

Digital signal processing.

124 Microtherm GmbH

Taeschenwaldstrasse 3
D-75134 Pforzheim

Tel: +49 7231 787-0
Fax: +49 7231 787155
E-mail:

Affiliation:
Contact: H Ossmann, Mktg Mgr

Organisation information
No of employees
Turnover (ECU*)
Sensors technology activities information
No of employees
Turnover (ECU*)
*1ECU=DM 1.89,FFr 6.30,£0.78,Lira 1909,Peseta 159.1

Organisation activities
Research and Development
Testing, consulting, design
Manufacture of finished sensors

Application \ Signal Sensed	Radiation	Optical	Electrical	Magnetic	Thermal	Chemical	Mechanical
Automotive					•		
Consumer products					•		
Process industries, mining, energy							
Manufacturing, machinery					•		
Medicine, health care					•		
HVAC, fire control, security					•		
Military, aerospace							
Telecommunications,office equip.					•		
Meteorology, environment					•		
Other *							

Main Activities
Manufacturing of thermal cutouts, thermal cut-offs, temperature and current limiters.

Specific Expertise

Development and manufacture of thermal sensors.

125 Mikro-Mess-Vertriebsgesellschaft mbH

Am Südende 15
D-31375 Lehrte

Tel: +49 5136 85494
Fax: +49 5136 85852
E-mail:

Affiliation:
Contact: C-K Varchim, MD

Organisation information
No of employees 11-25
Turnover (ECU*)
Sensors technology activities information
No of employees 11-25
Turnover (ECU*)
*1ECU=DM 1.89,FFr 6.30,£0.78,Lira 1909,Peseta 159.1

Organisation activities
Testing, consulting, design
Manufacture of finished sensors

Application \ Signal Sensed	Radiation	Optical	Electrical	Magnetic	Thermal	Chemical	Mechanical
Automotive							
Consumer products							
Process industries, mining, energy					•	•	•
Manufacturing, machinery							
Medicine, health care							
HVAC, fire control, security					•	•	•
Military, aerospace							
Telecommunications,office equip.							
Meteorology, environment					•	•	•
Other *							

Main Activities

Electrical measurement of physical quantities - pressure, rotation rate, humidity level, power, temperature, direction. Controllers and limit switches; customised measurement systems; development, manufacturing, sales.

126 Murata Europe Management GmbH

Holbeinstraße 21-23
D-90441 Nürnberg

Tel: +49 911 6687120
Fax: +49 911 6687270
E-mail:

Affiliation: Murata Japan
Contact: W Gottsachling, Dipl Ing
Marketing (Europe)

Organisation information
No of employees >100
Turnover (ECU*) >5m

Sensor technology activities information
No of employees >100
Turnover (ECU*) >5m

*1ECU=DM 1.89,FFr 6.30,£0.78,Lira 1909,Peseta 159.1

Organisation activities
Research and development.
Manufacture of finished sensors.

Application \ Signal Sensed	Radiation	Optical	Electrical	Magnetic	Thermal	Chemical	Mechanical
Automotive			●	●	●		●
Consumer products			●	●	●		
Process industries,mining energy			●	●	●		●
Manufacturing, machinery			●		●		●
Medicine, health care					●		
HVAC, fire control, security			●	●	●		
Military, aerospace					●		●
Telecommunications,office equip.			●	●	●		
Meteorology, environment			●	●	●		
Other *							

Main Activities

Matrix Management: the Murata Group is composed of more than 40 foreign and domestic companies, with Murata Manufacturing Co. Ltd at the nucleus. These companies engage in a variety of business activities, namely the production and sale of electronic components developed by the head Office Development Group or the Product Development departments.
Under Murata's unique global Matrix Management, each related company can display its worth as an independent entity creating products and processes. Respective business divisions also control product management within this frame work, united by Murata Manufacturing's operational staff. Murata has built networks to consolidate and augment the strength of the group. Production and sales bases are located in Europe, North and South America, and South East Asia. In Europe, production or sales bases function in Germany, France, Italy, UK and the Netherlands.

Specific Expertise

Thin film temperature sensors; thin film pyroelectric sensors; thin film piezoelectric sensors.

Collaboration Offered

Special design for special purposes; strong cooperation with customer.

127 Novocontrol GmbH

Obererbacher Straße 9
D-56414 Hundsangen

Tel: +49 6435 96230
Fax: +49 6435 962333
E-mail: 100257.263@compuserve.com

Affiliation:
Contact: Gerard den Dulk, MD
NIcole Endler, Product Manager

Organisation information
No of employees 1-25
Turnover (ECU*) 1m-5m
Sensor technology activities information
No of employees 1-10
Turnover (ECU*) <0.5m
*1ECU=DM 1.89,FFr 6.30,£0.78,Lira 1909,Peseta 159.1

Organisation activities
Testing, consulting, design

Application \ Signal Sensed	Radiation	Optical	Electrical	Magnetic	Thermal	Chemical	Mechanical
Automotive							
Consumer products							
Process industries,mining energy					•		
Manufacturing, machinery					•		
Medicine, health care							
HVAC, fire control, security							
Military, aerospace							
Telecommunications,office equip.							
Meteorology environment					•		
Other							

Main Activities

Since 1975, Novocontrol has built a reputation based on superior product quality and excellent consulting. The company has specialised in temperature sensors with a complementary line of related components and accessories. The growing importance of a high quality standard is highlighted in all Novocontrol products. To meet the demand of the ISO 9000 standards, Novocontrol can supply resistance temperature detectors of close tolerance inwire-wound, thick- and thin-film technology. All sensors are individually tested in order to maintain the high quality standard.

Specific Expertise

The company specialises in selecting component specifications to customer requirements and helps to determine what temperature sensors will suit best for specific applications.

Collaboration Offered

Technical consulting for temperature measurement and control applications.

Collaboration Sought

Collaboration with manufacturers of platinum resistance detectors and thermocouples.

128 PacTech Packaging Technologies GmbH

Reinickestr 8
D-14612 Falkensee

Tel: +49 30 46403 233
Fax: +49 30 46403 133
E-mail: ostm12706@szrz.zrz.tu-berlin.de

Affiliation:
Contact: Gerald Motulla, Development Engineer
Andreas Ostmann, FC-assembly

Organisation information

No of employees 1-25
Turnover (ECU*) 1m-5m

Sensor technology activities information

No of employees 1-10
Turnover (ECU*) <0.5m

*1ECU=DM 1.89,FFr 6.30,£0.78,Lira 1909,Peseta 159.1

Organisation activities

Research and development on sensors sensor materials or applications
Supply of raw materials or semi-finished products used to make sensors

Application \ Signal Sensed	Radiation	Optical	Electrical	Magnetic	Thermal	Chemical	Mechanical
Automotive							
Consumer products							
Process industries,mining energy							
Manufacturing, machinery							
Medicine, health care							
HVAC, fire control, security							
Military, aerospace							
Telecommunications,office equip.							
Meteorology environment							
Other (specify)		•	•	•			

* Consulting and service

Main Activities

We are not a supplier of sensors. We offer development, consulting and service of (sensor) wafer metallisation for FC-assembly.

Specific Expertise

Electroless Ni/Au metallisation of Al-pads on wafers; solder application; Flipchip assembly (service, consulting and development); seminars and tutorials on FC-assembly.

Collaboration Offered

Contract service for metallisation on FC-assembly; consulting and development service.

129 Parker Hannifin GmbH

Am Metallwerk 9
D-33659 Bielefeld

Tel: +49 521 4048 319
Fax: +49 521 4048 384
E-mail:

Affiliation:
Contact: U Baumann, Product Manager

Organisation information
No of employees 101-1000
Turnover (ECU*) >50m
Sensor technology activities information
No of employees 11-25
Turnover (ECU*) 0.5m-1m
*1ECU=DM 1.89,FFr 6.30,£0.78,Lira 1909,Peseta 159.1

Organisation activities
Research and development on sensors, sensor materials or applications.
Testing, consulting, design.
Supply of raw materials or semi-finished products used to make sensors.
Manufacture of systems used in the manufacture, testing or development of sensors.
Manufacture of finished sensors.
Manufacture of sensor based products.

Application \ Signal Sensed	Radiation	Optical	Electrical	Magnetic	Thermal	Chemical	Mechanical
Automotive							
Consumer products							
Process industries,mining energy		•	•		•		
Manufacturing, machinery		•	•		•		
Medicine, health care							
HVAC, fire control, security							
Military, aerospace							
Telecommunications,office equip.							
Meteorology, environment		•	•		•		
Other *							

Main Activities

Parker Hannifin is a worldwide organisation of hydraulic and pneumatic components.

Specific Expertise

Optical, electrical and thermal sensors.

Collaboration Offered

Electronic development of hand held instruments.

Collaboration Sought

Ground technology to manufacture a pressure sensor.

130 Parsytec Computer GmbH

Auf der Hùls 183
D-52068 Aachen

Tel: +49 24 19 69 6 0
Fax: +49 24 19 69 600
E-mail: www@parsytec.de

Affiliation:
Contact: T Glodek, PR Manager

Organisation information
No of employees 101-1000
Turnover (ECU*) 5m-50m

Sensor technology activities information
No of employees 11-25
Turnover (ECU*) 1m-5m
*1ECU=DM 1.89,FFr 6.30,£0.78,Lira 1909,Peseta 159.1

Organisation activities
Research and development on sensors, sensor materials or applications
Manufacture of sensor-based products

Application \ Signal Sensed	Radiation	Optical	Electrical	Magnetic	Thermal	Chemical	Mechanical
Automotive							•
Consumer products							
Process industries,mining energy							•
Manufacturing, machinery							•
Medicine, health care							
HVAC, fire control, security							
Military, aerospace							
Telecommunications,office equip.							
Meteorology environment							
Other							

Main Activities

Parsytec is concentrating work on products using parallel processing, and on real-time applications, particularly the realisation of complete, turn-key industrial systems containing interfaces, pre-processing and pattern recognition.

Specific Expertise

Specific expertise has been built up at Parsytec in the area of image processing and complex signal processing for measurement and control purposes as for example semi-conductor quality control, surface inspection, documented management and medical imaging. Image processing techniques have been successfully used for contactless spatial dimensional measurement or the detection of motion.

131 PME Meßsysteme H. Zimmer GmbH

Lise-Meitner-Str. 20
D-63457 Hanau

Tel: +49 6181 95760
Fax: +49 6181 957620
E-mail:

Affiliation:
Contact: H Zimmer, MD

Organisation information
No of employees 11-25
Turnover (ECU*) 0.5m-1m
Sensors technology activities information
No of employees 11-25
Turnover (ECU*) <0.5
*1ECU=DM 1.89,FFr 6.30,£0.78,Lira 1909,Peseta 159.1

Organisation activities
Research and Development

Application \ Signal Sensed	Radiation	Optical	Electrical	Magnetic	Thermal	Chemical	Mechanical
Automotive			•		•		
Consumer products							
Process industries, mining, energy			•		•		
Manufacturing, machinery							
Medicine, health care							
HVAC, fire control, security							
Military, aerospace			•		•		
Telecommunications,office equip.							
Meteorology, environment							
Other *							

Main Activities

The manufacture of signal conditioning equipment for thermal or electrical sensors.

Collaboration Offered

Datacase-24/system. Autodat-24/system. LV-Pt100-3L/4L.

Collaboration Sought

Great Britain, France.

132 Polytec GmbH

Polytec- Platz 1-7
D-76337 Waldbronn

Tel: +49 7243 6040
Fax: +49 7243 69944
E-mail:

Affiliation:
Contact: H Selbach, Director

Organisation information
No of employees >100
Turnover (ECU*) >5m
Sensors technology activities information
No of employees 11-25
Turnover (ECU*) 1m-5m
*1ECU=DM 1.89,FFr 6.30,£0.78,Lira 1909,Peseta 159.1

Organisation activities
Manufacturer of sensor-based products

Application \ Signal Sensed	Radiation	Optical	Electrical	Magnetic	Thermal	Chemical	Mechanical
Automotive							•
Consumer products							
Process industries, mining, energy							
Manufacturing, machinery							•
Medicine, health care							
HVAC, fire control, security							•
Military, aerospace							
Telecommunications,office equip.							•
Meteorology, environment							
Other *							

Main Activities

Interferometric measurement sensors and sensors based on the laser Doppler principle.

Collaboration Offered

Research and manufacture.

Collaboration Sought

Development capacity.

133 Precitec

Draisstr 1
D-76571 Gaggenau-Bad Ratenfeld

Tel: +49 7225 684 330
Fax: +49 7225 684 900
E-mail: 72258840-0001@t-online.de

Affiliation:
Contact: Dr U Zoske, Sales & Mkting
Dr S Biermann, Development

Organisation information
No of employees 26-100
Turnover (ECU*) 5m-50m
Sensor technology activities information
No of employees 26-100
Turnover (ECU*) 1m-5m
*1ECU=DM 1.89,FFr 6.30,£0.78,Lira 1909,Peseta 159.1

Organisation activities
Manufacture of finished sensors
Manufacture of sensor-based products

Application \ Signal Sensed	Radiation	Optical	Electrical	Magnetic	Thermal	Chemical	Mechanical
Automotive			•				•
Consumer products							
Process industries,mining energy							
Manufacturing, machinery			•				•
Medicine, health care							
HVAC, fire control, security							
Military, aerospace							
Telecommunications,office equip.							
Meteorology environment							
Other							

Main Activities

Development and manufacturing of sensors for rough surroundings in industrial applications.

Specific Expertise

Height sensors and focusing optics for laser tooling; seam - following systems for welding applications; optical triangulators for high reflective surfaces; distance sensors for high temperatures inside power plants; height sensors for water cutting systems.

134 Reten Electronic GmbH & Co

Höhenstr. 23
D-65520 Bad Camberg

Tel: +49 6434 240
Fax: +49 6434 2480
E-mail:

Affiliation:
Contact: R Theess, MD

Organisation information
No of employees >100
Turnover (ECU*) >5m
Sensors technology activities information
No of employees 11-25
Turnover (ECU*) 0.5m-1m
*1ECU=DM 1.89,FFr 6.30,£0.78,Lira 1909,Peseta 159.1

Organisation activities
Research and Development
Manufacturer of sensor-based products
Manufacture of finished sensors

Application \ Signal Sensed	Radiation	Optical	Electrical	Magnetic	Thermal	Chemical	Mechanical
Automotive							
Consumer products							
Process industries, mining, energy							•
Manufacturing, machinery							•
Medicine, health care							
HVAC, fire control, security							
Military, aerospace							•
Telecommunications,office equip.							
Meteorology, environment							•
Other *							

Main Activities

Development, manufacture and sales of the following products: acoustical instrumentation for applications in environment, machinery, underwater etc; instrumentation for the water industry for the applications of leak detection, network control, monitoring, pipe tracing etc; optical inspection equipment for the sewage industry.

Specific Expertise

The company specialises in piezoelectric and ultrasonic sensors in a variety of configurations, with or without integrated electronics.

Collaboration Offered

Expertise in vibration sensors for extreme conditions.

Collaboration Sought

The company is looking for co-operation with another manufacturer of ultrasonic sensors.

135 RWTH - Aachen

WZL - Laboratorium für Werkzeugmaschinen und Betriebslehre
Institüt für Werkzeugmaschinen
Steinbachstr. 53 B
D-52056 Aachen

Tel: +49 241 80-7407
Fax: +49 241 8888293
E-mail:

Affiliation:
Contact: Prof F Klocke, Chairholder
Prof T Pfeifer

Organisation information
No of employees >100
Turnover (ECU*) >5m

Sensors technology activities information
No of employees 11-25
Turnover (ECU*)
*1ECU=DM 1.89,FFr 6.30,£0.78,Lira 1909,Peseta 159.1

Organisation activities
Research and Development
Testing, consulting, design

Application \ Signal Sensed	Radiation	Optical	Electrical	Magnetic	Thermal	Chemical	Mechanical
Automotive							
Consumer products							
Process industries, mining, energy							
Manufacturing, machinery	•		•				•
Medicine, health care							
HVAC, fire control, security							
Military, aerospace							
Telecommunications,office equip.							
Meteorology, environment							
Other *							

Main Activities

Research activities: machine tool design; dynamical properties of machine tools; drive systems and gears; CNC control systems; tool and machine monitoring; CIM-system; robotics.

Specific Expertise

Machine-integrated force measurement system based on piezoelectric sensors; sensors for high resoluted measurements of torque directly on rotating stiff machine tool spindles; measurement data transfer system for rotating spindles, especially developed for piezoelectric sensors applicated on the spindle and for the integration in machine tools or gears; sensor systems for robot grippers which include measurement evaluating; orientation sensor systems for easy programming of robot movements, using high frequency electromagnetic fields, gravitation, the earth magnetic field or inertial systems.

136 Schiele Industriewerke GmbH

D-78132 Hornberg

Tel: +49 7833 78368
Fax: +49 7833 78318
E-mail:

Affiliation: MBB
Contact: FH Muchowski

Organisation information
No of employees >100
Turnover (ECU*) >5m
Sensors technology activities information
No of employees 1-10
Turnover (ECU*) <0.5
*1ECU=DM 1.89,FFr 6.30,£0.78,Lira 1909,Peseta 159.1

Organisation activities
Research and Development
Manufacturer of sensor-based products
Testing, consulting, design
Manufacture of finished sensors

Application \ Signal Sensed	Radiation	Optical	Electrical	Magnetic	Thermal	Chemical	Mechanical
Automotive							•
Consumer products							
Process industries, mining, energy	•						•
Manufacturing, machinery	•						•
Medicine, health care							
HVAC, fire control, security							
Military, aerospace	•						•
Telecommunications,office equip.							
Meteorology, environment							
Other *							

Main Activities

Schiele designs, manufactures and distributes control products and comprehensive industrial control packages including: programmable controllers; contactors and control relays; thermal overload relays; residual current operated circuit breakers; electronic timers; measuring and monitoring relays; fault indicating systems; industrial automation systems; electric control and switch boards; intelligent sensor systems.

Specific Expertise

Profile detection based on the triangulation principle. Non-contact measurement and profile detection using laser technology. Measuring range between 150 mm and 350 mm from sensor. Fast infrared temperature measurement. Non-contact temperature measurement is via detection of infrared radiation. The intelligent microprocessor controlled measuring system displays the temperature values measured which are availabe for evaluation at the RS 485 serial port.

137 Schmidt Feintechnik GmbH

Feldbergstrasse 1
D-78112 St Georgen

Tel: +49 7724 899-0
Fax: +49 7724 899101
E-mail:

Affiliation:
Contact: F Eigel, Sales & Mark Mgr

Organisation information
No of employees 101-1000
Turnover (ECU*) 5m-50m
Sensor technology activities information
No of employees 26-100
Turnover (ECU*) 0.5m-1m
*1ECU=DM 1.89,FFr 6.30,£0.78,Lira 1909,Peseta 159.1

Organisation activities
Manufacture of finished sensors.
Manufacture of sensor-based sensors.

Application \ Signal Sensed	Radiation	Optical	Electrical	Magnetic	Thermal	Chemical	Mechanical
Automotive							•
Consumer products							
Process industries,mining energy					•		•
Manufacturing, machinery					•		•
Medicine, health care							
HVAC, fire control, security					•		•
Military, aerospace							
Telecommunications,office equip.							
Meteorology, environment							
Other *							

Main Activities

Developing and manufacturing of: acceleration sensors, vibration controls for machine monitoring; safety related sensors in automotive application; airflow sensors for climate control, cleanroom applications, process control measurement.

Specific Expertise

Piecoelectrical acceleration sensors and mechanical precision switching sensors for acceleration and tilt. Heat transfer (calorimetric) flow sensors.

Collaboration Offered

Complete development and manufacturing capacity for high volume OEM product.

Collaboration Sought

Sales and marketing corporation in France.

138 sgt Sensorberatung Dr Guido Tschulena

Reichenberger Str 5
D-61273 Wehrheim

Tel: +49 6081 56168
Fax: +49 6081 57222
E-mail: tschulena.sgt@t-online.dc

Affiliation:
Contact: Dr G R Tschulena

Organisation information
No of employees 1-25
Turnover (ECU*) <1m
Sensor technology activities information
No of employees 1-10
Turnover (ECU*) <0.5m
*1ECU=DM 1.89,FFr 6.30,£0.78,Lira 1909,Peseta 159.1

Organisation activities
Testing, consulting, design.

Application \ Signal Sensed	Radiation	Optical	Electrical	Magnetic	Thermal	Chemical	Mechanical
Automotive							
Consumer products							
Process industries,mining energy							
Manufacturing, machinery							
Medicine, health care							
HVAC, fire control, security							
Military, aerospace							
Telecommunications,office equip.							
Meteorology, environment							
Other *	•	•	•	•	•	•	•

* Consulting

Main Activities

Techno-economic consulting in the area of sensors and micro-system technology. Market investigations for specific products and technologies. Sector and branches analyses. International cooperation. seminars and workshops.

Specific Expertise

Micro-system techonology. Market investigation.

139 Siemens AG

Dept: HL OCD MFH
Wesuerwirlestr 2
D-93049 Regensburg

Tel: +49 941 202 2342
Fax: +49 941 202 3786
E-mail:

Affiliation:
Contact: M Kutzner, Marketing Manager

Organisation information
No of employees No comment
Turnover (ECU*) No comment
Sensor technology activities information
No of employees No comment
Turnover (ECU*) No comment
*1ECU=DM 1.89,FFr 6.30,£0.78,Lira 1909,Peseta 159.1

Organisation activities
Research and development on sensors, sensor materials or applications.
Manufacture of packaged semi conductor deoices used in the manufacture, testing or development of sensors.

Application \ Signal Sensed	Radiation	Optical	Electrical	Magnetic	Thermal	Chemical	Mechanical
Automotive				•			
Consumer products							
Process industries,mining energy							
Manufacturing, machinery				•			
Medicine, health care							
HVAC, fire control, security							
Military, aerospace				•			
Telecommunications,office equip.							
Meteorology, environment							
Other *							

Main Activities

Overall: Siemens AG: general electric products.
Semiconductor Group: communication YC'S; memories, discrete and power devices; opto semiconductors.
Opto Semiconductors: laser; LED; couples; displays; infrared, detectors, magnetic sensors.

Specific Expertise

Magnetic sensors: magnetoresistors, based on InSb; Hallsensor, based on GaHS. All Analogue devices. Specific advantages for high accuracy, high temperature applications and high lifetime stability.

Collaboration Offered

Development and production of customers specific devices, if volume is sufficient.

140 STW - Sensor-Technik Wiedemann GmbH

Am Bärenwald 6
D-87600 Kaufbeuren

Tel: +49 8341 95050
Fax: +49 8341 950555
E-mail:

Affiliation:
Contact: W Wiedmann, MD

Organisation information
No of employees 26-100
Turnover (ECU*) 1m-5m

Sensors technology activities information
No of employees 26-100
Turnover (ECU*) 1m-5m
*1ECU=DM 1.89,FFr 6.30,£0.78,Lira 1909,Peseta 159.1

Organisation activities
Research and Development
Manufacturer of sensor-based products
Testing, consulting, design
Manufacture of finished sensors

Application \ Signal Sensed	Radiation	Optical	Electrical	Magnetic	Thermal	Chemical	Mechanical
Automotive					●		●
Consumer products					●		●
Process industries, mining, energy					●		●
Manufacturing, machinery					●		●
Medicine, health care							
HVAC, fire control, security							
Military, aerospace							
Telecommunications,office equip.							
Meteorology, environment					●		●
Other *							

Main Activities

Development, production and sales of pressure and temperature sensors; building of test equipment; specific customer applications.

141 tec5 Sensorik und Systemtechnik GmbH

Industriestrasse 23
D-61449 Steinbach

Tel: +49 6171 9758 15
Fax: +49 6171 9758 50
E-mail: 100451.2401@compuserve.com

Affiliation:
Contact: P Huber, Marketing
M Kother, CEO

Organisation information
No of employees 1-25
Turnover (ECU*) <1m
Sensor technology activities information
No of employees 1-10
Turnover (ECU*) <0.5m
*1ECU=DM 1.89,FFr 6.30,£0.78,Lira 1909,Peseta 159.1

Organisation activities
Research and development on sensors, sensor materials or applications.
Manufacture of sensor-based products.

Application \ Signal Sensed	Radiation	Optical	Electrical	Magnetic	Thermal	Chemical	Mechanical
Automotive							
Consumer products							
Process industries,mining energy		•	•				
Manufacturing, machinery		•	•	•			•
Medicine, health care			•				
HVAC, fire control, security							
Military, aerospace							
Telecommunications,office equip.							
Meteorology, environment		•					
Other *							

Main Activities

Product/electronics engineering and manufacturing; system integration; engineering expertise.

Specific Expertise

Electronics assemblies for spectral sensors; data acquisition and processing; optical inspection technologies; systems integration.

Collaboration Offered

Customer specific systems; adaptation of products.

142 Temic Microsystems Sensors

Wolff-Hirth-str 7
D-73230 Kirchheim/Teck

Tel: +49 7027 989 595
Fax: +49 7021 989 599
E-mail: kurtradtke@temickirchheim.ccmail.compuserve.com

Affiliation:
Contact: Kurt Radtke, Manager Sales

Organisation information

No of employees	101-1000
Turnover (ECU*)	5m-50m

Sensor technology activities information

No of employees	>100
Turnover (ECU*)	>5m

*1ECU=DM 1.89,FFr 6.30,£0.78,Lira 1909,Peseta 159.1

Organisation activities
Research and development on sensors sensor materials or applications
Manufacture of finished sensors

Application \ Signal Sensed	Radiation	Optical	Electrical	Magnetic	Thermal	Chemical	Mechanical
Automotive						•	•
Consumer products							•
Process industries,mining energy						•	•
Manufacturing, machinery							
Medicine, health care							
HVAC, fire control, security							
Military, aerospace							•
Telecommunications,office equip.							•
Meteorology environment							
Other (specify)							

Main Activities

Development of sensor systems for automobile industry and other industrial applications. High volume production of piezoelectrical and micromachined sensor modules.

Specific Expertise

Piezoelectric and quartz based accelerometer, YAW-late sensors; micromalised accelerometers; oil-level- and oil-quality-sensors.

143 TU-Berlin

Technologien der Mikropherik
Gustav-Meyer-Allee25
D-13355 Berlin

Tel: +49 30 314 72769
Fax: +49 30 314 72603
E-mail: oberm@mat.ee.tu-berlin.de

Affiliation:
Contact: Prof. E Obermeier
P Krause, Technology
K Hilgendorf, Simulation (FEA)

Organisation information
No of employees >1000
Turnover (ECU*)

Sensor technology activities information
No of employees 26-100
Turnover (ECU*) 1m-5m
*1ECU=DM 1.89,FFr 6.30,£0.78,Lira 1909,Peseta 159.1

Organisation activities
Research and development on sensors sensor materials or applications
Testing, consulting,design

Application \ Signal Sensed	Radiation	Optical	Electrical	Magnetic	Thermal	Chemical	Mechanical
Automotive					•	•	•
Consumer products	•				•	•	•
Process industries,mining energy							
Manufacturing, machinery							
Medicine, health care							
HVAC, fire control, security	•				•	•	
Military, aerospace							
Telecommunications,office equip.							
Meteorology environment					•	•	
Other							

Main Activities

Research and development in the field of microsensors and microactuators.

Specific Expertise

MEMS, micromachining (bulk and surface) anodic bonding, simulation, high temperature sensors, sensors for pressure, force, flow, radiation, acceleration, gases etc, silicon and thin film technology, microactuators, smart sensors.

Collaboration Offered

Silicon and thin film technology, development of processes and sensors, low volume production of sensors, FEA, anodic bonding service.

Collaboration Sought

Design of circuits for smart sensors.

144 Technical University of Chemnitz-Zwickau

Dept of Microsystem & Device Technology
D-09107 Chemnitz

Tel: +49 371 531 3264
Fax: +49 371 531 3259
E-mail: doetzel@infotech.tu-chemnitz.de

Affiliation:
Contact: Prof Dr.-ing W Doetzel

Organisation information

No of employees 1-25
Turnover (ECU*) <1m

Sensor technology activities information

No of employees 1-10
Turnover (ECU*) <0.5m

*1ECU=DM 1.89,FFr 6.30,£0.78,Lira 1909,Peseta 159.1

Organisation activities

Research and development on sensors sensor materials or applications
Testing, consulting, design

Application \ Signal Sensed	Radiation	Optical	Electrical	Magnetic	Thermal	Chemical	Mechanical
Automotive		●					●
Consumer products		●					●
Process industries,mining energy							
Manufacturing, machinery							
Medicine, health care							●
HVAC, fire control, security							
Military, aerospace							●
Telecommunications,office equip.		●					
Meteorology environment							
Other							

Main Activities

Research work at the department of Microsystem and Device Technology is aimed at functional design, modelling, simulation, and optimisation as well as experimental analysis of silicon-based micromechanical systems. This focus corresponds to both the high potential of success of the silicon-micromechanic and to the experiences made at the department. It is as well in accordance with the close cooperation with the department of microtechnologies and with the outstanding possibilities given at the Centre of Microtechnologies of the Chemnitz-Zwickau-University for producing micromechanical silicon components. This allows us to produce functional samples and pilot production and to use results of experimental analysis for on-linesimulation input.

145 TKM Microelectronic GmbH

Computertechnik, BTX
Auguste-Viktoria-Str. 63
D-14199 Berlin

Tel: +49 30 8251001
Fax: +49 30 8264018
E-mail:

Affiliation:
Contact: W Richardt, Gen Mgr

Organisation information
No of employees 1-10
Turnover (ECU*) 1m-5m
Sensors technology activities information
No of employees 1-10
Turnover (ECU*) 0.5m-1m
*1ECU=DM 1.89,FFr 6.30,£0.78,Lira 1909,Peseta 159.1

Organisation activities
Research and Development
Manufacture of finished sensors

Application \ Signal Sensed	Radiation	Optical	Electrical	Magnetic	Thermal	Chemical	Mechanical
Automotive							•
Consumer products							
Process industries, mining, energy							•
Manufacturing, machinery							•
Medicine, health care							
HVAC, fire control, security							
Military, aerospace							•
Telecommunications,office equip.							•
Meteorology, environment							
Other *							

Main Activities
Development of sensors, special stoke transducers LVDT, ball bearing gauge heads, linear displacement transducers, linear motion transducers. Measuring and control electronics for these sensors and tranducers. Software development for sensor applications, control, measuring in combination with MS-DOS/IBM compatible. Production and sales of the above.

Specific Expertise
The company specialises in sensors, and Main Activities are as given: development and production of special custom made LVDT stroke transducers, gauge heads, linear displacement transducers and motion transducers. Software for measuring and control of sensor applications, especially electronic measurement of mechanical values.

Collaboration Offered
High quality standard and custom made products.

Collaboration Sought
Sales and distribution nationally/worldwide.

146 Technische Universität Hamburg-Harburg

Optik und Messtechnik
Eissendorfer Str. 40
D-21073 Hamburg

Tel: +49 40 77183214
Fax: +49 40 77182860
E-mail:

Affiliation:
Contact: Prof R Ulrich, Fibre Optics Sensors

Organisation information
No of employees >100
Turnover (ECU*) >5m

Sensors technology activities information
No of employees 1-10
Turnover (ECU*) <0.5m
*1ECU=DM 1.89,FFr 6.30,£0.78,Lira 1909,Peseta 159.1

Organisation activities
Research and Development
Testing, consulting, design

Application \ Signal Sensed	Radiation	Optical	Electrical	Magnetic	Thermal	Chemical	Mechanical
Automotive			•				•
Consumer products	•		•			•	•
Process industries, mining, energy	•		•			•	•
Manufacturing, machinery	•		•			•	•
Medicine, health care	•		•			•	•
HVAC, fire control, security	•		•			•	•
Military, aerospace							
Telecommunications,office equip.	•		•			•	•
Meteorology, environment	•		•			•	•
Other *							

Main Activities

The Optics and Instrumentation Division within the Technical University, is concerned with fundamental research on optic and fibre optic sensors, from simple shutter-type devices to sophisticated interferometers, from the development of special sensor fibres to the evaluation of various encoding principles.

Specific Expertise

Fundamental research into the principles and problems of fibre optic sensors; development and testing of prototypes; using single-mode and multi-mode fibres for applications in industry, medicine, science, and sewage treatment; instrumentation for the evaluation of optical fibres and of integrated optical waveguides.

Collaboration Offered

Fibre optic sensors. Integrated optics analysis.

147 Technische Universität Hamburg-Harburg

Institut für Halbleitertechnologie
Eissendorfer Str. 42
D-21073 Hamburg

Tel: +49 40 77183029
Fax: +49 40 77182396
E-mail:

Affiliation:
Contact: Prof J Müller, Head of Dept

Organisation information
No of employees 25-100
Turnover (ECU*) 1m-5m

Sensors technology activities information
No of employees 11-25
Turnover (ECU*) 0.5m-1m
*1ECU=DM 1.89,FFr 6.30,£0.78,Lira 1909,Peseta 159.1

Organisation activities
Research and Development
Testing, consulting, design

Application \ Signal Sensed	Radiation	Optical	Electrical	Magnetic	Thermal	Chemical	Mechanical
Automotive					•		•
Consumer products							
Process industries, mining, energy						•	•
Manufacturing, machinery							
Medicine, health care							•
HVAC, fire control, security							
Military, aerospace							
Telecommunications,office equip.							
Meteorology, environment						•	
Other *							

Main Activities

Development of thin film materials and applications for sensors, integrated optics, passivation and isolation, high temperature superconductors, selective membranes. Development of deposition processes such as reactive sputtering, LP-, PE-, and laser CVD for transparent isolating and passivating layers using halogenids and organics on planar and non-planar substrates.

Specific Expertise

Pt temperature sensors for automotive applications; TiON piezoresistive pressure sensors for high temperature (500°C) and medical applications; ZnO chemical sensors for CO, SO_2, and NO_x detection including enhancement of selectivity by monolayer catalytic metals. Integrated optical circuits on the base of SiON-SiO_2-layers on silicon substrates including detectors, micromachining and electronics in the substrate for motion, distance, temperature and pressure/ force sensing.

Collaboration Offered

Development of all kinds of new sensor materials and applications also of non-planar transducer bodies including technology and stabilisation procedures, analysis of microstructure and composition of thin film technology.

Collaboration Sought

New problems and applications, development of housing and electronics, exchange and complementary technology and analysis.

148 TU München

Lehrstuhl für Elektrische Messtechnik
D-80290 München

Tel: +49 89 28 92 83 48
Fax: +49 89 28 92 33 48
E-mail: emt@e-technik.tu-munchen.de

Affiliation:
Contact: H Hammerschmidt, Director

Organisation information
No of employees >1000
Turnover (ECU*)

Sensor technology activities information
No of employees 11-25
Turnover (ECU*) 1m-5m
*1ECU=DM 1.89,FFr 6.30,£0.78,Lira 1909,Peseta 159.1

Organisation activities
Research and development on sensors, sensor materials or applications
Testing, consulting, design

Application \ Signal Sensed	Radiation	Optical	Electrical	Magnetic	Thermal	Chemical	Mechanical
Automotive							
Consumer products			•				
Process industries,mining energy			•			•	•
Manufacturing, machinery			•				•
Medicine, health care							
HVAC, fire control, security			•				
Military, aerospace							
Telecommunications,office equip.							
Meteorology environment		•	•			•	
Other (specify)							

Main Activities

Lehrstuhl für Elektrische Messtechnik der Technisdien Universität is involved in the development of new sensors and signal analysis.

Specific Expertise

- Environmental measurements (with Fourier Transform Spectrometers FTS)
- SAW sensors for different quantities and with different possibilities to apply (energy and signal transmitted by radio waves)
- VIMP = distributed intelligent sensor systems for the private home

Collaboration Offered

Our company offers consulting and prototypes.

Collaboration Sought

- Applications
- Production

149 Ultrakust Electronic GmbH

Schulstrasse 30
Postfach 63
D-94239 Gotteszell

Tel: +49 9929 3010
Fax: +49 9929 301112
E-mail:

Affiliation:
Contact: Mr Zimmermann, Proj Mgr

Organisation information
No of employees >100
Turnover (ECU*) >5m

Sensors technology activities information
No of employees 11-25
Turnover (ECU*) 1m-5m
*1ECU=DM 1.89,FFr 6.30,£0.78,Lira 1909,Peseta 159.1

Organisation activities
Manufacturer of sensor-based products
Manufacture of finished sensors

Application \ Signal Sensed	Radiation	Optical	Electrical	Magnetic	Thermal	Chemical	Mechanical
Automotive	•				•	•	
Consumer products							
Process industries, mining, energy	•				•	•	
Manufacturing, machinery	•				•	•	
Medicine, health care					•		
HVAC, fire control, security					•	•	
Military, aerospace							
Telecommunications,office equip.							
Meteorology, environment	•				•	•	
Other *							

Main Activities

Temperature and humidity measurement.

Specific Expertise

Temperature sensors to connect to portable measuring instruments or to highly sophisticated laboratory measuring equipment; executed as thermocouples, resistance thermometers, semiconductor probes. Temperature probes for stationary use, with all traditional sensor types available; special models for special applications are available on request. Physchometric probe for humidity and temperature with capacitive sensors; Hygrophil-h, the impact jet psychrometer for humidity measurement under extreme conditions. Pressure transducer to measure static and dynamic pressure. Sensors for the fast determination of pH values.

Collaboration Offered

As detailed in Specific Expertise.

Collaboration Sought

Low cost moisture sensor.

150 Unitronic GmbH

Mündelheimer Weg 9
D-40472 Düsseldorf

Tel: +49 211 9511 0
Fax: +49 211 9511 111
E-mail:

Affiliation:
Contact: R Breider, Dipl. Ing., President

Organisation information
No of employees 26-100
Turnover (ECU*) 5m-50m

Sensor technology activities information
No of employees 1-10
Turnover (ECU*) 1m-5m
*1ECU=DM 1.89,FFr 6.30,£0.78,Lira 1909,Peseta 159.1

Organisation activities
Testing, consulting, design.

Application \ Signal Sensed	Radiation	Optical	Electrical	Magnetic	Thermal	Chemical	Mechanical
Automotive				•		•	
Consumer products		•	•	•		•	
Process industries,mining energy							
Manufacturing, machinery							
Medicine, health care						•	
HVAC, fire control, security			•			•	
Military, aerospace							
Telecommunications,office equip.			•				
Meteorology, environment			•	•		•	
Other *							

Main Activities

Unitronic GmbH develop and manufacture optical,electrical, magnetic and chemical sensors for application in different areas.

Collaboration Offered

Diploma work and studies.

151 Universität der Bundeswehr

Institut fur Mess-und Automatisierungstechnik
Postfach 8014
D-85577 Neubiberg

Tel: +49 896004 374041
Fax: +49 896004 2557
E-mail:

Affiliation:
Contact: Mr Horn, Director

Organisation information
No of employees
Turnover (ECU*)
Sensors technology activities information
No of employees
Turnover (ECU*)
*1ECU=DM 1.89,FFr 6.30,£0.78,Lira 1909,Peseta 159.1

Organisation activities
Research and Development
Testing, consulting, design

Application \ Signal Sensed	Radiation	Optical	Electrical	Magnetic	Thermal	Chemical	Mechanical
Automotive						•	•
Consumer products						•	•
Process industries, mining, energy						•	
Manufacturing, machinery							
Medicine, health care							
HVAC, fire control, security							
Military, aerospace							
Telecommunications,office equip.							
Meteorology, environment						•	
Other *							

Main Activities
Low cost sensor and microcomputer-based signal processing for intelligent measuring systems with high accuracy. Mathematical methods for correction of influences on measurement (eg. temperature, creep, hysteresis). Gas calibration technique and calibration of gas sensors.

Specific Expertise
Type of signal: force, pressure, displacement, acceleration, gas flow, gas concentration.
Sensor principles: eddy current, Hall effect, oxidation of gas.
Personnel: mechanical signal - 5, chemical signal - 5.

Collaboration Offered
Gas calibration in practice and theory. Know-how in sensor-specific signal. Pressure calibration. Flow sensors.

Collaboration Sought
Integration of sensor signal processing hardware and a microcomputer in an ASIC. Contact with and information from gas sensor manufacturers. Contact and information on gas calibration technique.

152 Universität Erlangen-Nürnberg

Lehrstuhl für Technische Thermodynamik
Am Weichselgarten 8
D-91058 Erlangen

Tel: +49 9131 859900
Fax: +49 9131 859901
E-mail: sw@ltt.uni-erlangen.de

Affiliation:
Contact: Dr Stefan Will, Head of Dept
"In-Process Measurement Technology"

Organisation information
No of employees 26-100
Turnover (ECU*) 1m-5m

Sensor technology activities information
No of employees 11-25
Turnover (ECU*) 0.5m-1m
*1ECU=DM 1.89,FFr 6.30,£0.78,Lira 1909,Peseta 159.1

Organisation activities
Research and development on sensors, sensor materials or applications
Testing, consulting, design

Application \ Signal Sensed	Radiation	Optical	Electrical	Magnetic	Thermal	Chemical	Mechanical
Automotive		•			•	•	
Consumer products							
Process industries,mining energy		•			•	•	•
Manufacturing, machinery							
Medicine, health care							
HVAC, fire control, security							
Military, aerospace							
Telecommunications,office equip.							
Meteorology environment						•	•
Other		•			•	•	•

Main Activities

Applied research for combustion, determination of thermophysical properties, environmental analysis, particle measurement, heat transfer, process engineering.
Development of optical, mainly laser-based sensors and measurement systems.

Specific Expertise

Optical, mainly laser-based sensors and systems for temperature, particle size, turbidity, colour, species.

Collaboration Offered

Custom-developed optical sensors and systems for temperature, particle size, turbidity, colour, species.

Collaboration Sought

Light sources (esp. blue laser diodes) and detectors (e.g. ICCD cameras). Interesting applications for sensors developed.

153 Universität Kaiserslautern

Lehrstuhl Feinwerktechnik
Postfach 3049
D-67653 Kaiserslautern

Tel: +49 631 2052835
Fax: +49 631 2052871
E-mail:

Affiliation:
Contact: Prof R Haberland

Organisation information
No of employees
Turnover (ECU*)
Sensors technology activities information
No of employees
Turnover (ECU*)
*1ECU=DM 1.89,FFr 6.30,£0.78,Lira 1909,Peseta 159.1

Organisation activities
Research and Development
Manufacturer of sensor-based products
Testing, consulting, design
Manufacture of finished sensors

Application \ Signal Sensed	Radiation	Optical	Electrical	Magnetic	Thermal	Chemical	Mechanical
Automotive							•
Consumer products							
Process industries, mining, energy	•						•
Manufacturing, machinery	•						•
Medicine, health care							
HVAC, fire control, security							
Military, aerospace	•						•
Telecommunications,office equip.							
Meteorology, environment							
Other *							

Specific Expertise

Elasto-mechanic devices for sensors, optimised structures. Capacitive sensors for any type of measurement eg. acceleration, distance, thickness, structure, lubrication. High quality electronics for capacitive, inductive, resistive and light-position signals. Sensor specialities for automation and control.

Collaboration Offered

Accelerometers. Tilt Sensors. Distance Sensors. Light-point location sensors and related products. Lubricant film measurement systems. Specialities for automation and control. Measurements and design on ball and gas bearings. Diamond turning. Accuracy improvement of grinding milling and turning machines.

Collaboration Sought

Co-operation in either development or research with laboratories or companies.

154 Universität Kassel

Fachbereich Physik
Gesamthochschule Kassel
Heinrich-Plett-Str.40
D-34109 Kassel

Tel: +49 561 8044532
Fax: +49 561 8044136
E-mail:

Affiliation:
Contact: Prof R Kassing, Head of Inst

Organisation information
No of employees
Turnover (ECU*)
Sensors technology activities information
No of employees
Turnover (ECU*)
*1ECU=DM 1.89,FFr 6.30,£0.78,Lira 1909,Peseta 159.1

Organisation activities
Research and Development
Testing, consulting, design
Manufacture of finished sensors

Application \ Signal Sensed	Radiation	Optical	Electrical	Magnetic	Thermal	Chemical	Mechanical
Automotive	●	●	●	●	●	●	●
Consumer products	●	●	●	●	●	●	●
Process industries, mining, energy	●	●	●	●	●	●	●
Manufacturing, machinery							
Medicine, health care	●	●	●	●	●	●	●
HVAC, fire control, security	●	●	●	●	●	●	●
Military, aerospace							
Telecommunications,office equip.							
Meteorology, environment							
Other *							

Main Activities
Microsystem technology: integration of sensors, microelectronics and actuators on chip. Microstructuring, deposition of thin films, ASICs, plasma and material analytics, process simulation.

Specific Expertise
Development of fundamental sensor and actuator principles. Realising microsystems ie. the combination of sensors and actuators with microelectronics. Microstructure technology (e-beam lithography and dry etching), development of new materials by plasma and ion beam assisted deposition techniques, plasma and materials characterisation and process modelling.

Collaboration Offered
Basic research, development of microsystems and production.

Collaboration Sought
Finance.

155 Universität Rostock

Fachbereich Elektrotechnik
Albert-Einstein-Str 2
D-18051 Rostock

Tel: +49 381 498 3491
Fax: +49 381 498 3479
E-mail:

Affiliation:
Contact: Dr J Gätke, Professor

Organisation information
No of employees 1-25
Turnover (ECU*)
Sensor technology activities information
No of employees 1-10
Turnover (ECU*)
*1ECU=DM 1.89,FFr 6.30,£0.78,Lira 1909,Peseta 159.1

Organisation activities
Research and development on sensors, sensor materials or applications.

Application \ Signal Sensed	Radiation	Optical	Electrical	Magnetic	Thermal	Chemical	Mechanical
Automotive							
Consumer products							
Process industries,mining energy							
Manufacturing, machinery							
Medicine, health care							
HVAC, fire control, security							
Military, aerospace							
Telecommunications,office equip.							
Meteorology, environment		•	•			•	•
Other *							

Main Activities

In the field of research and development: current and flow measurement techniques in acoustical (that means especially clamp-on-performance) and optical way (that means LDA and PDA and CCD-technique), even for particle size measurements; environmental monitoring systems; microanalysis methods of mass-spectrometry and gas-chromatography.

Collaboration Offered

Research and development by employer; measurements; expert evidence and deliberation.

156 Universität GH Siegen

Institut fur Messtechnik
Holderlingstr 3
D-57068 Siegen

Tel: +49 271 7404419
Fax: +49 271 7402396
E-mail:

Affiliation:
Contact: Prof KW Bonfig, Director

Organisation information
No of employees
Turnover (ECU*)
Sensors technology activities information
No of employees
Turnover (ECU*)
*1ECU=DM 1.89,FFr 6.30,£0.78,Lira 1909,Peseta 159.1

Organisation activities
Research and Development

Application \ Signal Sensed	Radiation	Optical	Electrical	Magnetic	Thermal	Chemical	Mechanical
Automotive	•	•	•	•	•	•	•
Consumer products	•	•	•	•	•	•	•
Process industries, mining, energy	•	•	•	•	•	•	•
Manufacturing, machinery	•	•	•	•	•	•	•
Medicine, health care							
HVAC, fire control, security							
Military, aerospace							
Telecommunications,office equip.							
Meteorology, environment	•	•	•	•	•	•	•
Other *							

Main Activities

The Institute for Measuring Techniques belongs to the University GH Siegen. It is responsible for teaching measuring techniques in the basic course and in the main study courses I and II. Work and research are concerned with the development of sensors, smart sensors, sensor systems and methods of sensor processing. Fields of interest are electrical measuring of electrical and non-electrical quantities, particularly the acquisition and processing of measured values. Current projects include continuous flow measurement of fluids and gases, radiometric measurement of density, turbidimetry, intelligent measurement principles for determination of process parameters and measurement of capacitance, resistance and inductance by correlation techniques.
Professor Dr Bonfig is the editor and director of the Technical-Scientific Council of the international journal "Messen Prufen Automatisieren".

Collaboration Offered

Flow measurement. Density measurement. Turbidimetry. Pressure measurement. Force measurement. Torque measurement. Fault detection.

Collaboration Sought

In all the areas specified in Collaboration Offered.

157 Universität Tübingen

Institute of Physical and Theoretical Chemistry
Auf der Morgenstelle 8
D-72076 Tübingen

Tel: +49 7071 29 77633
Fax: +49 7071 29 6910
E-mail: jlm@ipc.uni-tuebingen.de

Affiliation: Universität Tubingen
Contact: Prof W Göpel, Head of Dept
J Mitrovics, Database Update

Organisation information
No of employees 26-100
Turnover (ECU*)

Sensor technology activities information
No of employees 26-100
Turnover (ECU*) 1m-5m
*1ECU=DM 1.89,FFr 6.30,£0.78,Lira 1909,Peseta 159.1

Organisation activities
Research and development on sensors sensor materials or applications
Testing, consulting,design
Supply of raw materials or semi-finished products used to make sensors
Manufacture of systems used in the manufacture, testing or development of sensors

Application \ Signal Sensed	Radiation	Optical	Electrical	Magnetic	Thermal	Chemical	Mechanical
Automotive						•	
Consumer products							
Process industries,mining energy						•	
Manufacturing, machinery							
Medicine, health care						•	
HVAC, fire control, security						•	
Military, aerospace							
Telecommunications,office equip.							
Meteorology environment						•	
Other (specify)							

Main Activities

Sensors, molecular electronics, catalysis, ion conductors, biosensors and biocompatibility, surface and interface analysis.

Specific Expertise

Gas sensors, new materials and transducers.
Materials: metals - Pt, Pd, Ni, Ag, Au, Sb, Rh; semi conductors - Si, GaAs, InP; ionic compounds - electron conductors (SnO_2, TiO_2, Ta_2O_5, ...); phathalocyanine, ziolites, calixarene, ...; polymers; biomolecular materials

Collaboration Offered

Cooperation based on national and international research projects.

158 Vitronic GmbH

Bildverarbeitungssysteme
Hasengartenstr. 14a
D-6200 Wiesbaden

Tel: +49 611 71520
Fax: +49 611 715233
E-mail:

Affiliation:
Contact: KM Bücher, Sales&Mktg

Organisation information
No of employees 26-100
Turnover (ECU*) 1m-5m
Sensors technology activities information
No of employees 26-100
Turnover (ECU*) 1m-5m
*1ECU=DM 1.89,FFr 6.30,£0.78,Lira 1909,Peseta 159.1

Organisation activities
Manufacturer of sensor-based products

Application \ Signal Sensed	Radiation	Optical	Electrical	Magnetic	Thermal	Chemical	Mechanical
Automotive	●						
Consumer products	●						
Process industries, mining, energy							
Manufacturing, machinery	●						
Medicine, health care							
HVAC, fire control, security							
Military, aerospace							
Telecommunications,office equip.							
Meteorology, environment							
Other *							

Main Activities
Vitronic is a well known manufacturer of 2D and 3D grey level image processing systems for a large variety of industrial applications such as: barcode auto-ID systems, character recognition, robot vision, direct parts identification, optical quality assurance systems for on-line and off-line parts inspection. One example is an optical navigation system for automatically guided vehicles (AGV).

Specific Expertise
Vitronic develops image vision systems to customers' specifications and is specialised in the field of industrial vision systems for automated processes. 100% of the software is written in-house, parts of the hardware are made outside, and the configuration is arranged in-house. Furthermore, the company is involved in various Resource and Development projects, sponsored by the German Government. More than 100 different vision systems have been installed so far to their customers' full satisfaction.

Collaboration Offered
A high degree of specific know-how for image vision processes, based on industrial experience. Standard system solutions for a large variety of industrial applications.

Collaboration Sought
Market knowledge and an operating sales force and service. Related products.

159 Weber Sensors GmbH

Strohdeich 32
D-25377 Kollmar

Tel: +49 4128 591
Fax: +49 4128 593
E-mail:

Affiliation:
Contact: W S Krause, Mark. Director

Organisation information

No of employees	26-100
Turnover (ECU*)	1m-5m

Sensor technology activities information

No of employees	26-100
Turnover (ECU*)	<0.5m

*1ECU=DM 1.89,FFr 6.30,£0.78,Lira 1909,Peseta 159.1

Organisation activities
Research and development on sensors, sensor materials or applications.
Testing, consulting, design.
Manufacture of finished sensors.

Application \ Signal Sensed	Radiation	Optical	Electrical	Magnetic	Thermal	Chemical	Mechanical
Automotive		•			•		•
Consumer products					•		•
Process industries,mining energy					•		•
Manufacturing, machinery							
Medicine, health care							
HVAC, fire control, security					•		•
Military, aerospace							
Telecommunications,office equip.							
Meteorology, environment					•		•
Other *		•			•		

*Steel

Main Activities

Weber Sensors is a global operating company with partners in most industrialised countries in the world which has been successfully developing, marketing and selling special sensors for the industry for more than 25 years. Focus areas are hot metal detection systems, flow detection sensors for liquids and gas/air and proximity sensors for large distances.

Specific Expertise

Sensors for detection of hot metal - photo captors. Proximity sensors for larger distances - proxi captors. Laser edge detection sensors - laser captors. Flow detection of liquids - flow captors. Flow detection of gas/air - vent captors.

Collaboration Offered

Weber sensors can help to engineer and develop speciality sensors, designed for specific applications.

Collaboration Sought

Marketing and sales support.

160 Weinlich GmbH & Co

Industriestr 6
D-68799 Reilingen

Tel: +49 6205 4025
Fax: +49 6205 17744
E-mail:

Affiliation:
Contact: Weinlich, GM, Owner

Organisation information
No of employees 1-25
Turnover (ECU*) <1m
Sensor technology activities information
No of employees 1-10
Turnover (ECU*) <0.5m
*1ECU=DM 1.89,FFr 6.30,£0.78,Lira 1909,Peseta 159.1

Organisation activities
Manufacture of sensor-based products.

Application \ Signal Sensed	Radiation	Optical	Electrical	Magnetic	Thermal	Chemical	Mechanical
Automotive							•
Consumer products		•					
Process industries,mining energy							
Manufacturing, machinery							
Medicine, health care							
HVAC, fire control, security		•					
Military, aerospace		•					
Telecommunications,office equip.							
Meteorology, environment							
Other *							•

* Training

Main Activities

Weinlich GmbH & Co is a small company developing, producing and selling: numerical controls; performance test stands for motors and other rotating machines; brake testers for vehicles; bullet velocity-meters for small arms; acoustic rate-or-fire meters.

Specific Expertise

Numeric controls - applications: robots and special applications outside of the area of machine tools;- used sensors: for position, partly for speed. Performance test:- stands - applications: combustion engines, electrical rotation machines, pumps etc, - sensing technologies: strain gauges, Hall sensors, shaft encoders, inductive sensors. Bullet speedmeters - applications: projectiles of all kinds of small arms and other flying small parts or linearly moving parts; sensing technology: light barriers. Rate-of-fire meters - applications: mechanisms whose speed cannot be measured by tachometers or stroboscopes,eg. because of inconstant frequency, - sensing technology: microphones.

Collaboration Offered

Regenerative engine dynamometer.

161 Wenglor Sensoric GmbH

Narzissenstrasse 3
D-88069 Tettnang

Tel: +49 75 4253990
Fax: +49 75 42539988
E-mail:

Affiliation:
Contact: P Pramann, Gen Mgr

Organisation information
No of employees 26-100
Turnover (ECU*) >5m
Sensors technology activities information
No of employees 26-100
Turnover (ECU*) >5m
*1ECU=DM 1.89,FFr 6.30,£0.78,Lira 1909,Peseta 159.1

Organisation activities
Research and Development
Manufacturer of sensor-based products
Testing, consulting, design
Manufacture of finished sensors

Application \ Signal Sensed	Radiation	Optical	Electrical	Magnetic	Thermal	Chemical	Mechanical
Automotive			●				
Consumer products							
Process industries, mining, energy			●				
Manufacturing, machinery			●				
Medicine, health care							
HVAC, fire control, security			●				
Military, aerospace							
Telecommunications,office equip.							
Meteorology, environment							
Other *							

Main Activities

Wenglor Sensoric GmbH is a young, flexible and dynamic high-tech company, which develops and produces equipment which can sense and recognise materials of all types, sizes and colours, without touching them. This greatly assists in problem-free automation of operating procedures in an extensive variety of machines and processes.

Specific Expertise

Opto-electronic sensors and light barriers, mostly working with infr ared light, are insensitive to secondary light and are fitted with a contamination warning. This warning is in the form of a flashing LED indicator and a separate output which enables the customer to intervene as soon as the sensing defect emerges. The LED also serves as an alignment aid.
A further special feature of high switching frequency of 20 KHZ is available on certain models. The company has a series of opto-electronic switches giving an analogue output varying with range from target. Grid sensing incorporating a multiplexed system and a number of opto-electronic light beams for accurate grid sensing over an area. Retro-reflective light barriers can be supplied incorporating polarisation filters for definite recognition of shiny objects.

162 WIKA Alexander Wiegand GmbH

Alexander-Wiegand-Straße
D-63911 Klingenberg

Tel: +49 9372 132 0
Fax: +49 9372 132 406
E-mail:

Affiliation:
Contact: Horst Landwehr, Marketing

Organisation information

No of employees	>1000
Turnover (ECU*)	>50m

Sensor technology activities information

No of employees	>100
Turnover (ECU*)	>5m

*1ECU=DM 1.89,FFr 6.30,£0.78,Lira 1909,Peseta 159.1

Organisation activities
Research and development on sensors sensor materials or applications
Manufacture of finished sensors
Manufacture of sensor-based products

Application \ Signal Sensed	Radiation	Optical	Electrical	Magnetic	Thermal	Chemical	Mechanical
Automotive							
Consumer products							
Process industries,mining energy					●		●
Manufacturing, machinery					●		●
Medicine, health care					●		●
HVAC, fire control, security							
Military, aerospace					●		●
Telecommunications,office equip.							
Meteorology environment					●		●
Other (specify)							

Main Activities

For 50 years WIKA has been a compentent producer of pressure and temperature measurement instruments. Our products are divided into four major groups: mechanical pressure gauges; chemical seals; electronic pressure measurement instruments; electronic and mechanical temperature measurement instruments.

Specific Expertise

Electronic pressure measurement instruments: pressure transducers and transmitters for the measurement of absolute and differential pressure and vacuum with piezoresistive, inductive and hall-effect sensors, as well as thin-film sensors. Pressure ranges covering --0.6mbar to 0-7000bar. Standardised current or voltage signal output. Digital and analog indicating and recording instruments. Pressure calibration equipment.

Collaboration Offered

Development of low-pressure sensor elements (up to 25 bar)

Collaboration Sought

Also development of low-pressure sensor elements (up to 25 bar).

163 Zirox Sensoren & Elektronik GmbH

Brandbeichstr. 19
D-17498 Greifswald

Tel: +49 13834 550 116
Fax: +49 13834 550 222
E-mail: zirox@t-online.de

Affiliation:
Contact: W Klinguer, Manager

Organisation information

No of employees 1-25
Turnover (ECU*) <1m

Sensor technology activities information

No of employees 1-10
Turnover (ECU*) <0.5m

*1ECU=DM 1.89,FFr 6.30,£0.78,Lira 1909,Peseta 159.1

Organisation activities

Research and development on sensors sensor materials or applications
Manufacture of sensor-based products

Application \ Signal Sensed	Radiation	Optical	Electrical	Magnetic	Thermal	Chemical	Mechanical
Automotive							
Consumer products							
Process industries,mining energy						•	
Manufacturing, machinery							
Medicine, health care							
HVAC, fire control, security							
Military, aerospace							
Telecommunications,office equip.							
Meteorology environment						•	
Other							

Main Activities

Development and production of gas sensor systems working on the base of solid electroytes. There are sensors for oxygen, carbon dioxide and hydrocarbons developed and produced. Development and production of precision calorimeter for the investigation of chemical reactions in liquid phases (isoperibol and adiabatic operation).

Additional Short Form Entries

ADUS Druckmesstechnik GmbH
Rosenbergstr.109
D-70193 Stuttgart

Contact	Rudolf Heyschmidt
Tel:	+49 711 6366835
Fax:	+49 711 631710

Sensors, sensor systems and components for gas and liquid mechanics, development.

AKTIV-ELECTRONIC GmbH
Griembergweg 34A
D-12305 Berlin

Contact	Hartmut Bloch
Tel:	+49 30 74703-0
Tel:	+49 30 7465736

Piezo resistive pressure sensors from sensor chip up to transmitters with complete evaluation electronic.

AMS THERMOTECH GmbH
Fraunhoferstr.22
D-82152 Martinsried/München

Contact	Sonja Mayer
Tel:	+49 89 89577-0
Fax:	+49 89 89577-199

AMS Thermotech GmbH represents Cambridge AccuSense Inc. A US manufacturer of airflow sensing devices utilising anemometer technology. Their product range includes airflow-switches, sensors and multi point measurement systems. The sensors can be calibrated to both, airflow (0 to 500 Std. cm/min) and differential pressure (0 to 1250 Pa). Also available are Precision Temperature switches and Thermocouple Monitor Systems.

Automation W + R GmbH
Romanstrasse 41
D-80639 München

Contact	Albin Weingartner
Tel:	+49 89 179199-0
Fax:	+49 89 179199-91

Industrial image processing, inspection systems for completeness, surfaces, dimensions, colours, optical character recognition.

BARTEC Componenten und Systeme GmbH
Dieselstrasse 6
D-97980 Bad Mergentheim

Contact	
Tel:	+49 7931 597-0
Fax:	+49 7931 597 119

Sensors for thermal and other quantities, sensor systems for thermal and other quantities as well as for gas and liquid mechanics.

Baumer Electric GmbH
Pfingstweide 28
D-61169 Friedberg

Contact	Dr Jürgen Nehler
Tel:	+49 6031 6007-0
Fax:	+49 6031 6007-70

Sensors and sensor systems for geometrical quantities, mech. Quantities, dynamic quantities, gas and liquid mechanics, optical quantities, sensor systems for: identity or presence, calibration and test systems for geometrical quantities, mechanical quantities, sensor components, industrial research: development, research, application-oriented tests.

Ernst Befort Wetzlar GmbH
Braunfelser Str.26-30
D-35578 Wetzlar

Contact	Ernst Befort
Tel:	+49 6441 22081 83
Fax:	+49 6441 24818

Sensors components for optical quantities, high tech-job shop, thin film technique.

BESTEC GmbH
Rudower Chaussee 6
D-12484 Berlin

Contact	Dr Hammerschmidt
Tel:	+49 30 6774376
Fax:	+49 30 6775718

High automatised systems for basic research in LASER technique and UHV technique on customer's request. PLD-UHV system, UHV-STM system, fs-LASER system, spectrometer, fast I/O boards, PC-OMS, energy measurement system, pulse delay generator, heater controller, bakeout controller.

betacontrol gmbH mess-und regeltechnik
Am Weidekamp 10
D-57258 Freudenberg

Contact	H G Schuster-Baumer
Tel:	+49 2734 468-0
Fax:	+49 2734 7711

Measurement and control systems for basis weight, thickness, density, coating thickness, moisture.

Franz Binder GmbH & Co
Elektrische Bauelemente KG
Rötelstr. 27
D-74172 Neckarsulm

Contact	W Fröhlich
Tel:	+49 7132 325-0
Fax:	+49 7132 325-150

Round connectors for sensors and sensor-bus systems, industrial versions with VDE-, UL-, GL- and SEV-approval, customer specified connector systems for industrial and medical applications, own design and engineering shop (CAD).

BMC Dr. Schetter GmbH
Boschstrasse 12
D-82178 Puchheim

Contact	
Tel:	+49 89 8002403
Fax:	+49 89 8002956

Sensors for pressure, acceleration, displacement, torque, measurement amplifiers, data logger-systems.

Bonn Sensorik
Hafendamm 6
D-47495 Rheinberg

Contact	Udo Bonn
Tel:	+49 02844 656
Fax:	+49 02844 1477

Optical laser sensors and sensor systems for thickness distance, gap, contour, 3-D surface scan, level, opacity, deformation, width, height, length, position, vibration, and more.

Bourdon Sedeme GmbH
Steinfurt 24
D-52222 Stolberg

Contact	B H Lenz
Tel:	+49 2402 9550-0
Fax:	+49 2402 9550-6

Sensor elements and transmitters for gas- and liquid-mechanics and force, transmitters for temperature. Calibration- and test systems for gas- and liquid mechanics.

Brose Systeme GmbH
Kienitzer Strasse 98
D-12049 Berlin

Contact:	Torsten Brose
Tel:	+49 30 62709193
Fax:	+49 30 62709195

Sensors for thermal quantities, sensor systems for geometrical, mechanical, dynamic, thermal, climatic, and other quantities and for gas and liquid mechanics, development consulting.

CIS Centrum für intelligente Sensorik Erfurt e.V
Haarbergstrasse 61
D-99097 Erfurt

Contact	Dipl.Ing. Roy Lischke
Tel:	+49 361 420510
Fax.:	+49 361 4205113

Sensor systems for humidity, liquid and length measurements, for climatic, chemical and other quantities.

CompuMess Elektronik GmbH
Lise-Meitner-Strasse 1
D-85716 Unterschleissheim

Contact
Tel: +49 89 321501-0
Fax: +49 89 321501-11

FFT-Analysators, digital, oscilloscopes, calibrators, impedance bridges, pulse generators, measuring transmitter.

Corrsys Korrelatorsysteme GmbH
Siegmund-Hiepe-Strasse 24
D-35578 Wetzlar

Contact Rainer Rottstock
Tel: +49 6441 9282-14
Fax: +49 6441-9282-17

Correvit (trade mark) sensors and systems for non-contact and non-clip measurement for length (distance) speed and drift for the dynamics of vehicle movement and process control of industrial strip material eg textiles, paper, steel plates and rubber with directional detection, customer specific sensor technology.

Datron-Messtechnik GmbH
Steinstrasse 5
D-35641 Schöffengrund

Contact Herr Köhler
Tel: +49 6445 608-0
Fax: +49 6445 7097

Correlation optical sensors (v-sensors and DLS-sensors), sensors and sensor-systems for non-contact, optical length and speed measurement at vehicles, distance and speed sensors (non-contact, optical multi-axle).

Delta OHM
Am Weimer 6
D-35410 Hungen

Contact Dr Th. Brokmann
Tel: +49 6402 9817
Fax: +49 6402 9602

Portable measuring instruments and transmitter for temperature, humidity, sound level, pressure, lux pH meters, conductivity meters, water activity.

Delta Regeltechnik GmbH
Türkenstr.11
D-80333 München

Contact Matthias Geller
Tel: +49 89 2820-43
Fax: +49 89 2850-41

Sensors for temperature (thermistors), pressure, force and acceleration.

Disynet GmbH
Holtschneiderweg 36
D-41379 Brüggen

Contact Hans A Holtkamp
Tel: +49 2157 8799-0
Fax: +49 2157 8799-22

Sensors and transducers for pressure, force, acceleration and torque based on foil and semiconductor gauges as well as micro-machined structures, customised transducers.

DISYS Mess-und Testsysteme GmbH
Auf der Grefenfurth 1-3
D-51503 Rösrath

Contact
Tel: +49 02205 84019
Fax; +49 02205 85244

Customised test systems for quality assurance, test and measuring systems.

DMT Druck-Messtechnik GmbH
Birkenstr.15
D-59514 Welver

Contact G. Olthuis
Tel: +49 2384 3037
Fax: +49 2384 497

Sensors and sensor systems for gas and liquid mechanics and for thermal quantities, calibration and test systems for gas and liquid mechanics and thermal quantities.

Döhrer Elektrooptik GmbH i.Gr
Ettlinger Str.5
D-76307 Karlsbad

Contact	Dipl-Ing. E Döhrer
Tel:	+49 7202 1535
Fax:	+49 7202 5616

Sensors elements for optical quantities, optical components.

Dosch Messapparate GmbH
Wiener Str.10
D-10999 Berlin

Contact	Diop.-Ing.R. Dosch
Tel:	+49 30 6124014
Fax:	+49 30 6124061

Sensors and sensor systems for gas and liquid mechanics and thermal quantities, application in machine engineering, production technology, building, heating, ventilation, air conditioning, environmental protection.

Drägerwerk AG
Moislinger Allee 53/55
D-23542 Lübeck

Contact	Burkard Dillig
Tel:	+49 451 882-0
Fax:	+49 451 2080

Sensors, sensor calibration and test-systems, sensor elements for chemical quantities, sensor systems for other quantities, R&D application-oriented tests, consulting, application in chemical industry, process technology, building, environmental protection, safety and security.

Driesen + Kern GmbH
Am Hasselt 25
D-24576 Bad Bramstedt

Contact	Dipl.Ing. H H Driesen
Tel:	+49 4192 9814
Fax:	+49 4192 7321

Instruments for: relative humidity, temperature, velocity, flow rate. Co2 + Co, pH-Value, depth, pressure. Data logger, miniature-data logger.

Druck & Temperatur Leitenberger GmbH
Bahnhofstrasse 33
D-72138 Kirchentellinsfurt

Contact	Erich Coulon
Tel:	+49 7121 60291
Fax:	+49 7121 600959

Pressure gauges, pressure transmitters, accessories, thermometers, pressure and temperature recorders, thermo hygrographs, deadweight tests, electronic calibrators, hand held test pumps, hand held tachometers.

ELKA-Elektronik GmbH
Talstr.104
D-58515 Lüdenscheid

Contact	
Tel:	+49 2351 176-0
Fax:	+49 2351 176 490

Sensors and sensor systems for pressure and humidity/temperature-acquisition. Portable meters and registration systems. Development and series production (OEM). Applications: MSR, climatic, building and bus-systems.

ELKUTEC Electronic GmbH
Erfurter Str.23
D-85386 Eching

Contact	
Tel:	+49 89 3192077
Fax:	+49 89 3195361

Sensor systems for thermal, climatic and chemical quantities and for gas and liquid mechanics.

EUROPAVIA-Deutschland GmbH
Meckenheimer Allee 67-69
D-53115 Bonn

Contact	
Tel:	+49 228 60430-0
Fax:	+49 228 60430-30

Electric measurement of mechanical quantities: strain gauges, load cells for force, pressure, displacement, acceleration, measurement amplifiers.

FOS Messtechnik GmbH
Rütgersstrasse 40
D-24790 Schacht-Audorf

Contact	Dr E Giese
Tel:	+49 4331 9065
Fax:	+49 4331 9066

Pressure sensors (fibre optical) for temperatures up to 800°C without cooling, cylinder pressure sensors for engine and motor application.

Foundry Control
Mess-und Regeltechnik GmbH
Robert-Bosch Strasse 5
D-56566 Neuwied

Contact	
Tel:	+49 2631 964000
Fax:	+49 2631 57867

Optical (infra-red) humidity measuring.

FPM Space Sensor GmbH
Am St. Niclas Schacht 13
D-09599 Freiberg/Sa

Contact	C Hamann
Tel:	+49 3731 781 216
Fax:	+49 3731 781 226

Contact less velocity sensors, length measuring device, sensors for flow measurement in gas and fluid, micro light barriers, real-time signal processor for frequency analysers.

GELMA Industrieelektronik GmbH
Mainzerstrasse 36-52
D-53179 Bonn

Contact	H Racke
Tel:	+49 228 8554-0
Fax:	+49 228 8554-237

Sensors and senors systems for geometrical and dynamic quantities for application in machine engineering and production technology.

Haenni-Mess - Systeme GmbH
Hofener Str.122
D-70335 Stuttgart

Contact	
Tel:	+49 711 5035-0
Fax:	+49 711 5035 175

Transducers and transmitters for gas and liquid mechanics as well as for thermal and climatic parameters. Systems for weight measurements. Calibration systems for pressure. Calibration services for pressure, temperature and humidity.

Erwin Halstrup Multur GmbH
Albert-Schweitzer-Str.2
D-79199 Kirchzarten

Contact	Herr Sura
Tel:	+49 7661 3963-0
Fax:	+49 7661 3963-99

Differential pressure transmitters for aeriform- and gaseous media up to 100 kPA. Digital manometer, calibration equipment, barometer, volume flow measuring equipment.

Hamamatsu Photonics Deutschland GmbH
Arzbergerstr.10
D-82211 Herrsching

Contact	Frau K Reutter
Tel:	+49 8152 375-0
Fax:	+49 8152 2658

Sensors for thermal, optical an other quantities, sensor systems for geometrical quantities, sensor elements for optical and thermal quantities, consulting.

Hamlin Electronics GmbH
Friedberger Str.124
D-61118 Bad Vilbel

Contact	
Tel:	+49 6101 7029
Fax:	+49 6101 2423

Product range Hamlin: Reed-switch products, magnetic position sensors, shock sensors and reed relays.

Heckendorf Engineering GmbH
Mess- und Regeltechnik Industriestr.
D-77729 Willstätt

Contact

Tel:	+49 7852 97340
Fax:	+49 7852 97341

Digital controllers with graphical programming, cycle-times up to 0.5 ms max. 18 MB RAM for data-logging, partly with ISDN-connection for remote controlling, ISDN-products, development of processor-systems and ISDN applications.

HEITRONICS Infrarot Messtechnik GmbH
Kreuzberger Ring 40
D-65205 Wiesbaden

Contact

Tel:	+49 611 97393-0
Fax:	+49 611 97393-26

Radiation pyrometers and systems for non-contact temperature measuring.

Hentschel System GmbH
Carl-Buderus-Strasse 5
D-30455 Hannover

Contact

Tel:	+49 511 94992-10
Fax:	+49 511 494581

Measuring and control technique for high security, data acquisition for crash tests.

Höntzsch GmbH
Robert Bosch Str.8
71334 Waiblingen

Contact	D Stutzke
Tel:	+49 7151 1716-0
Fax:	+49 7151 58402

Sensors and sensor systems for gas and liquid mechanics, development, consulting.

Honeywell AG, Sensorik
Kaiserleistrasse 39
D-63067 Offenbach

Contact

Tel:	+49 69 8064-0
Fax:	+49 69 8064 442

Pressure, airflow, level position, current and magnetic field sensors. Photoelectric controls, micro and limit switches. Optoelectronic and fibre optic components. Bus systems, bus sensors and valves.

IBR Prozessautomation GmbH & Co.
Messtechnik KG
Kirchstr.20
D-36166 Haunetal - Rhina

Contact	Dipl.-Ing.Robert Rohrbach
Tel:	+49 6673 919180
Fax:	+49 6673 919185

Sensors and sensor systems for displacement and angle measurements. Customs-specific development, miniaturisaiton and integration.

ifm electronic gmbh
Teichstrasse 4
D-45127 Essen

Contact	Thomas Nolde
Tel:	+49 201 242 20
Fax:	+49 201 2 42 22 00

Inductive and capacitive proximity switches, photoelectric sensors, electronic flow sensors, encoders, electronic pressure sensors, controller and evaluation systems, industrial communication system AS-interface.

IMM Messtechnologie GmbH
Carl-von-Linde-Str.40
D-85716 Unterschleissheim

Contact	Friedrich Raith
Tel:	+49 89 321412-0
Fax:	+49 89 321412-11

Sensors on optoelectronic or laser basis for measuring of geometric and mechanic parameter and colours.

IMOS Gubela GmbH Retroreflektoren
Schleifweg 2
D-77871 Renchen

Contact	H.E Gubela
Tel:	+49 7843 84181
Fax:	+49 7843 563

Retro reflectors, reflex foils, microstructures, laser reflex systems, production, research and development.

InfraTec GmbH
Infrarotsensorik und Messtechnik
Gostritzer Str.61-63
D-01217 Dresden

Contact	Dr M Krauss
Tel:	+49 351 4734 8620
Fax:	+49 351 4734 8727

Infra red imaging systems, thermographic services, Contact less temperature measurement (pyrometry). IR calibration sources, single and multi-element IR detectors, OEM components for IR technique.

INOR Transmitter GmbH
Postfach 1410
D-64836 Münster

Contact	Karl A Herdt
Tel:	+49 6071 612008 u.612009
Fax:	+49 6071 612007

Transmitters for temperature and physical quantities, 2-wire transmitters for in-head and DIN-rail mounting, 4-wire transmitters for DIN-rail and rack mounting, signal isolators, alarm units, loop indicators.

KAISER sensoren + systeme
Paartalstrasse 24
D-86316 Friedberg

Contact	H J Kaiser
Tel:	+49 821 269900
Fax:	+49 821 2699011

Sensors and systems for pressure measuring.

Klipp & Zonen Deutschland GmbH
Obere Dammstrasse 10
D-42653 Solingen

Contact	
Tel:	+49 212 587579
Fax:	+49 212 587599

Mess data acquisition, environmental measurement.

Kistler Instrumente GmbH
Daimlerstrasse 6
D-73760

Contact	Klaus H Deppler
Tel:	+49 711 3407-0
Fax:	+49 711 3707-159

Measuring equipment for electrical measurement, quartz-crystal-transducers for pressure, force, acceleration, tension and strain as well as load amplifiers, piezo esistive pressure transducer and amplifiers.

KOBOLD Messring GmbH
Nordring 22-24
D-65719 Hofheim/Ts

Contact	Thomas Budian
Tel:	+49 6191 299-0
Fax:	+49 6192 23398

Production and export of measuring, check and control instruments for flow, pressure, level and temperature.

Krick Messtechnik & Partner KG
Kinzigstr.100
D-63505 Langenselbold

Contact	Dip.-Ing. Herbert Krick
Tel:	+49 6184 62550
Fax:	+49 6184 4386

Small but flexible firm which specialists on customer demands in the field of thermometry. Resistance thermometer, thermocouples and service.

Dr. Kubelik Industrievertretungen
Auf dem Guggelensberg 14
D-76227 Karlsruhe

Contact	Dr. Ute Kubelik
Tel:	+49 721 405552
Fax:	+49 721 408385

Sensors and sensor systems for pressure, tension, torque, temperature.

Kugler GmbH
Heiligenberger Str.100
D-88682 Salem

Contact	Lothar Kugler
Tel:	+49 7553 9200-0
Fax:	+49 7553 9200-5

Optical sensor systems for geometrical quantities, application in machine engineering and production technology.

Kulite Semi-Conductor GmbH
Niederhofheimer Str.56
D-65719 Hofheim/Ts

Contact	Helge Petri
Tel:	+49 6192 7065
Fax:	+49 6192 24182

Sensors and sensor systems for gas and liquid mechanics, mechanical and dynamic quantities.

LAP GmbH Laser Applikationen
Zeppelinstr.23
D-21337 Lüneburg

Contact	Dr Armin Röckseisen
Tel:	+49 4131 9511 95
Fax:	+49 4131 951196

Laser for alignment and positioning, non-contract measurement systems for distance, length, width, thickness, diameter, velocity. Sensors, mechanics software.

Lenord, Bauer & Co GmbH
Dohlenstrasse 32
D-46145 Oberhausen

Contact	Hans-Georg Wilk
Tel:	+49 208 9963-0
Fax:	+48 208 676292

Controls for the automation of motion sequences: positioning controllers, synchronisation, controls, camshaft gears. Sensors for the detection of geometric and dynamic dimensions: incremental encoders, absolute encoders, linear scales, built-in pulse generators.

Leuze electronic GmbH & Co
In der Braike 1
D-73277 Owen-Teck

Contact	Rudi Dölfel
Tel:	+49 7021 5730
Fax:	+49 7021 573199

Sensors for geometrical quantities, sensor systems for other quantities.

Lorenz Messtechnik GmbH
Obere Schloßstrasse 131
D-73553 Alfdorf

Contact	Dipl.-Ing.Greiner
Tel:	+49 7172 3675 32268
Fax:	+49 7172 32172

Torque-force-weight-sensors, required supply and evaluation electronics, pressure sensors, extension sensors, strain gauges-contract application platform weighers, container weighing.

LUST-Oriel GmbH
Laser Optik Technologie
Im Tiefen See 58
D-07629 Hermsdorf

Contact	D Schönherr
Tel:	+49 36601 61103
Fax:	+49 3660 64282

Development and manufacturing of customer specified sensors and sensor systems in thick-film-technology for geometrical, optical and dynamic quantities.

MESA Systemtechnik GmbH
Schiffstr.6
D-78464 Konstanz

Contact Heinrich Bercher
Tel: +49 7531 9371-0
Fax: +49 7531 9371-71

Sensors and sensor systems for gas and liquid mechanics, geometrical, dynamic, thermal, climatic and other quantities.

Messko A. Hauser GmbH & Co. KG
Gablonzer Str.25-27
D-61440 Oberursel

Contact Werner Krüger
Tel: +49 6171 6398-0
Fax; +49 6171 6398-98

Sensors for thermal quantities, application in machine engineering, production and process technology, chemical industry, medical technology, environmental protection.

MICRO-SENSYS GmbH
Juri-Gagarin-Ring 27
D-99084 Erfurt

Contact Jurisch
Tel: +49 361 598740
Fax: +49 361 598747

RD-ID systems, rf - transponder, telemetry modules, minimised data logger for temperature and shock.

Natec Schultheiss GmbH
Niels-Bohr Str.11
Postfach 15 18
D-85748 Garching

Contact
Tel: +49 89 3291086
Fax: +49 89 3206297

Sensors and sensor systems for gas and liquid mechanics, calibration and test systems.

National Semiconductor GmbH
Livry-Gargan-Strasse 10
D-82256 Fürstenfeldbruck

Contact John Owsley
Tel: +49 180 530 8585
Fax: +49 180 530 8586

Design, manufacturing and marketing of high performance semiconductor components worldwide. Leading provider of integrated circuits (Ics) system solutions and technologies for communication, personal systems, industrial and consumer markets.

Newport Electronics GmbH
Daimlerstrasse 26
D-75392 Deckenpfronn

Contact Siegfried Stahl
Tel: +49 7056 3017
Fax: +49 7056 8540

Sensors and systems for thermal, climatic and mechanical quantities.

OEC GmbH
Vogelbergstrasse 20
D-86441 Zusmarshausen

Contact H-G Juraschek
Tel: +49 8291 9536 8347
Fax: +49 08291 8348

UV-VIS-NIR- IR-Detectors, optics, optical materials, VIS- NIR-laser diode modules, pump cavities, HeNe, krypton flash-lamps, amps TE-controllers, laser drive, customer manufacturing, pcb's, ceramics, thick- thin-film, SMD moulded plastic packages, complete systems.

OPTOLAB GmbH
Warnstedtstr. 57-59
D-22525 Hamburg

Contact Heinz-G Franz
Tel: +49 40 54744 200
Fax: +49 40 54744-201

Sensors for geometrical, dynamic, climatic, optical and other quantities, sensor systems for IR and X-ray radiation, identity and pattern recognition.

Penny & Giles GmbH
Mauthstr.9
D-85049 Ingolstadt

Contact	Dipl.-Ing Gustav Kramer
Tel:	+49 841 93503-0
Fax:	+49 841 93503-31

Sensorsystems for length and angle, inclination and pressure sensors.

Pfaudler-Werke GmbH
Pfaudlerstrasse
D-68723 Schwetzingen

Contact	B. Stolz
Tel:	+49 6202 85-0
Fax:	+49 6202 2241 26151 22487

Sensors for thermal and chemical quantities, for gas and liquid mechanics.

Philips Thermocoax
Meiendorfer Str.205
D-22145 Hamburg

Contact	W Weithoner
Tel:	+49 40 67960 274
Fax:	+49 40 67960 333

Sensors for temperature, fluxmeter, force, weighing (in motion) speed, vibration, sensor elements for nuclear quantities and thermal qualities.

Physik Instrumente (PI) GmbH & Co.
Polytec-Platz 1
D-76337 Waldbronn

Contact	Dr Karl Spanner
Tel:	+49 7243 604 100
Fax:	+49 7243 604 145

Capacitive displacement sensors and measuring systems with 0, 1 Nanometre resolution for distance measurement and foil thickness measurement. Range 15μm to 0, 3mm. Bandwidth: 10kHz. Piezo r ceramics for acceleration, vibration and level measurement, ultrasonic and actuation applications by daughter company "PI Ceramic GmbH".

PI Ceramic GmbH
Keramische Technologien und Bauelemente
Lindenstrasse
D-07589 Lederhose

Contact	Frank Moller
Tel:	+49 36604 882-0
Fax:	+49 36604 88225

Development and manufacturing of piezo r ceramic components for sensors and micro positioning, sensors and sensor systems for mechanical and dynamic quantities, gas and liquid mechanics, piezoelectric actuators for micro positioning application, job shop.

Pier-Electronic GmbH
Nassaustrasse 35
D-65719 Hofheim-Wallau

Contact	
Tel:	+49 6122 20654
Fax:	+49 6122 16533

Sensors for relative liquid, dewpoint, absolute humidity (in gases, liquids and solid states), measurement equipment.

piezosystem jena
Wildenbruchstr.15
D-07745 Jena

Contact	
Tel:	+49 3641 675 390
Fax:	+49 3641 675 393

Piezoelectric actuators and positioning systems up to 400μm movement for one and more axes, fibre positioning systems, piezoelectric driven optical slit, power supplies, integrated measurement systems, special developments.

PIL Sensoren GmbH
Eugen Kaiserstr.1
D-63486 Bruchköbel

Contact	Klaus Isele
Tel:	+49 6181 71085
Fax:	+49 6181 78824

Ultrasonic-distance-sensors for distance-measurement, presence detection, level measurements.

PRÜFTECHNIK AG
Oskar-Messter-Str.19-21
D-85737 Ismaning

Contact	
Tel:	+49 89 99616-0
Fax:	+49 89 99616-200

Sensor systems for geometrical and dynamic quantities, laser optical measurement systems, application in production and process technology.

Pulsotronic Merten GmbH & Co KG
Fritz-Kotz-Str.8
D-51674 Wiehl-Bomig

Contact	Harald Bassler
Tel:	+49 2261 702 02
Fax:	+49 2261 702 144

Sensors and sensor systems for the investment industry, metal detection and separation systems for the plastics and food industry. Image processing systems for the automated quality control in all lines of business.

Rheintacho GmbH & Co.
Mess-und Regeltechnik
Walterhofener Str.1
D-79111 Freiburg.i.Br

Contact	Herr Wangler
Tel:	+49 761 4513-0
Fax:	+49 761 44527-4

Sensors and sensor systems for dynamic quantities.

RJM Rheinmetall Jena Image Technology GmbH
Löbstedter Str.107-109
D-07749 Jena

Contact	Barbara Tietz
Tel:	+49 3641 6765 (0)18
Fax:	+49 3641 6765-41

Sensors and sensor systems for geometrical quantities and identification, development and consulting.

Rössel Messtechnik GmbH & Co
Lohstr.2
D-59368 Werne

Contact	Leopold Rössel
Tel:	+49 2389 409 01
Fax:	+49 2389 409 80

Sensors, sensor - calibration and test systems, sensor elements for thermal quantities, electrical measurement amplifiers, connections, development, application-oriented tests.

ROLEC
Mess & Systemtechnik
Ing. Buro Robert Lechner
Kirchstr.5
D-82211 Herrsching-Breitbrunn

Contact	Dip.Ing. . Lechner
Tel:	+49 8152 40178
Fax:	+49 8152 925107

Processor controlled indicators and strobing systems for geometrical, mechanical, climatic, thermal and other quantities, prototyping and series production, programmable logic, system integration, customer specific development, consulting production.

Klaus Schäfer Gesellschaft für Verfahrenstechnik mbH
Mörfelder Landstr.33
Postfach 1660
D-63225 Langen

Contact	Klaus Schäfer
Tel:	+49 6103 79085
Fax:	+49 6103 71799

Sensors and sensor systems for gas and liquid mechanics.

Schäfter & Kirchhoff
Celsiusweg 15
D-22761 Hamburg

Contact	Herr Federau
Tel:	+49 40 8512562
Fax:	+49 40 8403137

CCD line scan cameras, interfaces and software for PC and IPC, system components for optoelectronic measuring and sensing techniques, laser collimator systems with beam shaping and coupling optics for lacier diodes 514-1550nm.

Schönbuch Electronic Hanesch GmbH & Co KG
Robert-Bosch Str. 1
D-71154 Nufringen

Contact	David Sanders
Tel:	+49 7032 9468 10
Fax:	+49 7032 9468-49

Sensors for geometrical, dynamic and optical quantities, sensor systems for colour recognition, applications in machine engineering, production technology, chemical industry, process and medical technology.

Dr. Seitner System-Sensortechnik GmbH
Gewerbestr.17
D-82211 Herrsching

Contact	Peter Enzinger
Tel:	+49 8152 9324-0
Fax:	+49 8152 932499

Sensors and sensor systems for geometrical, mechanical, and dynamic quantities, application in machine engineering, production technology, building, heating, ventilation, air condition, aerospace industry, medical technology, safety and security, environmental and sewage.

Sensortechnics GmbH
Aubinger Weg 27
D-82178 Puchheim.b. München

Contact	Friedbert Werner
Tel:	+49 89 800830
Fax:	+49 89 8008333

Sensors and sensor systems for gas and liquid mechanics, calibration and test systems, humidity sensors.

Sentronic GmbH- Gesellschaft für Optische Meßsysteme
Gostritzer Str.61-63
D-01217 Dresden

Contact	Uwe Kirschner
Tel:	+49 351 473 48653
Fax:	+49 351 473 48448

Fibre optic sensors and US-VIS-NIR miniature spectrometers for chemical and physical measurements, fir for online process control.

SIE Sensorik Industrie-Elektronik GmbH
Edisonstr 2
D-68519 Viernheim

Contact	Herr Kubasta
Tel:	+49 6204 79058 59
Fax:	+49 6204 791 44

Sensors for geometrical and dynamic quantities. Sensors and sensor systems for gas and liquid mechanics, systems for electrical quantities, electrical measurement amplifiers, development

Siemens AG
EC B W M/V
Siemensstr. 13
D-67343 Speyer

Contact	H Iden
Tel:	+49 6232 30 2366
Fax:	+49 6232 30 2152

Magnetic sensors for the contact less measurement of magnetic field strength, position, rotation, displacement and angular. Magnetic proximity switches. Standard and customer specific sensors.

Steinbichler Optoelektronik GmbH
Am Bauhof 4
D-83115 Neubeuern

Contact Dr Hans Steinbichler
Tel: +49 8035 8704-0
Fax: +49 8035 1010

Sensor systems for dynamic and other quantities, optical measuring and test system (ESPI, shearing) contour measurement systems, optical components, application in automotive technics, machine engineering, production technology, aerospace industry.

Entwicklungsbüro Dipl.-Phys. Gerhard Stresing
Reinholdstre.5-6
D-12051 Berlin

Contact Gerhard Stresing
Tel: +49 30 6854506
Fax: +49 30 6851256

CCD-line scan cameras with simple computer interface for all kinds of optical measuring tasks, like object recognition and quality control, position and distance measurement, laser beam diagnostics or spectroscopy systems (also cooled systems for IR or UV)

Telemeter Electronic
Josef-Gänslet-Str.10
D-86609 Donauwörth

Contact F Weber
Tel: +49 906 70693 47
Fax: +49 906 70693-50

Sensors and sensor systems for dynamic and thermal quantities.

TESTEM Gesellschaft für Mess- und Datentechnik mbH
Reismühlerstr.65
D-81477 München

Contact Veronika Rösch
Tel: +49 89 748242-0
Fax: +49 89 748242-42

Sensors and sensor systems for geometrical, mechanical, dynamic, thermal and climatic quantities and for gas and liquid mechanics.

TSI GmbH
Zieglerstr.1
D-52078 Aachen

Contact
Tel: +49 241 523 03 0
Fax: +49 241 523 03 49

Sensor systems for gas and liquid mechanics, for length and speed for industrial production control, particle and aerosol measurement , industrial hygiene.

TWK-Elektronik GmbH
Heinrichstr.85
D-40239 Düsseldorf

Contact T Kessler
Tel: +49 211 632 067
Fax: +49 211 637 705

Analogue and digital transducers for rotary and linear displacement (potentiometer, inductive, electro optical, absolute, incremental, magnetostrictive).

Tylan General GmbH
Kirchhoffstr.8
D-85386 Eching

Contact Dr. Ulrich Breitmeier
Tel: +49 8165 9511-9
Fax: +49 8165 61399

Mass flow in gases. Pressure: capacity-membrane-vacuummeter.

Vaisala GmbH
Marlowring 21
D-22502 Hamburg

Contact	Rüdiger Ludewig
Tel:	+49 40 858 029
Fax:	+49 40 850 8444

Measurement instruments for relative humdiity, temperature, dewpoint, absolute humidity, CO_2, pressure, meteorological and traffic safety systems.

WACHENDORFF PROZESSTECHNIK GmbH & Co.KG
Industriestrasse 7
D-65360 Geisenheim

Contact	Andreas Kühn
Tel:	+49 6722 9965-0
Fax:	+49 6722 996543

Product range: counter, preset counter, rate indicator, timers, digital panelmeter for DMS, ana.log signals, PT100, thermocouples, PID-controller, IPC's, operator panels and message centers for rough industrial applications.

Wallace & Tiernan GmbH
Auf der Weide 10
D-89312 Günzburg

Contact	
Tel:	+49 8221 904-0
Fax:	+49 8221 904 203

Sensor systems for gas and liquid mechanics, calibration and test systems.

Hungary

Entry 164

164 Hungary Academy of Sciences

ATKI Research Institute for Materials Science
PO Box 49
H-1525 Budapest

Tel: +36 1 169 5165
Fax: +36 1 155 0694
E-mail: barsony@ra.atki.kfki.hu

Affiliation: Hungarian Academy of Science
Contact: István Bársony PhD, MD

Organisation information
No of employees 26-100
Turnover (ECU*) 1m-5m

Sensor technology activities information
No of employees 11-25
Turnover (ECU*) <0.5m
*1ECU=DM 1.89,FFr 6.30,£0.78,Lira 1909,Peseta 159.1

Organisation activities
Research and development on sensors sensor materials or applications
Testing, consulting, design
Supply of raw materials or semi-finished products used to make sensors
Manufacture of finished sensors

Application \ Signal Sensed	Radiation	Optical	Electrical	Magnetic	Thermal	Chemical	Mechanical
Automotive							•
Consumer products							
Process industries,mining energy		•		•			•
Manufacturing, machinery							•
Medicine, health care		•					
HVAC, fire control, security						•	
Military, aerospace				•			
Telecommunications,office equip.		•		•		•	•
Meteorology environment							
Other							

Main Activities

The institute is primarily committed to carry out basic and applied research in the field of materials science including materials technology (formation and modification of thin films), materials analysis (ion-beam analysis, X-ray analysis, STM) and materials characterisation (optical, magnetic and electrical). Besides the research activities the institute is also involved in the exploration of the studied basic phenomena in practical applications. This is resulting in the development of transferable know-how, and even in prototyping of the developed equipment or small scale manufacturing of R&D products.

Specific Expertise

Silicon piezoresistive-type pressure sensors, silicon micromachining (bulk and surface), combustion type chemical sensors, grid coupled integrated optical sensors, photosensors, materials for nonlinear optics (YIG, GGG), magnetic field sensors, thin-films for sensors.

Collaboration Offered

Microtechnology for sensor development, contract research on different sensing principles.

Collaboration Sought

System integration, marketing, large scale realisation.

Ireland

Entries 165 to 171

165 Dublin City University

School of Chemical Sciences
Glasnevin
IRL-Dublin 9

Tel: +353 1 7045404
Fax: +353 1 7045503
E-mail:

Affiliation:
Contact: Dr Diamond, Research Dir

Organisation information
No of employees >100
Turnover (ECU*)
Sensors technology activities information
No of employees 1-10
Turnover (ECU*) <0.5m
*1ECU=DM 1.89,FFr 6.30,£0.78,Lira 1909,Peseta 159.1

Organisation activities
Research and Development
Testing, consulting, design

Application \ Signal Sensed	Radiation	Optical	Electrical	Magnetic	Thermal	Chemical	Mechanical
Automotive							
Consumer products							
Process industries, mining, energy							
Manufacturing, machinery							
Medicine, health care			•				
HVAC, fire control, security							
Military, aerospace							
Telecommunications,office equip.							
Meteorology, environment			•				
Other *							

Main Activities

Teaching and research in chemistry on B.Sc. and Ph.D. level. Degree courses: B.Sc. analytical science; B.Sc. applied science; B.Sc. applied chemistry (1992); B.Sc. chemistry and languages (1992); M.Sc. instrumental analysis.

Specific Expertise

Potentiometric sensors for application in vivo, medical applications and general ion selective electrodes for Na, K, Li etc.; electrochemical biosensors, based on enzymes, antibodies; sensors based on polymer modified electrodes for the detection of selected analytes, including metals, anions.

Collaboration Offered

A series of novel sensors ready for commercialisation.

Collaboration Sought

Production and design of sensor unit.

166 Dublin City University

School of Physical Sciences
Optical Sensors Laboratory
Glasnevin
IRL-Dublin 9

Tel: +353 1 7045299
Fax: +353 1 7045384
E-mail:

Affiliation:
Contact: B MacCraith
V Ruddy

Organisation information
No of employees >100
Turnover (ECU*)

Sensors technology activities information
No of employees 11-25
Turnover (ECU*)
*1ECU=DM 1.89,FFr 6.30,£0.78,Lira 1909,Peseta 159.1

Organisation activities
Research and Development

Application \ Signal Sensed	Radiation	Optical	Electrical	Magnetic	Thermal	Chemical	Mechanical
Automotive							
Consumer products							
Process industries, mining, energy						•	
Manufacturing, machinery							
Medicine, health care						•	
HVAC, fire control, security						•	
Military, aerospace							
Telecommunications,office equip.							
Meteorology, environment						•	
Other *							

Main Activities
Organisation: Technological University; Department: Development of sensors for industrial, medical, environmental applications using: optical fibre and waveguides technology. Activities cover UV, visible, IR transmitting fibres.

Specific Expertise
1. Sensors based on evanescent wave interactions on optical fibres and waveguides eg. evanescent absoption on fluoride fibres for monitoring flammable gases.
2. Monitoring water quality by remote detection of fluorescence/absorbance.
3. Sensors based on sol-gel entrapment of chemically sensitive dyes.

Collaboration Offered
Basic and applied research. Sensor development to pre-prototype stage.

Collaboration Sought
Complementary research. Development of product from research environment to production stage.

167 Nitron Europe Limited

1 Brews Hill
Navan
Co Heath
Ireland

Tel: +353 46 71 333
Fax: +353 46 71 331
E-mail:

Affiliation: G Blakenby, MD
Contact:

Organisation information

No of employees 1-25
Turnover (ECU*) 1m-5m

Sensor technology activities information

No of employees 11-25
Turnover (ECU*) <0.5m

*1ECU=DM 1.89,FFr 6.30,£0.78,Lira 1909,Peseta 159.1

Organisation activities

Manufacture of systems used in the manufacture, testing or development of sensors.
Manufacture of finished sensors.

Application \ Signal Sensed	Radiation	Optical	Electrical	Magnetic	Thermal	Chemical	Mechanical
Automotive		●					
Consumer products							
Process industries,mining energy		●				●	
Manufacturing, machinery							
Medicine, health care							
HVAC, fire control, security		●				●	
Military, aerospace							
Telecommunications,office equip.		●					
Meteorology, environment	●					●	
Other *							

Main Activities

Gas detection in general; refrigerant identifiers; halon identifiers.

168 Panametrics Limited

Shannon Airport
Co Clare
Ireland

Tel: +353 61 47 1377
Fax: +353 61 47 1359
E-mail:

Affiliation:
Contact: James Gibson, General Manager

Organisation information

No of employees	26-100
Turnover (ECU*)	5m-50m

Sensor technology activities information

No of employees	26-100
Turnover (ECU*)	1m-5m

*1ECU=DM 1.89,FFr 6.30,£0.78,Lira 1909,Peseta 159.1

Organisation activities
Research and development on sensors sensor materials or applications
Manufacture of finished sensors
Manufacture of sensor-based products

Application \ Signal Sensed	Radiation	Optical	Electrical	Magnetic	Thermal	Chemical	Mechanical
Automotive				•			
Consumer products							
Process industries,mining energy			•	•			
Manufacturing, machinery			•	•			
Medicine, health care							
HVAC, fire control, security			•				
Military, aerospace			•	•			
Telecommunications,office equip.			•				
Meteorology environment			•	•			
Other							

Main Activities

Panametrics specialises in sensors and instrumentation for moisture and oxygen measurement. Panametric is also a manufacturer of transducers and it uses these in flowmetering instrumentation.

Specific Expertise

Thin film aluminium oxide sensors; zirconium oxide sensors; polymer sensors; piezoelectric transducers.

169 Teagasc

Moorepark
Fermoy
Co Cork
Ireland

Tel: +353 25 31422
Fax: +353 25 32563
E-mail: djocallaghan@dpc.teagasc.ie

Affiliation:
Contact: D O'Callaghan, Engineer

Organisation information
No of employees >1000
Turnover (ECU*) 5m-50m
Sensor technology activities information
No of employees 1-10
Turnover (ECU*) <0.5m
*1ECU=DM 1.89,FFr 6.30,£0.78,Lira 1909,Peseta 159.1

Organisation activities
Research and development on sensors, sensor materials or applications.

Application \ Signal Sensed	Radiation	Optical	Electrical	Magnetic	Thermal	Chemical	Mechanical
Automotive							
Consumer products							
Process industries,mining energy		•			•		
Manufacturing, machinery							
Medicine, health care							
HVAC, fire control, security							
Military, aerospace							
Telecommunications,office equip.							
Meteorology, environment							
Other *							

Main Activities

Research and development for food processing.

Specific Expertise

Sensors for on-line monitoring of rheological properties, and detection of changes in rheological properties, in food processing plants, including cheese manufacture and milk powder manufacture.

Collaboration Offered

Facilities for evaluation of on-line sensors in food processing plants.

Collaboration Sought

On-line technology for food processing lines.

170 University College of Galway

UCG
Department of Physics
IRL-Galway

Tel: +353 91 524411
Fax: +353 91 525700
E-mail:

Affiliation:

Contact: Prof PW Walton, Dir Sensor Gr
M Lang, Ass Dir

Organisation information

No of employees 11-25
Turnover (ECU*)

Sensors technology activities information

No of employees 11-25
Turnover (ECU*)
*1ECU=DM 1.89,FFr 6.30,£0.78,Lira 1909,Peseta 159.1

Organisation activities

Research and Development
Testing, consulting, design

Application \ Signal Sensed	Radiation	Optical	Electrical	Magnetic	Thermal	Chemical	Mechanical
Automotive							
Consumer products							
Process industries, mining, energy	•	•	•	•	•	•	•
Manufacturing, machinery							
Medicine, health care	•	•	•	•	•	•	•
HVAC, fire control, security							
Military, aerospace							
Telecommunications,office equip.							
Meteorology, environment	•	•	•	•	•	•	•
Other *							

Main Activities

Experimental Physics Department: is responsible for teaching physics to the faculties of science, engineering and medicine in addition to providing four year honors degrees in experimental physics, applied physics & electronics, and physics & computing. Research groups include: optical spectroscopy, laser applications, astrophysics, sensors/instrumentation and atmospheric physics.

Specific Expertise

Biosensors (for use in medicine, food processing, environmental monitoring):-
Techniques used: surface acoustic waves; gravimetric; calorimetric; particle based systems.
Other biomedical sensors: sensor for digital radiography; peripheral blood flow measurement.

Collaboration Offered

Novel sensor technologies for commerical application.

Collaboration Sought

Funding. Supply of specialist piezoelectric materials.

171 University of Limerick

School of Engineering
Advanced Sensors Research Unit
Plassey Technological Park
IRL-Limerick

Tel: +353 61 333644
Fax: +353 61 330316
E-mail:

Affiliation:
Contact: V Casey, Dir ASRU
JB McMonagle

Organisation information
No of employees 11-25
Turnover (ECU*) <0.5m

Sensors technology activities information
No of employees 11-25
Turnover (ECU*) <0.5
*1ECU=DM 1.89,FFr 6.30,£0.78,Lira 1909,Peseta 159.1

Organisation activities
Research and Development
Testing, consulting, design

Application \ Signal Sensed	Radiation	Optical	Electrical	Magnetic	Thermal	Chemical	Mechanical
Automotive					●		
Consumer products				●			
Process industries, mining, energy							
Manufacturing, machinery							
Medicine, health care			●			●	
HVAC, fire control, security							
Military, aerospace							
Telecommunications,office equip.							
Meteorology, environment						●	
Other *							

Specific Expertise

Deposition and full electronic characterisation of semiconducting oxide thin films; Langmuir-Blodgett thin film deposition onto sensor substrates; Conducting polymer preparation, doping and characterisation; gas sensor testing; thin/thick film sensor patterning.

Collaboration Offered

Preparation and deposition of sensitised materials for incorporation in a wide range of sensors. Gas sensor characterisation.

Collaboration Sought

Provision of sensor test vehicles such as ISFET and SAW structures for sensor development.

Italy

Entries 172 to 179

172 Dibe University of Genoa

via Opera Pia 11A
I-16145 Genoa

Tel:
Fax: +39 10 35 32 777
E-mail: gzutta@dibe.umige.it

Affiliation:
Contact: Prof M Grattaroia

Organisation information
No of employees 26-100
Turnover (ECU*) 1m-5m
Sensors technology activities information
No of employees 1-10
Turnover (ECU*) <0.5m
*1ECU=DM 1.89,FFr 6.30,£0.78,Lira 1909,Peseta 159.1

Organisation activities
Research and Development
Testing, consulting, design

Application \ Signal Sensed	Radiation	Optical	Electrical	Magnetic	Thermal	Chemical	Mechanical
Automotive							
Consumer products							
Process industries, mining, energy							
Manufacturing, machinery							
Medicine, health care						•	
HVAC, fire control, security							
Military, aerospace							
Telecommunications,office equip.		•					
Meteorology, environment						•	
Other *							

Main Activities

Universita' Degli Studi di Genova is involved in teaching courses and research programmes:

- Bioelectromagnetics and Molecular Electronics
- Bioelectronics and Neurobioengineering
- Computer systems architecture and integrated VLSI systems
- Electromagnetism
- Electronic systems, multimedia systems, networking, telematics for education
- Image and signal processing and understanding
- Mathematical physics
- Microelectronics
- Nonlinear circuits, complex systems and neural networks
- Physical structures of perception and computation

Specific Expertise

- Characterisation by means of atomic force microscopy and acoustic microscopy
- Automated testing of performances
- Computer simulations

Collaboration Offered

Characterisation, testing, design

Collaboration Sought

Chemical sensors and biosensors to be integrated in a system
Biomedical and chemical know-how

173 Franco Corradi

Via Cornaggia 11
PO Box 124
I-20017 RHO Milano

Tel: +39 2 930 29 55
Fax: +39 2 931 69 312
E-mail:

Affiliation:
Contact: F Corradi, President

Organisation information
No of employees 1-25
Turnover (ECU*) <1m

Sensor technology activities information
No of employees 1-10
Turnover (ECU*) 0.5m-1m
*1ECU=DM 1.89,FFr 6.30,£0.78,Lira 1909,Peseta 159.1

Organisation activities
Testing, consulting, design.
Supply of raw materials or semi-finished products used to make sensors.
Manufacture of finished sensors.

Application \ Signal Sensed	Radiation	Optical	Electrical	Magnetic	Thermal	Chemical	Mechanical
Automotive							
Consumer products							
Process industries,mining energy		●			●		
Manufacturing, machinery							
Medicine, health care			●		●		
HVAC, fire control, security							
Military, aerospace		●	●		●		
Telecommunications,office equip.		●	●		●		
Meteorology, environment					●		
Other *							

Main Activities

Temperature measurement sensors and instruments for research and industrial application - precision temperature measurements from 0.05K with accuracy up to 0.01°C - cryogenic resistance elements and thermometers - metal sheathed thermocouple wires and thermocouples - microwelding systems - calibration baths - miniature precision tubes - sealed CO_2 laser tubes and systems - windows, substrates, prisms, rods, high purity metals - sputtering evaporation targets - cavitation and subcooled boiling detection and measurement - infrared pyrometer.

174 IROE-CNR

Via Panciaticmi
I-64-50123 Firenze

Tel: +39 55 42351
Fax: +39 55 410893
E-mail: rubino@iroe.fi.cnr.it

Affiliation: National Council of Research
Contact: N Rubino, Director
F Baldini, Researcher

Organisation information

No of employees 26-100
Turnover (ECU*) 1m-5m

Sensor technology activities information

No of employees 1-10
Turnover (ECU*) <0.5m

*1ECU=DM 1.89,FFr 6.30,£0.78,Lira 1909,Peseta 159.1

Organisation activities
Research and development on sensors sensor materials or applications

Application \ Signal Sensed	Radiation	Optical	Electrical	Magnetic	Thermal	Chemical	Mechanical
Automotive		●					
Consumer products							
Process industries,mining energy							
Manufacturing, machinery							
Medicine, health care		●					
HVAC, fire control, security							
Military, aerospace							
Telecommunications,office equip.							
Meteorology environment		●					
Other							

Main Activities

Our organisation is invovled in seven main areas: science of TME Atmosphere; optoelectronics and photonics; remote sensing; image sciences; signal processing; integration EM fields-matter; structure of matter and spectroscopy.

Specific Expertise

Optical fibre sensors for physical and chemical parameters for both biomedical and environmental applications. In particular prototype already developed for the detection of: position, vibration, temperature, particle size; pH, oxygen, bilirubin.

Collaboration Offered

Our knowledge and experience in conceiving, designing and developing optical fibre sensor prototypes.

Collaboration Sought

Since we are a Government Research Institute, we are seeking collaboration for testing our prototypes in real conditions.

175 Laser Optronic SRL

via Quaranta 55
I-20139 Milano

Tel: +39 2 57 465 1
Fax: +39 2 56 4101 27
E-mail: 101522.2747@compuserve.com

Affiliation:
Contact: G Galimberti, MD
A Marinelli, Product Manager

Organisation information

No of employees 26-100
Turnover (ECU*) 5m-50m

Sensor technology activities information

No of employees 1-10
Turnover (ECU*) 0.5m-1m
*1ECU=DM 1.89,FFr 6.30,£0.78,Lira 1909,Peseta 159.1

Organisation activities
Testing, consulting, design

Application \ Signal Sensed	Radiation	Optical	Electrical	Magnetic	Thermal	Chemical	Mechanical
Automotive		●					
Consumer products							
Process industries,mining energy					●		
Manufacturing, machinery		●			●		
Medicine, health care					●		
HVAC, fire control, security					●		
Military, aerospace		●			●		
Telecommunications,office equip.		●					
Meteorology environment					●		
Other (specify)							

Main Activities

Supply of electro-optic sensors for energy of power measurements and informed detectors.

176 Magneti Marelli

Div. Elettronica
via Trieste 67
I-27100 Pavia

Tel: +39 382 599 567
Fax: +39 382 599 496
E-mail: braschi@pavia.marelli.it

Affiliation:
Contact: Enrico Braschi, Hardware Engineering
Rino Rocchelli, Program Manager

Organisation information

No of employees	>1000
Turnover (ECU*)	>50m

Sensor technology activities information

No of employees	26-100
Turnover (ECU*)	1m-5m

*1ECU=DM 1.89,FFr 6.30,£0.78,Lira 1909,Peseta 159.1

Organisation activities
Research and development on sensors sensor materials or applications
Manufacture of finished sensors

Application \ Signal Sensed	Radiation	Optical	Electrical	Magnetic	Thermal	Chemical	Mechanical
Automotive							•
Consumer products							
Process industries,mining energy							
Manufacturing, machinery							
Medicine, health care							
HVAC, fire control, security							
Military, aerospace							
Telecommunications,office equip.							
Meteorology environment							
Other (specify)							

Main Activities

Our organisation is involved in all the electronics for the automotive markets.

Specific Expertise

Expertise of our Organisation is on thick film piezoelectric pressure sensors for automotive applications, in particular for pressure over 5bar and when aggressive fluids are involved.

Collaboration Offered

Good experience on: sensor electronic for automotive applications; medium range pressure sensor.

Collaboration Sought

To develop high pressure (>1,000 bar) sensors.

177 University of Modena

Physics Department
Via G Campi
I-2131A 41100 Modena

Tel: +39 59 58 60 30
Fax: +39 59 36 74 88
E-mail: prodew2vati@1moax1.unimo.ij

Affiliation:
Contact: Prof Maria Prudenziati, Group Leader
Prof Bruno Morten, Technology

Organisation information
No of employees 26-100
Turnover (ECU*) >50m

Sensor technology activities information
No of employees 1-10
Turnover (ECU*) >5m
*1ECU=DM 1.89,FFr 6.30,£0.78,Lira 1909,Peseta 159.1

Organisation activities
Research and development on sensors sensor materials or applications
Testing, consulting, design

Application \ Signal Sensed	Radiation	Optical	Electrical	Magnetic	Thermal	Chemical	Mechanical
Automotive				●			●
Consumer products				●	●		●
Process industries,mining energy							
Manufacturing, machinery				●			●
Medicine, health care							
HVAC, fire control, security	●						
Military, aerospace							●
Telecommunications,office equip.							
Meteorology environment							
Other (specify)							

Main Activities

General Scope: to design, implement and test new thick film materials and thick film sensors and actuators. Experience: Thick film piezoresistors - optimisation of pastes, and process, applications in pressure and acceleration sensors; thick film magnetoresistors - design of pastes, characterisation, implementation of sensor prototypes of proximity, linear or angular displacement, rotational speed, identification of application areas; development of PZT-based piezoelectric films and their applications in resonant sensors, and sensors for the measurement of acceleration, torque, fluid flow; thick film resistors for cryogenic thermometers; development of elastic resonant structures and composite ultrasonic transducers.

Specific Expertise

Development of novel material prepared with the thick film technology, i.e. screen printing of adequate pastes and firing processes, by means of: identification of adequate paste compositions (nature of the functional material, size and shape, additives, etc.); preparation and characterisation of thick films; outline of the films' properties as a function of the preparing conditions (composition, firing cycle, terminations, etc.); design of sensor/actuator structures; prototype tests for determining the performance characteristics.

Collaboration Offered

Experience in developing optimised paste compositions: experience in developing optimised process parameters (firing conditions, substrate, structures): experience in correlation between thick film composition, microstructure and functional properties.

Collaboration Sought

Identification of application areas; definition of specific peformances; tests of sensors in field; production and marketing.

178 Università di Pavia

Department of Electronics
via Ferrata 1
I-27100 Pavia

Tel: +39 382 505204
Fax: +39 382 422583
E-mail: donati@ipvsm6.unipv.it

Affiliation:
Contact: S Donati, Research Director

Organisation information

No of employees 26-100
Turnover (ECU*) 1m-5m

Sensor technology activities information

No of employees 1-10
Turnover (ECU*) <0.5m

*1ECU=DM 1.89,FFr 6.30,£0.78,Lira 1909,Peseta 159.1

Organisation activities
Research and development on sensors, sensor materials or applications.

Application \ Signal Sensed	Radiation	Optical	Electrical	Magnetic	Thermal	Chemical	Mechanical
Automotive							•
Consumer products							
Process industries,mining energy							
Manufacturing, machinery			•	•			
Medicine, health care						•	
HVAC, fire control, security							
Military, aerospace							•
Telecommunications,office equip.		•			•		
Meteorology, environment							
Other *							•

* Constructions

Main Activities

Research and development of sensors, mainly fibre-optics, and related circuits.

Specific Expertise

Optical fibre gyroscopes, current sensors, magnetic field sensors, smart sensors, thick-film sensors.

Collaboration Offered

Research and development up to laboratory prototype follow-up of engineering on prototypes. Facilities in a technological park (presently being built) for industrial spin-off.

Collaboration Sought

Specs of the sensors. Research and development contracts.

179 Università degli Studi di Firenze

Dip. di Sanità Pubblica Epid. e Chim. Analitica Ambientale
Via Gino Capponi 9
I-50121 Firenze

Tel: +39 55 2757274
Fax: +39 55 2476972
E-mail: mascini.cesitl.unifi.it

Affiliation: Dip. Sanita Pubbica, Epidem. e Chimica Analitica Ambientale
Contact: Prof. Marco Mascini, Biosensors Lab. Director

Organisation information
No of employees 1-25
Turnover (ECU*)
Sensor technology activities information
No of employees 11-25
Turnover (ECU*) <0.5m
*1ECU=DM 1.89,FFr 6.30,£0.78,Lira 1909,Peseta 159.1

Organisation activities
Research and development on sensors sensor materials or applications
Testing, consulting, design
Manufacture of finished sensors
Manufacture of sensor-based products

Application \ Signal Sensed	Radiation	Optical	Electrical	Magnetic	Thermal	Chemical	Mechanical
Automotive							
Consumer products							
Process industries,mining energy							
Manufacturing, machinery		•	•			•	
Medicine, health care							
HVAC, fire control, security							
Military, aerospace							
Telecommunications,office equip.		•	•			•	
Meteorology environment							
Other (specify)		•	•			•	

Main Activities

Biosensor Technology Development. Sensors for medicine, environment and food chemistry. Coupling of immobilised enzymes and electrochemical or optical sensors. Immobilisation of proteins on piezo-electric quartz for antibody recognition. Immunochemistry based biosensors. DNA probes, electrochemical or piezoelectrical detection. Immobilised enzyme stabilisation. Immobilised proteins or oligonucleotides studies.

Specific Expertise

Immobilisation of proteins by several procedure in order to obtain catalytic or affinity biosensors. Applications include medicine, environment and food chemistry. Immobilisation of enzymes, antibodies and oligonucleotides for specific detection of substrate in complex media like blood, urine, waste water, milk, extract of fish, etc.

Collaboration Offered

Experience on the above technology. Realisation of prototypes for the market. Patents on specific applications. Biosensors in the field of medicine (in vivo, ex vivo), environment (pesticides, bacteria, etc.), food process (freshness, quality, contamination).

Collaboration Sought

Problems to solve, marketing, ability to transfer products from prototype stage to consumer.

Luxembourg

Entry 180

180 IEE-International Electronics&Engineering

2b route de Trèves
L-2632 Luxembourg

Tel: +352 424737-1
Fax: +352 424737-200
E-mail:

Affiliation:
Contact: Herbet Naumer, General Manager

Organisation information

No of employees 101-1000
Turnover (ECU*) 5m-50m

Sensor technology activities information

No of employees >100
Turnover (ECU*) >5m

*1ECU=DM 1.89,FFr 6.30,£0.78,Lira 1909,Peseta 159.1

Organisation activities

Research and development on sensors sensor materials or applications
Manufacture of finished sensors

Application \ Signal Sensed	Radiation	Optical	Electrical	Magnetic	Thermal	Chemical	Mechanical
Automotive			●	●			
Consumer products			●				
Process industries,mining energy							
Manufacturing, machinery							
Medicine, health care			●				
HVAC, fire control, security							
Military, aerospace							
Telecommunications,office equip.							
Meteorology environment							
Other							

Main Activities

Force Sensing Resistors (FSR(TM)) to be used for Passenger Presence Detection systems and Child Seat Orientation and Detection systems for Air-Bag Deployment control in automobiles.

Collaboration Offered

Engineering know-how and FSR technology.

Collaboration Sought

Local marketing/sales support and investment support for local production engineering.

Netherlands

Entries 181 to 193

181 3T BV

Institutenweg 6
PO Box 3639
NL-7500 DP Enschede

Tel: +31 53 4 336633
Fax: +31 53 4 33 6869
E-mail: henk.leeuwis@3t.nl

Affiliation:
Contact: H Leeuwis, Mgr MST

Organisation information

No of employees	26-100
Turnover (ECU*)	1m-5m

Sensor technology activities information

No of employees	1-10
Turnover (ECU*)	0.5m-1m

*1ECU=DM 1.89,FFr 6.30,£0.78,Lira 1909,Peseta 159.1

Organisation activities
Research and development.
Manufacture of sensor-based products.

Application \ Signal Sensed	Radiation	Optical	Electrical	Magnetic	Thermal	Chemical	Mechanical
Automotive				•			•
Consumer products							
Process industries,mining energy							
Manufacturing, machinery				•			•
Medicine, health care						•	
HVAC, fire control, security							
Military, aerospace						•	
Telecommunications,office equip.							•
Meteorology, environment						•	
Other *							

Main Activities

Contract development of micro sensors, actuators and systems in close cooperation with the MESA Research Institute of the University of Twente. Project teams consist of 3T employees supported by university researchers. Development (and production) of sensor based systems by means of micro-electronics (M/W and S/W) and data processing.

Specific Expertise

Sensors and actuators based on micromechanics/micro machining (microphones, flow sensors, pumps, valves, nozzles); thin film magnetic sensors (permalloy); ion sensors (conductivity, ISFET); dissolved oxygen sensors; micro fluidic systems (fluid management, microanalysis systems).

Collaboration Offered

Technology for new products.

Collaboration Sought

Market and application know-how.

182 Bronkhorst High-Tech BV

Nijverheidsstraat 1a
NL-7261 AK Ruurlo

Tel: +31 573 45 88 00
Fax: +31 573 45 88 08
E-mail: sales@bht.nl

Affiliation:
Contact: W Jouwsing, Tech Director
T Bruggeman, Commercial Director

Organisation information
No of employees 26-100
Turnover (ECU*) 5m-50m
Sensor technology activities information
No of employees 26-100
Turnover (ECU*) 1m-5m
*1ECU=DM 1.89,FFr 6.30,£0.78,Lira 1909,Peseta 159.1

Organisation activities
Research and development on sensors, sensor materials or applications.
Testing, consulting, design.
Supply of raw materials or semi-finished products used to make sensors.
Manufacture of finished sensors.
Manufacture of sensor-based products

Application \ Signal Sensed	Radiation	Optical	Electrical	Magnetic	Thermal	Chemical	Mechanical
Automotive			•	•	•		
Consumer products							
Process industries,mining energy			•	•	•		
Manufacturing, machinery			•	•	•		
Medicine, health care			•	•	•		
HVAC, fire control, security			•	•	•		
Military, aerospace			•	•	•		
Telecommunications,office equip.			•				
Meteorology, environment			•	•	•		
Other *							

Main Activities

Bronkhorst High-Tech BV specialises in the field of mass flow measurement and control, and electronic pressure measurement and control systems for both gases and liquids based on modern sensor technology.

R&D, design, manufacturing and selling of sensors and actuators in the area of flow and pressure.

Specific Expertise

Flow measurement and control. Pressure measurement and control. Proportional electromagnetic control values. Thin film sensors based on increased temperature measurement.

Collaboration Offered

R&D in flow measurement and control.

Collaboration Sought

Applications in above mentioned areas.

183 Dewit Industrial Sensors BV

Frans Halsstraat 3
PO Box 202
NL-3440 AE Woerden

Tel: +31 3484 13154
Fax: +31 3484 22352
E-mail:

Affiliation:
Contact: H R Wieske, Director

Organisation information
No of employees 1-10
Turnover (ECU*) <0.5m

Sensors technology activities information
No of employees 1-10
Turnover (ECU*) <0.5m
*1ECU=DM 1.89,FFr 6.30,£0.78,Lira 1909,Peseta 159.1

Organisation activities
Research and Development
Manufacture of finished sensors

Application \ Signal Sensed	Radiation	Optical	Electrical	Magnetic	Thermal	Chemical	Mechanical
Automotive							
Consumer products							
Process industries, mining, energy							
Manufacturing, machinery	•			•			
Medicine, health care							
HVAC, fire control, security							
Military, aerospace							
Telecommunications,office equip.							
Meteorology, environment	•				•		
Other *							

Main Activities

Dewit Industrial Sensors develops and produces a range of industrial sensors based on a square compact housing, allowing easier, faster mounting, robustness and in most instances IP 67 or better protection. The present range comprises: induction proximity switches; infrared sensor; temperature and moisture sensors; proximity sensors with reed contact or Hall IC.

Specific Expertise

Inductive proximity switches: size 10x10x7 up to 40x40x25 mm, sensing range 1-18 mm; supply voltage 10 -30VDC/20 - 250 VAC; infrared sensor size transmitter/receiver 30x30x15 mm self-contained or with separate amplifier; maximum range 30 m dirt resistant, not sensitive to condensation and water.
Temperature sensors: size 20x20x12 - 30x30x15 or 40x40x25 mm, linear output 10 mV/K or 1 micro A/K; temperature range -40° to +120°C.
Humidity sensors: size 40x40x35 mm, sensing range 0 -100%RH; output 0 - 1V linear ±0,5%; temperature range 0 - 85°C; supply voltage 10 - 30 VDC.

Collaboration Offered

Production under label.

Collaboration Sought

Ultra-sonic sensors.

184 Hollandse Signaalapparaten BV

PO Box 42
NL-7550 GD Hengelo

Tel: +31 74 2439874
Fax: +31 74 483070/425936
E-mail:

Affiliation: Thomson - CSF
Contact: R G Van Scaik,
Head Commercial Dept
I Edwards, Communications

Organisation information
No of employees >100
Turnover (ECU*) >5m
Sensors technology activities information
No of employees >100
Turnover (ECU*) >5m
*1ECU=DM 1.89,FFr 6.30,£0.78,Lira 1909,Peseta 159.1

Organisation activities
Manufacturer of sensor-based products

Application \ Signal Sensed	Radiation	Optical	Electrical	Magnetic	Thermal	Chemical	Mechanical
Automotive							
Consumer products	•						
Process industries, mining, energy							
Manufacturing, machinery							
Medicine, health care							
HVAC, fire control, security	•						
Military, aerospace	•						
Telecommunications,office equip.							
Meteorology, environment							
Other *							

Main Activities

Defence electronics: surveillance systems eg. radar, infrared; fire control systems; command and control systems; integrated management of land-based and naval applications; air traffic control and vessel graffic management systems.
Situated in Eindhoven, Signaal-Usfa specialises in electronic and electro-optical equipment and has more than forty years experience in the development and manufacture of high technology products and systems. Its product range comprises; night vision equipment, cryogenic cooling engines, active and reserve lithium batteries and electronic fuses for ammunition purposes.

Specific Expertise

Medium-long range 2D surveillance radar; medium-long range 3D surveillance radar; multi function radar systems; fire control radar systems; data handling systems; infrared surveillance systems; optronic fire control systems. Main Activities are: ATC; naval vessels; battlefield/air defence.
Signaal-Usfa develops a range of electro-optical sensors which includes image intensifiers and thermal imagers covering both military and civil applications. The image intensifier range comprises; binocular night sights for mounting on platforms and weapons, observation sights for multi-purpose applications, self-powered hand-held night viewers for various applications, individual weapon night sights and night-driving periscopes.

185 Revere Transducers Europe BV

Ramshoorn 7
PO Box 67909
NL-4802 HX Breda

Tel: +31 76 54 80 700
Fax: +31 76 54 15 711
E-mail:

Affiliation: Revere Transducers Incorporated
Contact: A Freire, Sales Director
J Kersten, Product Manager

Organisation information

No of employees	26-100
Turnover (ECU*)	5m-50m

Sensor technology activities information

No of employees	26-100
Turnover (ECU*)	>5m

*1ECU=DM 1.89,FFr 6.30,£0.78,Lira 1909,Peseta 159.1

Organisation activities
Research and development on sensors, sensor materials or applications.
Manufacture of finished sensors.

Application \ Signal Sensed	Radiation	Optical	Electrical	Magnetic	Thermal	Chemical	Mechanical
Automotive							
Consumer products							
Process industries,mining energy							•
Manufacturing, machinery							•
Medicine, health care							
HVAC, fire control, security							
Military, aerospace							•
Telecommunications,office equip.							
Meteorology, environment							•
Other *							

Main Activities
Manufacture and sales of hermetically sealed stainless steel local cells for the weighing industry. Load cells are also metrologically approved as well as intrinsically approved.

Specific Expertise
Strain gauge technology and materials science know-how with respect to stainless steel.

Collaboration Offered
Dedicated facility ISO approved for the manufacture of approved (metrological) load cells.

Collaboration Sought
Specific know-how on intelligent sensors. Specific know-how on the chemical industry as related to sensors.

186 Sentron BV

PO Box 125
NL-9300 AC Roden

Tel: +31 5050 13800
Fax: +31 5050 16834
E-mail:

Affiliation:
Contact: H Vanhealen

Organisation information
No of employees 26-100
Turnover (ECU*)

Sensors technology activities information
No of employees 26-100
Turnover (ECU*)
*1ECU=DM 1.89,FFr 6.30,£0.78,Lira 1909,Peseta 159.1

Organisation activities
Research and Development
Manufacturer of sensor-based products
Testing, consulting, design
Manufacture of finished sensors

Application \ Signal Sensed	Radiation	Optical	Electrical	Magnetic	Thermal	Chemical	Mechanical
Automotive							
Consumer products							
Process industries, mining, energy						•	
Manufacturing, machinery							
Medicine, health care						•	•
HVAC, fire control, security							
Military, aerospace							
Telecommunications,office equip.							
Meteorology, environment						•	
Other *						•	

Main Activities

Medical sensors:- pressure sensor catheters for: cardiology, urology, gastroenterology, neurology. Chemical sensors:- pH sensors for: laboratory and industrial applications. Contract development and production for: silicon chip foundry and micro assembly.

Specific Expertise

Medical sensors: Sentron develops, produces and markets catheter tip sensors for the determination of intracorporeal pressure. Applications include the following fields: Cardiology, gastroenterology, urology and neurology.
Chemical sensors: Sentron develops, produces and markets chemical sensors based on the ion sensitive field effect transistor (ISFET). The ISFET offers unique characteristics that allow applications in areas where glass electrodes or membrane electrodes fail.

187 Signaal USFA

Meerenakkerweg 1
NL-5652 AR Eindhoven

Tel: +31 40 250 3776
Fax: +31 40 250 3777
E-mail: zewa@usfa.nl

Affiliation: Hollandse Signaalapparaten
Thomson-CSF
Contact: M Wÿnheÿmer, Electro-Optics
D Verbeek, Cryogenics
M van Oostrom, Fuses
J theuns, Batteries

Organisation information
No of employees 101-1000
Turnover (ECU*) 5m-50m
Sensor technology activities information
No of employees 26-100
Turnover (ECU*) 1m-5m
*1ECU=DM 1.89,FFr 6.30,£0.78,Lira 1909,Peseta 159.1

Organisation activities
Manufacture of sensor-based products.

Application \ Signal Sensed	Radiation	Optical	Electrical	Magnetic	Thermal	Chemical	Mechanical
Automotive							
Consumer products							
Process industries,mining energy							
Manufacturing, machinery							
Medicine, health care							
HVAC, fire control, security							
Military, aerospace		•	•		•		
Telecommunications,office equip.							
Meteorology, environment							
Other *							

Main Activities

Signaal USFA was established over forty-five years ago as part of Philips' Gloeilampen Fabrieken N.V. in Eindhoven. Since then, the company has been developing and manufacturing high-technology electronic equipment particularly for defence applications.

Specific Expertise

Electro-optics.
Cryogenics.
Reserve batteries.
Naval proximity fuzes.

Collaboration Offered

Both direct and indirect compensation/offset programmes,especially in the field of electrooptical fine mechanics, are accepted.

188 TNO-FEL

Physics and Electronic Laboratory
PO Box 96864
NL-2509 JG The Hague

Tel: +31 70 3740000
Fax: +31 70 3280961
E-mail: bekman@fel.tno.nl

Affiliation: TNO
Contact: H Bekman, Researcher
J L Joppe, Section Leader

Organisation information
No of employees 101-1000
Turnover (ECU*) 1m-5m
Sensor technology activities information
No of employees 1-10
Turnover (ECU*) <0.5m
*1ECU=DM 1.89,FFr 6.30,£0.78,Lira 1909,Peseta 159.1

Organisation activities
Research and development on sensors sensor materials or applications
Testing, consulting, design

Application \ Signal Sensed	Radiation	Optical	Electrical	Magnetic	Thermal	Chemical	Mechanical
Automotive						•	•
Consumer products							
Process industries,mining energy							
Manufacturing, machinery							•
Medicine, health care							
HVAC, fire control, security					•		
Military, aerospace	•	•			•	•	•
Telecommunications,office equip.					•		
Meteorology environment						•	
Other (specify)							

Main Activities

We are involved in the development of a sensor platform based on hybrid integrated optic techniques. Various sensors can be developed with this platform as a starting point Our interest at the moment is for miniature sensitive optical pressure sensors.

Specific Expertise

Areas of expertise: integrated optics; growth of low loss optical waveguide materials; positioning with sub-micron precision of laserdiodes. Optical pressure sensors via membranes. Specific materials: silicon; zinc oxide; silicon dioxide; III-V laser diodes.

Collaboration Offered

Positioning/fixing laser diodes with sub-micron precision realising specialised integrated-optic stand-alone sensors where the partner has ideas for a specific optical measuring technique.

Collaboration Sought

Ideas for optical measuring techniques and related markets when implemented in our sensor platform.

189 TNO Nutrition & Food Research Institute

PO Box 360
NL-3700 AJ Zeist

Tel: +31 30 69 44 636
Fax: +31 30 69 44 901
E-mail: koopal@voeding.tno.nl

Affiliation: Netherlands Organisation for Applied Scientific Research
Contact: Dr C Koopal, Product Manager
Dr H Hofstra, Head of Dept

Organisation information
No of employees 101-1000
Turnover (ECU*) 5m-50m
Sensor technology activities information
No of employees 1-10
Turnover (ECU*) 0.5m-1m
*1ECU=DM 1.89,FFr 6.30,£0.78,Lira 1909,Peseta 159.1

Organisation activities
Research And development on sensors sensor materials or applications
Testing, consulting, design

Application \ Signal Sensed	Radiation	Optical	Electrical	Magnetic	Thermal	Chemical	Mechanical
Automotive							
Consumer products							
Process industries,mining energy							
Manufacturing, machinery							
Medicine, health care			●			●	
HVAC, fire control, security							
Military, aerospace							
Telecommunications,office equip.							
Meteorology environment		●			●		
Other (specify)							

Main Activities

TNO Nutrition and Food Research Institute's core activities include research, evaluation and advice in:
- Integrated chain management in the food industry, aimed at process innovation and quality control.
- Functionality of food and feed constituents, particularly proteins and carbohydrates.
- Gene technology for product, process and crop improvement.
- Packaging of food products with an emphasis on taste, shelf-life and safety.
- Favourable and unfavourable properties of human and veterinary drugs in health care.
- Chemical labour conditions in the chemical, agrochemical and food industries.
- Safety and quality of traditional and novel foods, raw materials and chemicals.
- Food consumption and digestive physiology in relation to health.

Collaboration Offered

Knowledge

Collaboration Sought

Enduser needs. Technology push from instrument manufacturers.

190 Universiteit Twente

MESA Research Institute
PO Box 217
NL-7500 AE Enschede

Tel: +31 55 4899111
Fax: +31 53 4892575
E-mail:

Affiliation:
Contact: Prof J Fluitman, Director

Organisation information
No of employees >100
Turnover (ECU*) >5m
Sensors technology activities information
No of employees 26-100
Turnover (ECU*) 1-5m
*1ECU=DM 1.89,FFr 6.30,£0.78,Lira 1909,Peseta 159.1

Organisation activities
Research and Development
Testing, consulting, design

Application \ Signal Sensed	Radiation	Optical	Electrical	Magnetic	Thermal	Chemical	Mechanical
Automotive				•			•
Consumer products	•		•	•			•
Process industries, mining, energy	•					•	•
Manufacturing, machinery				•	•		•
Medicine, health care	•		•			•	•
HVAC, fire control, security							
Military, aerospace							
Telecommunications,office equip.	•		•	•			
Meteorology, environment						•	
Other *	•					•	

Main Activities

The MESA Institute has nine divisions plus a cleanroom facility. MESA is an acronym for: Micro Electronics, Materials Engineering, Sensors & Actuators.
The divisions are: IC - process and device technology; biosensors; lightwave devices; optoelectronics; molecular engineering; micromechanical transducers; analogue, time discrete and adaptive circuits; testable design and testing of digital ICs; information storage technology.

Specific Expertise

Micromechanics in a wide area for sensors, actuators and systems. ISFET-based sensors (eg. CHEMFETS); immuno and enzyme sensors; optical wave guide sensors and switches (eg. glucose); surface plasmon sensors; magnetic sensors (angle & position).

Collaboration Offered

Possibilities for contract research co-operation.

Collaboration Sought

Input of any organisation or industry that can use the Institute's expertise.

191 University of Delft

Department of Electrical Engineering
Electronic Instrumentation Laboratory
PO Box 5031
NL-2600 GA Delft

Tel: +31 15 785951
Fax: +31 15 785755
E-mail:

Affiliation:
Contact: H Leeuwis, Project Mgr, Sensor Tech

Organisation information
No of employees
Turnover (ECU*)
Sensor technology activities information
No of employees
Turnover (ECU*)
*1ECU=DM 1.89,FFr 6.30,£0.78,Lira 1909,Peseta 159.1

Organisation activities
Research or development on sensors, sensor materials or applications.
Testing, consulting, design

Application \ Signal Sensed	Radiation	Optical	Electrical	Magnetic	Thermal	Chemical	Mechanical
Automotive							•
Consumer products	•			•	•		
Process industries,mining energy						•	•
Manufacturing, machinery				•	•		•
Medicine, health care	•		•			•	
HVAC, fire control, security							•
Military, aerospace						•	
Telecommunications,office equip.	•						
Meteorology, environment							•
Other *							

Main Activities

The Electronic Instrumentation Laboratory is one of the 16 laboratories of the Department of Electrical Engineering, and has one full professor, five scientific staff members, ten technical and administrative staff members, fifteen PhD students and 25 MS students. The research carried out at the Laboratory consists of topics ranging from silicon technology, through sensors and signal conditioning to digital instrumentation systems. Most of the projects use the services of the new Delft Institute of Microelectronics and Sub-Micron Technology.
In the last two decades, progress in silicon planar technology has exceeded the most daring predictions, leading to a shortage of suitable sensors, which is why the Electronic Instrumentation Laboratory started a large project on silicon sensors many years ago.

Specific Expertise

Silicon sensors are made with silicon planar technology, which permits batch-fabrication of the sensors and also the integration of the sensing element and electronic circuitry on the same chip leading to “smart sensors”. The Laboratory is active in all five signal domains and was successful in designing silicon sensors for measuring nuclear particles, soft X-rays, colour, light intensity, light images, infrared, flow, vacuum, strain, tilt, acceleration, temperature differences, magnetic field and direction, humidity and gas concentration. Further, much experience was gained in interface electronics for sensors, like OpAmps, SAW components, bridge-frequency conversion, offset correction, bus systems, etc.

Collaboration Offered

Scientific cooperation. Design of prototypes. In general, interested in joining the Common-market programmes.

192 VAF Instruments BV

Vierlinghstraat 24
NL-3316 EL Dordrecht

Tel: +31 78 6183100
Fax: +31 78 6177068
E-mail:

Affiliation:
Contact: P Klouwens,
Dir Business Dev & Sales
A Van der Put, MD
D Snel, Manager Research & Dev

Organisation information
No of employees 26-100
Turnover (ECU*) >5m
Sensors technology activities information
No of employees
Turnover (ECU*)
*1ECU=DM 1.89,FFr 6.30,£0.78,Lira 1909,Peseta 159.1

Organisation activities
Research and Development
Manufacture of finished sensors

Application \ Signal Sensed	Radiation	Optical	Electrical	Magnetic	Thermal	Chemical	Mechanical
Automotive							
Consumer products							
Process industries, mining, energy							•
Manufacturing, machinery							
Medicine, health care							
HVAC, fire control, security							
Military, aerospace							
Telecommunications,office equip.							
Meteorology, environment	•						
Other *							

Main Activities

Manufacture, sales and service of the following sensor and control systems worldwide: Viscosity measurement and control; flow metering, flow control for batch and blending applications; oil concentration measurement in water for effluent monitoring.

Specific Expertise

Flowmeters:- Manufacture, sales and service of the following types of flow meters: Positive displacement meters; magnetic flow meters; corriolis massflow meters. Viscosity sensors and systems: Manufacture, sales and service of viscosity sensors and viscosity control systems for heavy fuel oil in power plants and marine applications. Oil concentration measurement for pollution prevention: Manufacture, sales and service of this equipment used for waste water monitoring and boiler feed water monitoring.

Collaboration Offered

Technology of light scatter for detection of suspensions and concentration measurement.

Collaboration Sought

Co-operation with manufacturer of flow meters to extend range of flow meter sales in Holland and/or worldwide.

193 Xensor Integration BV

PO Box 3233
NL-2601 de Delft

Tel: +31 15 2697 696
Fax: +31 15 2697 748
E-mail:

Affiliation:
Contact: Dr A W van Herwaarden, Director
P J M Aalberse, Mkting & Sales

Organisation information
No of employees 1-25
Turnover (ECU*) <1m
Sensor technology activities information
No of employees 1-10
Turnover (ECU*) <0.5m
*1ECU=DM 1.89,FFr 6.30,£0.78,Lira 1909,Peseta 159.1

Organisation activities
Research and development on sensors, sensor materials or applications
Testing, consulting, design
Manufacture of finished sensors

Application \ Signal Sensed	Radiation	Optical	Electrical	Magnetic	Thermal	Chemical	Mechanical
Automotive							
Consumer products					•		
Process industries,mining energy					•	•	
Manufacturing, machinery					•		
Medicine, health care					•	•	
HVAC, fire control, security							
Military, aerospace	•				•		
Telecommunications,office equip.							
Meteorology environment	•				•	•	
Other							

Main Activities

Xensor Integration BV was founded in 1988 in Delft, the Netherlands. Since then XI has built a vast experience in the development and production of Application Specific Integrated Sensors (ASIS) and silicon micro structures.

Specific Expertise

ASIS development and production, micro structure development and production thermal sensors, micromachining, anodic bonding, sensor consultancy.

Collaboration Offered

Production facilities, expertise on thermal sensors, integrated sensors and micromachining.

Norway

Entries 194 to 197

194 Datainstruments AS

Nye Sandviksvei 56A
PO Box 1561
N-5035 Bergen

Tel: +47 55 311415
Fax: +47 55 323396
E-mail:

Affiliation:
Contact: K Faeraas, Senior Dev Engineer

Organisation information
No of employees 26-100
Turnover (ECU*) 1-5m
Sensors technology activities information
No of employees 11-25
Turnover (ECU*) 1-5m
*1ECU=DM 1.89,FFr 6.30,£0.78,Lira 1909,Peseta 159.1

Organisation activities
Research and Development
Manufacturer of sensor-based products
Testing, consulting, design
Manufacture of finished sensors

Application \ Signal Sensed	Radiation	Optical	Electrical	Magnetic	Thermal	Chemical	Mechanical
Automotive							•
Consumer products							
Process industries, mining, energy							•
Manufacturing, machinery							
Medicine, health care							
HVAC, fire control, security							
Military, aerospace							•
Telecommunications,office equip.							
Meteorology, environment							
Other *							

Main Activities

Design and international marketing of: Traffic analysers; automatic traffic control systems; dynamic vehicle weighing systems; road surface analysers; crane monitoring systems; weighing systems for front-wheel loaders.

Specific Expertise

Datainstrument AS manufacture and market and own patent rights to the S-sensor (TM). The sensor is particularly suited for the monitoring of forces in steel structures.
The sensor has been used in the offshore North Sea environment for 10 years and has an unprecedented record of reliability and stability.
The sensor is also being used by the company on weighing systems in the automotive industry.

Collaboration Offered

Datainstrument AS offer extensive knowledge in the application of the S-sensor for various monitoring requirements.

195 Geonor AS

PO Box 99 Roa
N-0701 Oslo 7

Tel: +47 67147550
Fax: +47 67145846
E-mail:

Affiliation:
Contact: T Levar, MD
G Caprona, Sensor Design

Organisation information
No of employees 11-25
Turnover (ECU*) 1-5m
Sensors technology activities information
No of employees 1-10
Turnover (ECU*) 0.5-1m
*1ECU=DM 1.89,FFr 6.30,£0.78,Lira 1909,Peseta 159.1

Organisation activities
Research and Development
Manufacturer of sensor-based products
Testing, consulting, design
Manufacture of finished sensors

Application \ Signal Sensed	Radiation	Optical	Electrical	Magnetic	Thermal	Chemical	Mechanical
Automotive							
Consumer products							
Process industries, mining, energy			•	•			
Manufacturing, machinery							
Medicine, health care							
HVAC, fire control, security							
Military, aerospace							
Telecommunications,office equip.							
Meteorology, environment				•			
Other *							

Main Activities

Manufacturing and marketing of test equipment/sensors to the construction industry.

Specific Expertise

Development and manufacturing of vibrating wire sensors and read out equipment.

Collaboration Offered

Custom designed sensor solution. Total installation quotations.

Collaboration Sought

Solid-state sensor development knowledge.

196 SensoNor AS

Knutsrodveien 7
PO Box 196
N-3191 Horten

Tel: +47 33 035000
Fax: +47 33 035105
E-mail:

Affiliation:
Contact: S Horntvedt, MD
H Jakobsen, Technology Manager
T Kvisteroy, Development Manager
S Sorensen, Marketing Manager

Organisation information
No of employees
Turnover (ECU*)
Sensors technology activities information
No of employees
Turnover (ECU*)
*1ECU=DM 1.89,FFr 6.30,£0.78,Lira 1909,Peseta 159.1

Organisation activities
Research and Development
Manufacturer of sensor-based products
Manufacture of finished sensors

Application \ Signal Sensed	Radiation	Optical	Electrical	Magnetic	Thermal	Chemical	Mechanical
Automotive							•
Consumer products							
Process industries, mining, energy							
Manufacturing, machinery							
Medicine, health care							•
HVAC, fire control, security							
Military, aerospace							•
Telecommunications,office equip.							
Meteorology, environment							
Other *							•

Main Activities

SensoNor takes on custom design tasks covering product definition, development, production of prototypes and series production, mainly of pressure and acceleration sensors and transducers.

Specific Expertise

SensoNor utilises the piezoresistive effect of silicon in pressure sensors and accelerometers, providing a signal from a wheatstone bridge proportional to the pressure or acceleration. Products serve in application areas such as: defence/aerospace; automotive; medical; research and development.

Collaboration Offered

Pressure sensors and transducers. Accelerometers.

197 Sintef Delab

N-7034 Trondheim

Tel: +47 7 3593000
Fax: +47 7 3591099
E-mail:

Affiliation: Sintef Group
Contact: F Bekkadal, Director of Research
O Pettersen, Acoustics, Ultrasound
R Ellingsen, Optics, Fibre Optics

Organisation information
No of employees >100
Turnover (ECU*) >5m

Sensors technology activities information
No of employees 11-25
Turnover (ECU*) 1-5m
*1ECU=DM 1.89,FFr 6.30,£0.78,Lira 1909,Peseta 159.1

Organisation activities
Research and Development
Testing, consulting, design

Application \ Signal Sensed	Radiation	Optical	Electrical	Magnetic	Thermal	Chemical	Mechanical
Automotive							
Consumer products							
Process industries, mining, energy	•		•				•
Manufacturing, machinery							
Medicine, health care	•				•		
HVAC, fire control, security	•		•				
Military, aerospace	•		•				
Telecommunications,office equip.	•		•				
Meteorology, environment							
Other *							

Main Activities

Sintef Delab is a research and development institute in the Sintef Group, specialising in research and development in information technology based on computer science, telecommunications, electronics and acoustics.

Specific Expertise

Microwaves and radar; acoustics, ultrasound and sonar; optics, fibre optics and lidar; solid-state electronics; signal processing.

Collaboration Offered

Expertise in research and development of sensors, sensor materials and design methods, signal processing and applications.

Collaboration Sought

Requirements for expertise given in 'Collaboration Offered' to perform research and development activities including design and applications of sensors.

Portugal

Entry 198

198 Universidad Nova de Lisboa

Faculdade de Ciências e Tecnologia
Quinta da Torre
P-2825 Monte da Caparica

Tel: +351 2954 464
Fax: +351 2954 461
E-mail: jnm@mail.fet.unl.pt

Affiliation:
Contact: Jose Marat-Mendes, Professor

Organisation information
No of employees 26-100
Turnover (ECU*) n/a
Sensor technology activities information
No of employees 11-25
Turnover (ECU*) n/a
*1ECU=DM 1.89,FFr 6.30,£0.78,Lira 1909,Peseta 159.1

Organisation activities
Research and development on sensors sensor materials or applications
Testing, consulting, design
Supply of raw materials or semi-finished products used to make sensors
Manufacture of systems used in the manufacture, testing or development of sensors
Manufacture of finished sensors
Manufacture of sensor-based products.

Application \ Signal Sensed	Radiation	Optical	Electrical	Magnetic	Thermal	Chemical	Mechanical
Automotive			•		•		•
Consumer products			•		•		•
Process industries,mining energy	•		•		•		•
Manufacturing, machinery	•		•		•		•
Medicine, health care			•		•		•
HVAC, fire control, security	•		•		•		•
Military, aerospace			•				•
Telecommunications,office equip.			•			•	
Meteorology environment	•		•			•	
Other (specify)							

Main Activities

The University specialises in sensors and transducers either at a graduate or a postgraduate level. Research into new sensor materials, the development of sensors and the application of sensors in instrumentation in various areas.

Specific Expertise

Main areas of expertise: piezo electric and pyro electric sensors; SMART sensors; electret microphones; weigh in motion sensors; vibration and acoustic sensors; radiation sensors; etc.
Specific materials: polymer electrets, teflon, aclar, PVDF; piezo electric and pyro electric polymers; SMART materials; piezo and pyro electric ceramics and composites; sol-gel produced ceramics; lead titanate ceramics.

Collaboration Offered

Production and characterisation of materials for sensors (as above); production of sensors (as above); development of systems with sensors (applications).

Collaboration Sought

Complimentary work either in research, or in applications and marketing.

Romania

Entries 199 to 200

199 BitNet-Research

Centre on Sensors & Systems Limited
Madach 67
ROM-3400 Cluj - Napoca

Tel: +40 64 195040
Fax: +40 64 195040
E-mail: bitnet@mail.cjnet.ro

Affiliation: Romanian Space Agency
Contact: Octavian Cristea, Director
Andrei Patriciu, Technology

Organisation information
No of employees 1-25
Turnover (ECU*) <1m

Sensor technology activities information
No of employees 1-10
Turnover (ECU*) <0.5m
*1ECU=DM 1.89,FFr 6.30,£0.78,Lira 1909,Peseta 159.1

Organisation activities
Research and development on sensors sensor materials or applications
Testing, consulting, design

Application \ Signal Sensed	Radiation	Optical	Electrical	Magnetic	Thermal	Chemical	Mechanical
Automotive							
Consumer products							
Process industries,mining energy							
Manufacturing, machinery							
Medicine, health care							
HVAC, fire control, security							
Military, aerospace		•	•		•	•	
Telecommunications,office equip.							
Meteorology environment		•	•		•	•	
Other (specify)							

Main Activities

BitNet Research Centre on Sensors & Systems Ltd is a contract R&D and project management organisation, based in Cluj, Romania. Registered as a private company in 1993.
Main R&D activities: contracts and grants with the Ministry of Science and Technology and the Romanian Academy.
Local project management: BitNet-RCSS is involved in several international projects concerning Earth Observation and Global Change.
Representation: BitNet-RCSS acts as representative in Transilvania for several R&D agencies, such as the Romanian Space Agency and the newly founded National Agency for Tech Transfer & Innovation.

Specific Expertise

R&D on: sensor applications development (vegetation and environment, ground truth for Earth Observation from space); field data loggers; networking applications.

Collaboration Offered

Sensor applications R&D. Tech transfer related activities (information management, technical demonstration, international cooperation opportunities, evaluation, etc.). Local project management and representation services.

Collaboration Sought

Exchange of ideas concerning: participation in international R&D projects; technical transfer related activities development; manufacture of sensor-based products.

200 Research Centre for Ecological Technologies

296 Splaiul Independentiec
PO Box 17-45
ROM-77700 Bucharest

Tel: +40 1 6430329
Fax: +40 1 3124180
E-mail: ccte@u1.ici.ro

Affiliation:
Contact: A Godeanu, Head of Laboratory

Organisation information
No of employees 26-100
Turnover (ECU*) 1m-5m
Sensor technology activities information
No of employees 11-25
Turnover (ECU*) 0.5m-1m
*1ECU=DM 1.89,FFr 6.30,£0.78,Lira 1909,Peseta 159.1

Organisation activities
Research and development on sensors sensor materials or applications
Testing, consulting, design
Manufacture of systems used in the manufacture, testing or development of sensors
Manufacture of finished sensors

Application \ Signal Sensed	Radiation	Optical	Electrical	Magnetic	Thermal	Chemical	Mechanical
Automotive							
Consumer products							
Process industries,mining energy		•	•	•	•	•	•
Manufacturing, machinery		•	•	•	•	•	•
Medicine, health care							
HVAC, fire control, security							
Military, aerospace			•				
Telecommunications,office equip.		•	•	•	•	•	•
Meteorology environment							
Other (specify)							

Main Activities

We are an ecological research institution dealing with problems concerning the ecological technologies (soil, water), ecological monitoring, ecological modelling, sensors, biosensors and electronic sensors for ecological and biological survey, control and process.

Specific Expertise

Biology and ecology, impact studies, research field, manufacture of products, coating new electronic sensors or adapting old ones.

Collaboration Offered

Sensors for humidity (already patented); installations for control survey and monitoring special systems (green houses, laboratories, plant cultures, cultures in special conditions).

Collaboration Sought

Joint ventures for building research and trade or only for promoting our products on the other markets.

Russia

Entries 201 to 208

201 CHROMDET Analytical Instruments

ul.K.Zetkin 18
Rus-Moscow 125299

Tel: +7 95 1545345
Fax: +7 95 4442275
E-mail: chdet@ineum.ac.ru

Affiliation:
Contact: Dr V Boudovitch, Director
Dr E Poloniouk, Sales & Mkting

Organisation information
No of employees 1-25
Turnover (ECU*) <1m
Sensor technology activities information
No of employees 1-10
Turnover (ECU*) <0.5m
*1ECU=DM 1.89,FFr 6.30,£0.78,Lira 1909,Peseta 159.1

Organisation activities
Research and development on sensors sensor materials or applications
Testing, consulting, design
Manufacture of finished sensors
Manufacture of sensor-based products

Application \ Signal Sensed	Radiation	Optical	Electrical	Magnetic	Thermal	Chemical	Mechanical
Automotive							
Consumer products							
Process industries,mining energy							
Manufacturing, machinery							
Medicine, health care							
HVAC, fire control, security							
Military, aerospace							
Telecommunications,office equip.		•	•				
Meteorology environment							
Other							

Main Activities

The main directions of CHROMDET activities are the following: research, development and manufacturing of photoionisation lamps and photoionisaition sensors; manufacturing of instruments based on photoionisation, including portable gas analysers, personal pocket-size monitors, GC and gas calibrators; manufacturing of the instruments based on the electrochemical sensors; manufacturing of multipoint permanent gas detection systems based on photoionisation and electrochemical sensors.

Specific Expertise

CHROMDET has developed photoionisation sensors providing high level of selectivity and more informative signals. This is achieved by means of photoionisation lamps emitting different UVU-spectra due to additional filters. The development of low-consuming photoionisation sensors design for application in personal PID.

Collaboration Offered

Different types of photoionisation lamps and sensors; manufacturing if PID based instruments; research and development of nonradioactive ionisation sources.

Collaboration Sought

Microprocessor part of portable instruments.

202 KWARTA

International Technopark Moskworetschje
Kaschkirskoje Chaussee 31, korp. 44 a
RUS-115409 Moscow

Tel: +7 095 323 9114
Fax: +7 095 524 5462
E-mail:

Affiliation:
Contact:

Organisation information
No of employees 63
Turnover (ECU*)

Advanced materials activities information
No of employees
Turnover (ECU*)
*1ECU=DM 1.89,FFr 6.30,£0.78,Lira 1909,Peseta 159.1

Organisation activities
Research and Development
Manufacturer of sensor-based products

Application \ Signal Sensed	Radiation	Optical	Electrical	Magnetic	Thermal	Chemical	Mechanical
Automotive							
Consumer products							
Process industries, mining, energy			•				
Manufacturing, machinery			•		•		•
Medicine, health care							
HVAC, fire control, security			•		•		
Military, aerospace							
Telecommunications,office equip.							
Meteorology, environment					•		
Other *							

Specific Expertise

Development of
- dosemeters
- pyrometers
- micrometers
- swirl-current-detecter
- acoustic leakage seeking tools.

All instruments are of high technical standard and of low costt.

Collaboration Offered

Contract manufacture and distribution

Collaboration Sought

International cooperation

203 Moscow University

Institute of Nuclear Physics
Leninskije gory
RUS-119899 Moscow

Tel: +7 095 939 2125
Fax: +7 095 939 0896
E-mail:

Affiliation:
Contact:

Organisation information
No of employees
Turnover (ECU*)
Advanced materials activities information
No of employees
Turnover (ECU*)
*1ECU=DM 1.89,FFr 6.30,£0.78,Lira 1909,Peseta 159.1

Organisation activities
Research and Development

Application \ Signal Sensed	Radiation	Optical	Electrical	Magnetic	Thermal	Chemical	Mechanical
Automotive							
Consumer products							
Process industries, mining, energy						•	
Manufacturing, machinery							
Medicine, health care							
HVAC, fire control, security						•	
Military, aerospace							
Telecommunications,office equip.							
Meteorology, environment						•	
Other *							

Main Activities

Instruments both to determine the concentration of radon, thoron and their dissociated constituents and to measure flow and spectrum of alpha-particles of little intensity. The work principle of this instrument is the use of a semiconductor detector for wide-area-radiation. The heavy alpha-particles, the light ß-particles, and the gamma-rays have different losses of energy, on this account they can easily be separated. The minimum of activity recorded by this instrument is 150 - 200 Bq/m^3. There are two modes of operation: real time analysis of radioactivity either of radon or of alpha-particles. Analyse in detail by connecting to a PC. Portable ferrosonde-magnetometer for precise size control and configuration control of native and technical magnetic fields.

Collaboration Offered

Sale of licences, contract manufacture.

Collaboration Sought

Cooperation in manufacture.

204 MTM

International Technopark Moskworetschje
ul. Ak. Kapizy 32-2-138
RUS-117647 Moscow

Tel: +7 095 429 6171
Fax: +7 095 420 6568
E-mail:

Affiliation:
Contact:

Organisation information
No of employees 35
Turnover (ECU*)

Advanced materials activities information
No of employees
Turnover (ECU*)
*1ECU=DM 1.89,FFr 6.30,£0.78,Lira 1909,Peseta 159.1

Organisation activities
Research and Development
Testing, consulting, design

Application \ Signal Sensed	Radiation	Optical	Electrical	Magnetic	Thermal	Chemical	Mechanical
Automotive							
Consumer products							
Process industries, mining, energy							
Manufacturing, machinery							
Medicine, health care	•						
HVAC, fire control, security	•						
Military, aerospace							
Telecommunications,office equip.							
Meteorology, environment							
Other *							

Main Activities

The main activities of the MTM company are research for modern sensors and design of testing of new ly developed sensors.

Specific Expertise

Developments:
- portable radon meter PRA-OIM
- sampling apparatus POU-2

The radon metre is used for rapid time measurement of air in quarters or in production rooms and of water.

Collaboration Offered

Cooperation in manufacture and distribution, contract manufacture

Collaboration Sought

A joint venture

205 REAKONT

Kaschkirskoje Chaussee 13 a
RUS-115201 Moscow

Tel: +7 095 111 1339
Fax: +7 095 111 3055
E-mail:

Affiliation:
Contact:

Organisation information
No of employees 68
Turnover (ECU*)
Advanced materials activities information
No of employees
Turnover (ECU*)
*1ECU=DM 1.89,FFr 6.30,£0.78,Lira 1909,Peseta 159.1

Organisation activities
Research and Development

Application \ Signal Sensed	Radiation	Optical	Electrical	Magnetic	Thermal	Chemical	Mechanical
Automotive							
Consumer products							
Process industries, mining, energy							•
Manufacturing, machinery							•
Medicine, health care							
HVAC, fire control, security							
Military, aerospace							•
Telecommunications,office equip.							
Meteorology, environment							
Other *							

Main Activities

The scientific and technical company REAKONT is preparing technical solutions in the areas of electronics and optics.

Specific Expertise

Portable supersonics-echotomoscope. Computer aided supersonics mammographer.
New algorithms to remove low acoustic signals in sounds were set up on basis of statistical theory and a new approach to solve tasks of back-resonance-solids to reconstruct their figures.

Collaboration Offered

Cooperation in R&D and manufacture, contract manufacture.

Collaboration Sought

International cooperation.

206 Russian Academy of Science

Institute of Automation and Electrometrics
Universiteskij pr. 1
RUS-630090 Nowosibirsk

Tel: +7 3832 354550
Fax: +7 3832 354851
E-mail:

Affiliation:
Contact:

Organisation information
No of employees 480
Turnover (ECU*)
Advanced materials activities information
No of employees
Turnover (ECU*)
*1ECU=DM 1.89,FFr 6.30,£0.78,Lira 1909,Peseta 159.1

Organisation activities
Research and Development

Application \ Signal Sensed	Radiation	Optical	Electrical	Magnetic	Thermal	Chemical	Mechanical
Automotive							
Consumer products							
Process industries, mining, energy							
Manufacturing, machinery		•	•				
Medicine, health care							
HVAC, fire control, security							
Military, aerospace							
Telecommunications,office equip.							
Meteorology, environment							
Other *							

Main Activities

The Institute of Automation of Electometrics focuses on R & D in the fields of electronics and optics.

Specific Expertise

Development of a laser-depiction generator. The work principle of this precision record system is the circular scanning of a focused argon laser. The maximum size of the substratum can be 300mm of a diameter, a gauge of substratum 1 - 20 mm.
The record method is based on thermochemical and thermostructural measurements released by laser in thin films.

Collaboration Offered

Cooperation in R&D, consultancy, contract manufacture

Collaboration Sought

International cooperation.

207 Russian Academy of Science

Institute of Physics of Semiconductors
Prospekt ak. Lawrentjewa 13
RUS-630090 Nowosibirsk

Tel: +7 3832 354286
Fax: +7 3832 357202
E-mail:

Affiliation:
Contact:

Organisation information
No of employees 550
Turnover (ECU*)
Advanced materials activities information
No of employees
Turnover (ECU*)
*1ECU=DM 1.89,FFr 6.30,£0.78,Lira 1909,Peseta 159.1

Organisation activities
Research and Development
Testing, consulting, design

Application \ Signal Sensed	Radiation	Optical	Electrical	Magnetic	Thermal	Chemical	Mechanical
Automotive							
Consumer products							
Process industries, mining, energy							
Manufacturing, machinery		•	•				
Medicine, health care		•	•				
HVAC, fire control, security		•	•				
Military, aerospace		•	•				
Telecommunications,office equip.							
Meteorology, environment		•	•				
Other *							

Main Activities

The Institute of Physics of Semiconductors is generally active in the fields of electronics and optics.

Specific Expertise

Development of hetero-structures for infrared matrix-cameras. The epitaxial structure with superior grids may be used to produce infrared matrix cameras for medicine and environmental monitoring. The superior grids are based on AIGa/GaAs and InGaAs/InAIAs.

Collaboration Offered

Scientific consultancy

Collaboration Sought

Cooperation in R&D and manufacture.

208 University of Transport

Department of Automation of Railway Traffic
pl. Opoltschenija 2
RUS-344017 Rostow

Tel: +7 8632 311 215
Fax:
E-mail:

Affiliation:
Contact:

Organisation information
No of employees 19
Turnover (ECU*)
Advanced materials activities information
No of employees
Turnover (ECU*)
*1ECU=DM 1.89,FFr 6.30,£0.78,Lira 1909,Peseta 159.1

Organisation activities
Research and Development
Testing, consulting, design

Application \ Signal Sensed	Radiation	Optical	Electrical	Magnetic	Thermal	Chemical	Mechanical
Automotive		•	•	•			
Consumer products							
Process industries, mining, energy							
Manufacturing, machinery							
Medicine, health care							
HVAC, fire control, security							
Military, aerospace							
Telecommunications,office equip.							
Mèteorology, environment							
Other *							

Main Activities

The Department of Automation of Railway Traffic is active in the fields of development and manufacture of magnetic, electronic and optical sensors.

Specific Expertise

Developments:
Magnetic sensor DM88; uninterrupted sensor SCHTANKE SCHMP93; differential-transformer-sensor SCHTANKE DP50; receiver for sensor signals. The sensors are used in the railway net to track down overheated points. These sensors are based on a new theory on reliable work of the sensors under the influence of electromagnetic interferences and difficult or instable conditions. Reliability, in use and over time, minimal maintenance, 20 - 25 per cent of world-market price.

Collaboration Offered

Cooperation in manufacture and R&D, order manufacture

Collaboration Sought

Joint ventures

Slovakia

Entries 209 to 210

209 Ploskon AT

Exnarova 15
Presov
SK-080 01 Slovakia

Tel: +42 91 732836
Fax: +42 91 734197
E-mail:

Affiliation:
Contact: S Ploskon, Director

Organisation information
No of employees 26-100
Turnover (ECU*) 1m-5m
Sensor technology activities information
No of employees 11-25
Turnover (ECU*) 0.5m-1m
*1ECU=DM 1.89,FFr 6.30,£0.78,Lira 1909,Peseta 159.1

Organisation activities
Research and development of sensors, sensor materials or applications.
Manufacture of finished sensors.
Manufacture of sensor-based products.

Application \ Signal Sensed	Radiation	Optical	Electrical	Magnetic	Thermal	Chemical	Mechanical
Automotive				•			•
Consumer products							
Process industries,mining energy		•		•			•
Manufacturing, machinery		•		•	•		•
Medicine, health care				•			•
HVAC, fire control, security							
Military, aerospace		•		•			•
Telecommunications,office equip.		•		•			•
Meteorology, environment		•		•	•		•
Other *							

Main Activities
Ploskon ATdeals with he manufacture of components for automated processes, the emphasis being laid on sensors. Moreover, the company manufactures chip and readers, card park barriers, management systems for health-care administration, swimming pool control systems, etc.

Specific Expertise
Ploskon AT deals with series production of sensors for industrial use. Thanks to constant development, the products have been constantly improved. In addition, there is the possibility to manufacture custom-made sensors based on the specifications of customers. Our processor-controlled sensors increase the quality-level of our products as well as the comfort of their applications.

Collaboration Offered
The company is in a position to manufacture sensors on the basis of documentation submitted by our partners or, sensors can be developed and manufactured based on in-house design. Sensors can frequently be used to complete the range of options for a comprehensive approach to a specific technical problem.

Collaboration Sought
To get involved in joint development projects concerning sensors, or aimed at extending our product range in view of our future plans. Furthermore, we would like to cooperate in the field of technologies.

210 Slovak Academy of Sciences

Institute of Electrical Engineering
Dubravská cesta 9
SK-842 39 Bratislava
Slovakia

Tel: +42 7 378 2755
Fax: +42 7 375 816
E-mail: dubecky @savba.sk

Affiliation: Slovak Academy of Sciences
Contact: Dr Frantisek Dubecky, Head of Dept
Dr Milan Majoras, Head of Dept

Organisation information

No of employees 101-1000
Turnover (ECU*) 1m-5m

Sensor technology activities information

No of employees 11-25
Turnover (ECU*) <0.5m

*1ECU=DM 1.89,FFr 6.30,£0.78,Lira 1909,Peseta 159.1

Organisation activities

Research and development on sensors sensor materials or applications
Testing, consulting, design
Supply of raw materials or semi-finished products used to make sensors
Manufacture of finished sensors

Application \ Signal Sensed	Radiation	Optical	Electrical	Magnetic	Thermal	Chemical	Mechanical
Automotive							
Consumer products							
Process industries,mining energy				•			
Manufacturing, machinery				•			
Medicine, health care	•						
HVAC, fire control, security							
Military, aerospace							
Telecommunications,office equip.		•					
Meteorology environment		•					
Other * Methology	•	•		•			

Main Activities

IEE SAS was founded in 1955. At present it has a scientific staff of about 50 and a technical and administrative staff of 60. The Institute has two basic scientific directions: superconductors and semiconductors, with four scientific departments each. The main activity in the field of semiconductors is directed to the basic applied research in the semiconductor physics/mostly III-V compounds, in the following departments: theory of semiconductor microstructures; optoelectronics; microelectronic structures and; semiconductor technology.
A part of research work is done with the collaboration with domestic as well as foreign institutes and universities.

Specific Expertise

Semiconductor detectors based on GaAs compound material, operated at room temperature, or up to 50°C with a Peltier cooler.
Form of detectors: single PAD detectors with different sensitive area 0.5-10mm2; array of detectors for imaging of planar distribution of irradiation in a form of pixel, strip, or microstrips, with different area, or pitch, down to a limit of 5μm for individual dimension.

Collaboration Offered

Irradiation sensors based on GaAs in different form, as semi-products, detector chips, or final detectors mounted in a holder; fast modification of detector topology within a few, 1-3 months, including photolithography masks proposal and production; testing of basic electrical and detection parameters of produced sensors.

Collaboration Sought

Proposal and fabrication of multichannel readout chip suitable for our detector array; proposal and fabrication of a complete instrumentation for individual applications in different fields.

Spain

Entries 211 to 214

211 Ciemat

Edificio 22
Avda Compluterise, 22
E-28040 Madrid

Tel: +34 1 346 60 72
Fax: +34 1 346 62 75
E-mail: josema@gtavxz.ciemat.es

Affiliation: Ministry of Industry
Contact: J Pascual, Head of Electronic Dept
J Perez, Researcher

Organisation information

No of employees >1000
Turnover (ECU*) <1m

Sensor technology activities information

No of employees 11-25
Turnover (ECU*) <0.5m

*1ECU=DM 1.89,FFr 6.30,£0.78,Lira 1909,Peseta 159.1

Organisation activities

Research and development on sensors sensor materials or applications
Manufacture of sensor -based products

Application \ Signal Sensed	Radiation	Optical	Electrical	Magnetic	Thermal	Chemical	Mechanical
Automotive							
Consumer products							
Process industries,mining energy	•						
Manufacturing, machinery							
Medicine, health care							
HVAC, fire control, security							
Military, aerospace	•						
Telecommunications,office equip.							
Meteorology environment							
Other							

Main Activities

Ciemat is a research and development laboratory for the study of technological aspects related with the production of energy. In particular, nuclear energy. The sensors we are interested in are radiation detectors.

Specific Expertise

Radiation room temperature solid state detectors.

Collaboration Offered

Experience in testing and construction of sensor based devices, mechanical and electronic facilities for the production of prototypes. Radioactive cells for testing and calibrating, all related with research programmes.

Collaboration Sought

Interest in participating in common projects related to the applications of new material or classical room temperature semiconductors.

212 Instituto de Bioingenieria - Fundacion CGM

Aveuida de la Marina 3,5
E-11007 Cadiz

Tel: +34 56 27 01 40
Fax: +34 56 28 99 80
E-mail:

Affiliation:
Contact: J Bustelo, Director

Organisation information
No of employees 1-25
Turnover (ECU*) <1m
Sensor technology activities information
No of employees 1-10
Turnover (ECU*) <0.5m
*1ECU=DM 1.89,FFr 6.30,£0.78,Lira 1909,Peseta 159.1

Organisation activities
Research and development on sensors, sensor materials or applications.
Manufacture of systems used in the manufacture, testing or development of sensors.

Application \ Signal Sensed	Radiation	Optical	Electrical	Magnetic	Thermal	Chemical	Mechanical
Automotive		•	•				•
Consumer products							
Process industries,mining energy							
Manufacturing, machinery							
Medicine, health care	•	•	•	•	•		•
HVAC, fire control, security							
Military, aerospace							
Telecommunications,office equip.							
Meteorology, environment							
Other *							

Main Activities

This Foundation is dedicated to bioengineering - iontophoresis, muscular sensors, optical sensors, electrical sensors, basic investigation, prototypes.

Specific Expertise

The Foundation's main dedication is in the field of the non-invasive biosensors, as a help to diagnosis and control of certain pathologies. Optical sensors using visible and infrared radiation and electrical sensors based in impedance measurements are being investigated. Present work is on blood glucose sensors. Future projects will be on cholesterol, bilirubbin and others

213 Universidad Complutense de Madrid

Dpto Optica-Facultad
CC-Fisicas-Ciudad Universitaria
E-28040 Madrid

Tel: +34 1 394 4555
Fax: +34 1 394 4674
E-mail: fiopt04@emducms1.sis.ucm.es

Affiliation:
Contact: E Bernabeu, Director

Organisation information
No of employees 26-100
Turnover (ECU*) 1m-5m
Sensor technology activities information
No of employees 1-10
Turnover (ECU*) <0.5m
*1ECU=DM 1.89,FFr 6.30,£0.78,Lira 1909,Peseta 159.1

Organisation activities
Research and development.
Testing, consulting,design.

Application \ Signal Sensed	Radiation	Optical	Electrical	Magnetic	Thermal	Chemical	Mechanical
Automotive		•					
Consumer products							
Process industries,mining energy		•					
Manufacturing, machinery		•					
Medicine, health care							
HVAC, fire control, security							
Military, aerospace		•					
Telecommunications,office equip.							
Meteorology, environment							
Other *							

Main Activities

The Optics Department of the Universidad Complutense maintains different active R&D lines in several fields of Optics. These lines may be selected in the following main areas: laser; optical technology - applied optics; optical technology - optical sensors and systems of inspection and control; analysis of light beams; lighting; optical spectroscopy of materials; vision.

Specific Expertise

Interferometric and diffractional control of materials; optical characterisation of materials; control of opthalmic optics by Moire deflectometry; optical encoders and tachometers; magneto-optical devices.

Collaboration Offered

Optical characterisation of materials; design and development of optical sensors for industrial applications (encoders, systems of inspection and control); calculation and design of optical systems.

Collaboration Sought

Partners for R&D international projects; sponsors for materials.

214 University of Cordoba

Faculty of Sciences
Analytical Chemistry Department
Arda San Alberto Magno, s/n
E-14004 Córdoba

Tel: +34 57 21 86 15
Fax: +34 57 21 86 06
E-mail:

Affiliation:
Contact: M D Lugue de Castro, Full Prof

Organisation information
No of employees 101-1000
Turnover (ECU*)
Sensor technology activities information
No of employees
Turnover (ECU*)
*1ECU=DM 1.89,FFr 6.30,£0.78,Lira 1909,Peseta 159.1

Organisation activities
Research and development on sensors, sensor materials or applications.

Application \ Signal Sensed	Radiation	Optical	Electrical	Magnetic	Thermal	Chemical	Mechanical
Automotive							
Consumer products							
Process industries,mining energy		•	•				
Manufacturing, machinery							
Medicine, health care							
HVAC, fire control, security							
Military, aerospace							
Telecommunications,office equip.							
Meteorology, environment		•	•				
Other *							

Main Activities
Design and development of flow-through chemical and biochemical sensors, development of analytical methods based on these sensors; development of robotic analytical methods; automation of solid sample treatment; automation of continuous chromatographic and non-chromatographic separation techniques.

Specific Expertise
Flow-through (bio)chemical sensors based on immobilisation of: chemicals; enzymes/co-factors; antibodies-antigens.

Collaboration Offered
Personnel specialised in the design and application of flow-through (bio)chemical sensors.

Collaboration Sought
Prototype development and commercialisation of flow-through (bio)chemical sensors.

Sweden

Entries 215 to 217

215 Institute of Optical Research

S-10044 Stockholm

Tel: +46 8 791 1300
Fax: +46 8 789 6672
E-mail: iof@optics.kth.se

Affiliation:

Contact: M Breidne, MD
U Persson, R&D Mgr
R Stubbe, Fibre Sensors
K Kleveby, Special Optical Fibres

Organisation information

No of employees 26-100
Turnover (ECU*) 1m-5m

Sensor technology activities information

No of employees 11-25
Turnover (ECU*) 1m-5m

*1ECU=DM 1.89,FFr 6.30,£0.78,Lira 1909,Peseta 159.1

Organisation activities

Research and development on sensors, sensor materials or applications.
Testing, consulting, design.
Supply of raw materials or semi-finished products used to make sensors.

Application \ Signal Sensed	Radiation	Optical	Electrical	Magnetic	Thermal	Chemical	Mechanical
Automotive							
Consumer products							
Process industries,mining energy		●					
Manufacturing, machinery							
Medicine, health care							
HVAC, fire control, security							
Military, aerospace		●					
Telecommunications,office equip.		●					
Meteorology, environment							
Other *							

Main Activities

The Institute sees its mission in filling industry's needs for R&D, gatekeeping and technology transfer, training and other services in modern optics. An important ingredient is collaboration - with universities, institutes, industry, in Sweden and internationally.

Specific Expertise

Special optical fibres (including special coatings), bragg gratings, distributed optical sensors for stain and temperature, TDM architecture.

216 National Defence Research Establishment

Department of Sensor Technology
P O Box 1165
S-58111 Linköping
Sweden

Tel: +46 13 31 80 00
Fax: +46 13 31 82 87
E-mail: 50073

Affiliation:
Contact: C Larsson, Head, Optical Tech
Prof S Ödman, Optoelectronics
Dr O Steinvall', Laser Technology
L Bohman, Optical Signatures
U Ringh, Electro-optical Reconnaissance

Organisation information
No of employees >5
Turnover (ECU*) >100

Sensor technology activities information
No of employees 1-5
Turnover (ECU*) 26-100
*1ECU=DM 1.89,FFr 6.30,£0.78,Lira 1909,Peseta 159.1

Organisation activities
Research and development on sensors, sensor materials or applications
Testing, consulting, design

Application \ Signal Sensed	Radiation	Optical	Electrical	Magnetic	Thermal	Chemical	Mechanical
Automotive							•
Consumer products							
Process industries,mining energy							
Manufacturing, machinery							
Medicine, health care						•	
HVAC, fire control, security							•
Military, aerospace	•			•		•	•
Telecommunications,office equip.							•
Meteorology environment	•					•	
Other (specify)							

Main Activities

Main research areas: Optoelectronics including biosensors; laser technology; optical signatures; electro-optical reconnaissance.

Specific Expertise

Ground based surveillance; biosensors for protection against explosives, chemical and biological weapons. Laser radar technology; coherent laser radars; laser detectors; laser scanners; signal processing of laser signal; laser technology (laser sources); laser eye protection; sensor protection from lasers. Infrared focal plane array technology; analysis and performance predictions; modelling, field evaluation of new sensor technology. Multi spectral reconnaissance using CCD cameras and IR; smart optical sensors for surveillance.

Collaboration Offered

New biosensor technology with supersensitivity. Participation in common development projects/field trials. Experience from multi-spectral video system using six different wavelengths between 450mm and 1100mm.

Collaboration Sought

Biotechnology tissue culturing, cleaning of receptors. Specific knowledge of areas in laser/laser radar technology. Experiences from multi-spectral video for reconnaissance.

217 SAAB Instruments AB

Box 1017
S-55111 Jonkoping

Tel: +46 36 194100
Fax: +46 36 194500
E-mail:

Affiliation:
Contact: C Dahlber, General Manager

Organisation information
No of employees
Turnover (ECU*)
Sensors technology activities information
No of employees
Turnover (ECU*)
*1ECU=DM 1.89,FFr 6.30,£0.78,Lira 1909,Peseta 159.1

Organisation activities
Research and Development
Manufacturer of sensor-based products
Testing, consulting, design
Manufacture of finished sensors

Application \ Signal Sensed	Radiation	Optical	Electrical	Magnetic	Thermal	Chemical	Mechanical
Automotive							
Consumer products							
Process industries, mining, energy							
Manufacturing, machinery							
Medicine, health care							
HVAC, fire control, security							
Military, aerospace							•
Telecommunications,office equip.							
Meteorology, environment							
Other *							

Main Activities

SAAB Instruments AB is the defence equipment branch of the SAAB Scania Combitech Group. Main product areas are: Aircraft and missile equipment (electrical, electronic, mechanical, hydraulic); aircraft and helicopter sights; submarine steering control systems; precision products (gyros, torque motors).

Specific Expertise

Rate Gyro Saab GR-65:- Electrically driven spring restrained rate gyros for aircraft, missile, torpedo and vehicle applications. Salient features are: High scale-factor accuracy over a wide temperature range; high temperature stability of null; high shock resistance (> 500g) for armoured fighting vehicle use; long operating life; low life cycle cost.
Rate Gyro Packages:- Single or multi-gyro packages with drive electronics (DC in/DC out).

Collaboration Offered

Standard line of rate gyros. Custom design of rate gyro. Versions and rate gyro packages (with or without electronics).

Switzerland

Entries 218 to 240

218 Battelle Geneva Research Centre

Carouge 1227
CH-1228 Geneva

Tel: +41 22 8272727
Fax: +41 22 8272310
E-mail:

Affiliation:
Contact: G Kotrotios, "M 3D"

Organisation information
No of employees
Turnover (ECU*)
Sensors technology activities information
No of employees
Turnover (ECU*)
*1ECU=DM 1.89,FFr 6.30,£0.78,Lira 1909,Peseta 159.1

Organisation activities
Research and Development
Testing, consulting, design

Application \ Signal Sensed	Radiation	Optical	Electrical	Magnetic	Thermal	Chemical	Mechanical
Automotive					•	•	•
Consumer products							
Process industries, mining, energy			•	•	•	•	•
Manufacturing, machinery					•		•
Medicine, health care						•	
HVAC, fire control, security	•				•	•	•
Military, aerospace							
Telecommunications,office equip.							
Meteorology, environment							
Other *							

Main Activities

Battelle is an independent research and development organisation, working under contract to industrial and government clients. The Department of Electronics and Optics at Battelle Geneva works in a range of disciplines for sensors research and development: optics; fibre optics; acoustics and electrical.

Specific Expertise

Optical, optoelectronic and fibre optic sensors; applied to industrial, automotive and biomedical areas. Specific sensors: Fibre optic distributed pressure sensing cable for intrusion sensing, pressure sensing, touch sensing; optoelectronic refractometer (flor lead acid battery density measurements, sugar solution concentration etc.); optical coagulameter; optical biomedical and immunoassay sensor; fibre optic strain gauge.

Collaboration Offered

Consultancy. Market and new business opportunities surveys. Research for new sensors (optical fibre sensors, optical sensors). Development of prototype sensors.

Collaboration Sought

Optoelectronic and fibre optic sensors.

219 Baumer Electric AG

Hummelstrasse 17
CH-8500 Frauenfeld

Tel: +41 52 7281122
Fax: +41 52 7281144
E-mail:

Affiliation:
Contact: B Furrer
S Schneider

Organisation information
No of employees
Turnover (ECU*)
Sensors technology activities information
No of employees
Turnover (ECU*)
*1ECU=DM 1.89,FFr 6.30,£0.78,Lira 1909,Peseta 159.1

Organisation activities
Manufacturer of sensor-based products
Manufacture of finished sensors

Application \ Signal Sensed	Radiation	Optical	Electrical	Magnetic	Thermal	Chemical	Mechanical
Automotive							•
Consumer products							•
Process industries, mining, energy							•
Manufacturing, machinery							•
Medicine, health care							
HVAC, fire control, security							•
Military, aerospace							•
Telecommunications,office equip.							
Meteorology, environment							
Other *							•

Main Activities

The policy of Baumer Electric is based on an intensive research and innovative development programme. The engineers and technicians continually update existing products and design new products by working closely with their own research and development department or using technology previously unexplored in the sensing market.

Specific Expertise

Sensors to detect position and motion, such as: Inductive linear sensors (IWA) which have an output voltage directly proportional to the object distance from the sensing face; Inductive two point sensors (IZRM) allowing independent setting of two switching points along the defined linear output signal of the IWA in question; Standard specialist sensors: Photo-electric proximity switches with background suppression for detection of objects against reflective backgrounds. Retroreflective sensors with security; Special intelligent sensors: Field plate sensors, security sensors for machine and tool protection.

Collaboration Offered

Proximity and motion sensing products. Solve customer application problems (Research and development facility). Photoelectronic sensors. Encoders. Inductive proximity switches.

220 Contelec AG

Portstrasse 38
Postfach 869
CH-2501 Biel-Bienne

Tel: +41 32 253225
Fax: +41 32 265601
E-mail:

Affiliation: Horst Siedle KG, Germany
Contact: Dr H J Richter, President
H P Zuili, Technical Engineer

Organisation information
No of employees >100
Turnover (ECU*) >5
Sensors technology activities information
No of employees >100
Turnover (ECU*) >5
*1ECU=DM 1.89,FFr 6.30,£0.78,Lira 1909,Peseta 159.1

Organisation activities
Research and Development
Manufacturer of sensor-based products
Testing, consulting, design
Manufacture of finished sensors

Application \ Signal Sensed	Radiation	Optical	Electrical	Magnetic	Thermal	Chemical	Mechanical
Automotive							•
Consumer products							
Process industries, mining, energy							•
Manufacturing, machinery							•
Medicine, health care							•
HVAC, fire control, security							•
Military, aerospace							•
Telecommunications,office equip.							•
Meteorology, environment							•
Other *							

Main Activities

Leading company in analogue, absolute sensors, potentiometric and optic principles.

Specific Expertise

Manufacturing Swiss quality-units to be used in measuring and control systems which demand the highest integrity.

221 CSEM - Centre Suisse d'Electronique

et de Microtechnique
Jaquet-droz 1
CH-2007 Neuchâtel

Tel: +41 32 7205 111
Fax: +41 32 7205 700
E-mail:

Affiliation:
Contact: F Rudolf,
Sensors & Actuators, Mkting

Organisation information
No of employees >100
Turnover (ECU*) >5m

Sensors technology activities information
No of employees 26-100
Turnover (ECU*) 1-5m
*1ECU=DM 1.89,FFr 6.30,£0.78,Lira 1909,Peseta 159.1

Organisation activities
Research and Development
Testing, consulting, design
Manufacture of finished sensors

Application \ Signal Sensed	Radiation	Optical	Electrical	Magnetic	Thermal	Chemical	Mechanical
Automotive							•
Consumer products							
Process industries, mining, energy			•	•	•		•
Manufacturing, machinery			•	•	•		•
Medicine, health care							•
HVAC, fire control, security					•		
Military, aerospace							•
Telecommunications,office equip.				•			
Meteorology, environment					•		
Other *							

Main Activities

CSEM carries out applied research and contract engineering in the fields of IC design, technologies for microelectronics and microstructures, optics, materials and special instrumentation. The "Sensors and Actuators" Division develops silicon sensors and actuators and microsystems containing such devices. The "Optics" Division carries out work in the field of optical sensors.

Specific Expertise

Micromachined silicon sensors and actuators:
Areas of work are the design and development of silicon sensors and actuators and of microsystems combining sensors and electronic circuitry. The technologies used include bulk silicon micromachining and the deposition of metal layers on silicon by electroplating. Typical applications and realisations are capacitive sensors for acceleration, pressure, force, microphones, electromagnetic actuators, valves, pumps, displays, microshutters.
Optical sensors:
Main fields of work are optical sensors based on optical fibres for point measurements and for linearly distributed measurements. Applications include sensors for electrical current, pressure and temperature.

Collaboration Offere

Contract engineering: feasibility studies; product developments; technology developments. Manufacturing of prototypes and small series.

Collaboration Sought

Basic research from universities. Know-how exchange through projects of the European Community.

222 Endress + Hauser (Int) Holding AG

Kägenstr. 7
CH-4153 Reinach

Tel: +41 61 7156590
Fax: +41 61 7110682
E-mail:

Affiliation:
Contact: D Schaudel,
Dir Technology &Technique

Organisation information
No of employees >100
Turnover (ECU*) >5m

Sensors technology activities information
No of employees >100
Turnover (ECU*) >5m
*1ECU=DM 1.89,FFr 6.30,£0.78,Lira 1909,Peseta 159.1

Organisation activities
Research and Development
Manufacturer of sensor-based products
Manufacture of finished sensors

Application \ Signal Sensed	Radiation	Optical	Electrical	Magnetic	Thermal	Chemical	Mechanical
Automotive							
Consumer products	•		•	•	•	•	•
Process industries, mining, energy	•		•	•	•	•	•
Manufacturing, machinery							
Medicine, health care					•	•	
HVAC, fire control, security							
Military, aerospace							
Telecommunications,office equip.	•		•	•			•
Meteorology, environment							
Other *							

Main Activities

Switzerland based subsidiary of the Endress + Hauser Group of companies. Main Activities: flow, pressure, bulk flow, temperature, moisture and conductivity measurement are the most important disciplines, followed by systems technology and involvement in control, recording and display systems, development and manufacturing of the company's own field sensors.

Specific Expertise

Application areas: chemical, food, cement, paper, fresh and waste water, power.
Sensor type involved: capacitive, hydrostatic, electromechanical, ultrasonic, radiometer, Karman's vortex shedding, Coriolis, impact plate, correlation, weighing, electromagnetic.

223 EPFL-Ecol Polytech Fédérale de Lausanne

Département d'Electricité
Laboratory of Metrology
CH-1015 Lausanne

Tel: +41 21 6932 649
Fax: +41 21 6932 614
E-mail:

Affiliation:
Contact: Prof P H Robert, Head of Laboratory

Organisation information
No of employees >100
Turnover (ECU*)
Sensors technology activities information
No of employees 11-25
Turnover (ECU*)
*1ECU=DM 1.89,FFr 6.30,£0.78,Lira 1909,Peseta 159.1

Organisation activities
Research and Development
Testing, consulting, design

Application \ Signal Sensed	Radiation	Optical	Electrical	Magnetic	Thermal	Chemical	Mechanical
Automotive							
Consumer products							
Process industries, mining, energy							
Manufacturing, machinery							
Medicine, health care							
HVAC, fire control, security	•		•		•	•	•
Military, aerospace							
Telecommunications,office equip.							
Meteorology, environment							•
Other *							

Main Activities

The Swiss Federal Institute of Technology of Lausanne (EPFL) has more than 3,500 students and about 2,400 staff. As a public service institution, the principle aims of the Institution are training of engineers and architects, basic and applied research, consultancy in engineering. The Laboratory of Metrology is in the Electrical Engineering Department of the EPFL.

Specific Expertise

Temperature sensors based on thin film Ni 1000 ohms (according to DIN 43760) and Pt 100 ohms (according to DIN/IEC 751) on glass substrate encapsulated in epoxy resin. Range of room, duct, piping and related temperature detectors incorporating these sensors; Nylon-band relative humidity sensors utilised in a range of electromechanical humidistats; Non-dispersive infrared spectroscopic sensing device for detection and measurement of gas concentrations (principallly CO_2) as an indicator of indoor air quality.

Collaboration Offered

Full range of products for the detection of temperature, humidity, pressure, water level and air quality. Principally for applications in the heating, ventilation and air conditioning fields of activity.

Collaboration Sought

Original equipment manufacturers (OEMs) for the products indicated as above in the following countries: Australia, Austria, Belgium, France, Great Britain, Hong Kong, Spain, Sweden, Singapore; Switzerland, USA, Germany and Ireland.

224 ETH Zürich (Centre for Chemical Sensors/Biosensors & Bioanalytical Chemistry)

Technoparkstr 1
CH 8005 Zürich

Tel: +41 1 445 12 31
Fax: +41 1 445 12 33
E-mail:

Affiliation: Swiss Federal Institute of Technology
Contact: U Spichiger, Head of Centre

Organisation information
No of employees 1-25
Turnover (ECU*)
Sensor technology activities information
No of employees 1-10
Turnover (ECU*) <01.5m
*1ECU=DM 1.89,FFr 6.30,£0.78,Lira 1909,Peseta 159.1

Organisation activities
Research and development.
Testing, consulting, design.
Supply of raw materials or semi-finished products used to make sensors.

Application \ Signal Sensed	Radiation	Optical	Electrical	Magnetic	Thermal	Chemical	Mechanical
Automotive							
Consumer products		•	•			•	
Process industries,mining energy							
Manufacturing, machinery		•	•			•	
Medicine, health care							
HVAC, fire control, security							
Military, aerospace							
Telecommunications,office equip.		•	•			•	
Meteorology, environment		•	•			•	
Other *							

Main Activities

The Centre develops optical,electrical and chemical sensors for different applications.
See Homepage : http://www.chemseus.pharma.ethz.ch

Specific Expertise

Host compounds, dyes, chemically active and selective layers. Specific sensor design and solutions for dedicated systems and analytical problems in biotechnology medical analyses, environmental analysis (air, water, etc.).

Collaboration Offered

Sensors for continuous monitoring in water, air, medical care. Sensor components (membranes). Development of a Fatal Analysis Sensor System.

Collaboration Sought

Support in electronics and physical hardware. True applications.

225 FR Sauter SA

Im Surinam 55
CH-4016 Basel

Tel: +41 61 6955555
Fax: +41 61 6955510
E-mail:

Affiliation:
Contact: C Müller, Sensor Product Manager

Organisation information
No of employees >100
Turnover (ECU*) >5m
Sensors technology activities information
No of employees 11-25
Turnover (ECU*) 1-5m
*1ECU=DM 1.89,FFr 6.30,£0.78,Lira 1909,Peseta 159.1

Organisation activities
Research and Development
Manufacturer of sensor-based products
Manufacture of finished sensors

Application \ Signal Sensed	Radiation	Optical	Electrical	Magnetic	Thermal	Chemical	Mechanical
Automotive			•				•
Consumer products							
Process industries, mining, energy			•	•	•		•
Manufacturing, machinery			•	•	•		•
Medicine, health care							•
HVAC, fire control, security						•	
Military, aerospace							
Telecommunications,office equip.							
Meteorology, environment							
Other *							

Main Activities

Fr Sauter SA is a family enterprise with three factories in Germany, France and Switzerland. Products are manufactured for building and water technology. The Controls Division is responsible for development and marketing of: amongst other devices, detectors for temperature, humidity, pressure, water levels, and air quality. The basis for these detectors, both solid state and electro-mechanical sensors, are developed in-house.

Specific Expertise

General: Fibre optic, optoelectronic and optical sensors. Application: Industrial, automotive area. Ergometric sensor for use in the biomedical area.
Specific sensors: Fibre optic current sensors using polarimetric and interferometric detection schemes, dedicated to high current under high voltage measurements. Fibre optic quasi-distrubuted sensors using FMAMCW technique for pressure/force and chemical sensing. Optical sensors for force measurement in high voltage environment. Ergometric sensor for measuring the energy expenditure due to the muscular activity of small mammals.

Collaboration Offered

Consultancy. Research and prototypes of new sensors (fibre optic optoelectronic and optical sensors).

226 Haenni & Co Limited

Bernstrasse
CH-3303 Jegenstorf

Tel: +41 31 764 99 11
Fax: +41 31 764 99 22
E-mail:

Affiliation: Zelinder Holding
Contact: Ernst Diggelmann, Manager
Peter Standacher, Mkting Director
F Diggelmann, Vice President - Electronics

Organisation information
No of employees 101-1000
Turnover (ECU*) 5m-50m
Sensor technology activities information
No of employees 26-100
Turnover (ECU*) >5m
*1ECU=DM 1.89,FFr 6.30,£0.78,Lira 1909,Peseta 159.1

Organisation activities
Manufacture of finished sensors
Manufacture of sensor-based products

Application \ Signal Sensed	Radiation	Optical	Electrical	Magnetic	Thermal	Chemical	Mechanical
Automotive							
Consumer products							
Process industries,mining energy					•		•
Manufacturing, machinery					•		•
Medicine, health care					•		•
HVAC, fire control, security					•		•
Military, aerospace							
Telecommunications,office equip.							
Meteorology environment					•		•
Other (specify)							•

Main Activities

Haenni develops, produces and sells products to measure pressure, temperature, level, humidity and weight. The product lines are organised in profit centres. Haenni has companies in Switzerland, Germany, France, and the USA with representatives all over the world.

Specific Expertise

Industrial packaging (stainless steel, Hastelloy, etc) of different raw sensors. Different technologies.

Collaboration Offered

Semi-finished and finished industrial products; worldwide market access.

Collaboration Sought

Technologies for raw sensors; raw sensors.

227 Kontron Instruments AG

Medical Sensors
Austrasse 25
CH-4051 Basel

Tel: +41 61 2788 111
Fax: +41 61 2788 181
E-mail:

Affiliation: Kontron Instruments Ltd (UK)
Contact: Dr P Eberhard

Organisation information

No of employees 1-25
Turnover (ECU*) 1m-5m

Sensor technology activities information

No of employees 11-25
Turnover (ECU*) 1m-5m

*1ECU=DM 1.89,FFr 6.30,£0.78,Lira 1909,Peseta 159.1

Organisation activities

Research and development on sensors and sensor materials or applications
Manufacture of finished sensors

Application \ Signal Sensed	Radiation	Optical	Electrical	Magnetic	Thermal	Chemical	Mechanical
Automotive							
Consumer products							
Process industries,mining energy							
Manufacturing, machinery		•	•			•	
Medicine, health care							
HVAC, fire control, security							
Military, aerospace							
Telecommunications,office equip.							
Meteorology environment							
Other (specify)							

Main Activities

Medical Sensors Department: development of method of measurement, sensor and instrument for the monitoring of blood and respiratory gases. Production of sensors.

Specific Expertise

Blood gas and respiratory gas sensors: measurement of PO_2 and F_1O_2, PCO_2, and pulse rate; measurement in the gas phase (F_1O_2), on the skin surface (transcutaneous PO_2 and PCO_2,). PO_2 and F_1O_2 sensor: Clark type electrochemical sensor. PCO_2 sensor; stow saveringhaus electrochemical sensor. light absorption sensors.

Collaboration Offered

Blood gas and respiratory gas sensor and front end electronic.

Collaboration Sought

Optical sensor expertise or product for blood gas and respiratory gas measurement.

228 KUBE Electronics Ltd

Glarnischstr. 59
CH-8712 Staefa

Tel: +41 19 26 43 62
Fax: +41 19 26 61 31
E-mail:

Affiliation:
Contact: HJ Keller, General Manager

Organisation information
No of employees 26-100
Turnover (ECU*) 5m-50m

Sensors technology activities information
No of employees 26-100
Turnover (ECU*) >5m
*1ECU=DM 1.89,FFr 6.30,£0.78,Lira 1909,Peseta 159.1

Organisation activities
Research and Development
Testing, consulting, design
Manufacture of finished sensors

Application \ Signal Sensed	Radiation	Optical	Electrical	Magnetic	Thermal	Chemical	Mechanical
Automotive							
Consumer products		•			•		
Process industries, mining, energy					•		
Manufacturing, machinery					•		
Medicine, health care					•	•	
HVAC, fire control, security		•			•		
Military, aerospace					•		
Telecommunications,office equip.							
Meteorology, environment					•		
Other *							

Main Activities

KUBE Electronics Ltd is involved in the development and manufacturing of:
- thermal infrared sensors
- pyroelectric detectors
- motion detectors
- presence detectors
- people sensors
- passive infrared sensors

Specific Expertise

The main area of expertise is:
- Passive infrared technology

Collaboration Offered

Components for presence detectors used in building automation/HVAC

229 Microsens SA

Rue Jacquet-Droz 1
CH-2007 Neuchatel

Tel: +41 38 7205 151
Fax: +41 38 7205 712
E-mail:

Affiliation:
Contact: Dr A Grisel, Director

Organisation information
No of employees 1-10
Turnover (ECU*) <0.5m
Sensors technology activities information
No of employees 1-10
Turnover (ECU*) <0.5m
*1ECU=DM 1.89,FFr 6.30,£0.78,Lira 1909,Peseta 159.1

Organisation activities
Research and Development
Manufacturer of sensor-based products
Testing, consulting, design
Manufacture of finished sensors

Application \ Signal Sensed	Radiation	Optical	Electrical	Magnetic	Thermal	Chemical	Mechanical
Automotive						•	•
Consumer products						•	
Process industries, mining, energy					•	•	
Manufacturing, machinery							•
Medicine, health care					•	•	•
HVAC, fire control, security					•	•	•
Military, aerospace						•	•
Telecommunications,office equip.							
Meteorology, environment					•	•	•
Other *							

Main Activities

Integrated sensors design and fabrication; development and application studies of integrated chemical and physical sensors; supply of integrated sensors and systems.

Specific Expertise

Types of sensors:- Integrated chemical sensors: ISFET (ion sensitive field effect transistors); pH-ISFET in CMOS technology, K^+, Ca^{++}, Mg^{++} ISFET type sensors. Gas sensors; amperometric integrated sensors (O_2, etc.), semiconductor gas sensor (H_2, CO, hydrocarbons, ethyl alcohol, etc.). Integrated Physical Sensors: Temperature sensors; capacitive pressure sensors; capacitive accelerometers.

Collaboration Offered

pH-ISFET, pH-meter with ISFET, pH-marker (pH-pen). CO, H_2, methane, ethyl alcohol, Integrated gas sensors, pressure sensors. Accelerometer. Applications studies. Measurement systems with integrated sensors.

Collaboration Sought

OEM for the development of sensor systems. Integrated sensor application studies. ISFET pH-meter for distribution. Gas sensors for distribution. Application studies, integrated sensor supply.

230 Reznicek & Hlach SA

CH-1728 Rossens

Tel: +41 26 4111 414
Fax: +41 26 4111 971
E-mail:

Affiliation:
Contact: Ing Reznicek, Chairman

Organisation information
No of employees 11-25
Turnover (ECU*) 0.5-1m
Sensors technology activities information
No of employees 11-25
Turnover (ECU*) 0.5-1m
*1ECU=DM 1.89,FFr 6.30,£0.78,Lira 1909,Peseta 159.1

Organisation activities
Research and Development
Manufacturer of sensor-based products
Testing, consulting, design
Manufacture of finished sensors

Application \ Signal Sensed	Radiation	Optical	Electrical	Magnetic	Thermal	Chemical	Mechanical
Automotive							
Consumer products							
Process industries, mining, energy							•
Manufacturing, machinery							•
Medicine, health care							
HVAC, fire control, security							
Military, aerospace							
Telecommunications,office equip.							
Meteorology, environment							
Other *							

Main Activities

Manufacturer of load cells and load measuring and controlling systems. Development and manufacture of strain gauge based transducers and systems for force measurement, detection and control.

Collaboration Offered

Customised applications of strain gauges.

Collaboration Sought

Software for applications.

231 Rotronic AG

Grindelstrasse 6
CH-8303 Bassersdorf

Tel: +41 1 8381111
Fax: +41 1 8364424
E-mail:

Affiliation:
Contact: Mr Juen, Head of Sensors Dept

Organisation information
No of employees >100
Turnover (ECU*) >5m

Sensors technology activities information
No of employees 26-100
Turnover (ECU*) >5m
*1ECU=DM 1.89,FFr 6.30,£0.78,Lira 1909,Peseta 159.1

Organisation activities
Research and Development
Testing, consulting, design
Manufacture of finished sensors

Application \ Signal Sensed	Radiation	Optical	Electrical	Magnetic	Thermal	Chemical	Mechanical
Automotive							
Consumer products							
Process industries, mining, energy						•	
Manufacturing, machinery						•	
Medicine, health care						•	
HVAC, fire control, security						•	
Military, aerospace						•	
Telecommunications,office equip.							
Meteorology, environment						•	
Other *							

Main Activities

Manufacture and sales of: Computer products; computer accessories; measuring equipment; humidity and temperature sensors; 19" - technology.

Specific Expertise

Humidity and temperature measuring and control systems as hand held units, transmitters, laboratory equipment, data logger and recorder.

232 Schiltknecht Messtechnik AG

Industriegebiet
CH-8625 Gossau (ZH)

Tel: +41 1 9352121
Fax: +41 1 9352471
E-mail:

Affiliation:
Contact: Dr F Doorenbosch,
President of Technology

Organisation information
No of employees 11-25
Turnover (ECU*) 1-5m
Sensors technology activities information
No of employees 11-25
Turnover (ECU*) 1-5m
*1ECU=DM 1.89,FFr 6.30,£0.78,Lira 1909,Peseta 159.1

Organisation activities
Research and Development
Manufacturer of sensor-based products
Manufacture of finished sensors

Application \ Signal Sensed	Radiation	Optical	Electrical	Magnetic	Thermal	Chemical	Mechanical
Automotive							
Consumer products							
Process industries, mining, energy					●		●
Manufacturing, machinery							
Medicine, health care					●		●
HVAC, fire control, security							
Military, aerospace							
Telecommunications,office equip.					●		●
Meteorology, environment							
Other *							

Main Activities

Schiltknecht Messtechnik focuses its activities on precision measuring equipment in three main areas: Flow of gases and liquids; differential pressure; meteorology. The programme covers standard products, customer specific equipment and customer oriented projects. The Schiltknecht company is known for Swiss precision and quality.

Specific Expertise

Sensors for flow, based on different physical phenomena; sensors for meteorological quantities; sensor-based systems and networks.

233 Scientific-Technical Consulting

Rütistrasse 55
CH-8032 Zürich

Tel: +41 1 2612 671
Fax: +41 1 2612 771
E-mail:

Affiliation:
Contact: Dr O Oehler, Consultant

Organisation information
No of employees 1-10
Turnover (ECU*) <0.5m
Sensors technology activities information
No of employees 1-10
Turnover (ECU*) <0.5m
*1ECU=DM 1.89,FFr 6.30,£0.78,Lira 1909,Peseta 159.1

Organisation activities
Research and Development
Testing, consulting, design

Application \ Signal Sensed	Radiation	Optical	Electrical	Magnetic	Thermal	Chemical	Mechanical
Automotive							
Consumer products							
Process industries, mining, energy						•	
Manufacturing, machinery							
Medicine, health care						•	
HVAC, fire control, security	•				•	•	
Military, aerospace							
Telecommunications,office equip.							
Meteorology, environment						•	
Other *							

Main Activities

Selective gas detection based on photoacoustic and photothermal spectroscopy; detection of oxygen on the basis of its paramagnetism and by the detuning of an ultrasonic resonator; calorimetry and colourimetry of microprobes by the thermal detuning of an ultrasonic resonator; fibre optical gas detection by using infrared fibres.

Specific Expertise

One area is the development of gas sensing methods, which are based on the photoacoustic and photothermal effect. Highly selective gas detection is obtained from the specific infrared spectra of gases. The heating of the gas on intensity modulated infrared absorption is measured either by gas pressure fluctuations (photoacoustics) or by the detuning of an ultrasonic resonator (photothermal method).
The latter method allows one additionally to measure oxygen by its accumulation in an ultrasonic resonator via an inhomogenous magnetic field.
Calorimetric and colourmetric measurements of microsamples can be carried out by measuring the temperature of the gas surrounding the sample by using an ultrasonic resonator.

Collaboration Offered

Know-how on the field of optical spectroscopy and gas sensing, as well as several international patents on these subjects. Close contact with the Swiss Federal Institute of Technology, Zurich.

Collaboration Sought

Production capabilities for optical gas sensing devices or microcalorimetric instruments.

234 SNT Sensortechnik AG

Bahnhofstrasse 25
CH-8153 Rümlang

Tel: +41 1 8172922
Fax: +41 1 8171083
E-mail:

Affiliation:
Contact: A Heim, MD, Research and Development

Organisation information
No of employees
Turnover (ECU*)

Sensors technology activities information
No of employees 11-25
Turnover (ECU*) 1-5m
*1ECU=DM 1.89,FFr 6.30,£0.78,Lira 1909,Peseta 159.1

Organisation activities
Research and Development
Manufacturer of sensor-based products
Testing, consulting, design
Manufacture of finished sensors

Application \ Signal Sensed	Radiation	Optical	Electrical	Magnetic	Thermal	Chemical	Mechanical
Automotive							•
Consumer products							
Process industries, mining, energy				•			•
Manufacturing, machinery				•			•
Medicine, health care							
HVAC, fire control, security							
Military, aerospace							
Telecommunications,office equip.							
Meteorology, environment							•
Other *							

Main Activities

Research and development of ultrasonic distance sensors in air, also fibre optic sensors.

235 STS Sensor Technik Sirnach AG

Buefelderstr 1
CH-8370 Sirnach

Tel: +41 71 9694 929
Fax: +41 71 9694 920
E-mail:

Affiliation:
Contact: H P Mueller, President
A Schneider, Production
M Von Ritter, Electronics

Organisation information
No of employees
Turnover (ECU*)
Sensors technology activities information
No of employees
Turnover (ECU*)
*1ECU=DM 1.89,FFr 6.30,£0.78,Lira 1909,Peseta 159.1

Organisation activities
Research and Development
Manufacturer of sensor-based products
Testing, consulting, design
Manufacture of finished sensors

Application \ Signal Sensed	Radiation	Optical	Electrical	Magnetic	Thermal	Chemical	Mechanical
Automotive							•
Consumer products							•
Process industries, mining, energy							•
Manufacturing, machinery							•
Medicine, health care							•
HVAC, fire control, security							
Military, aerospace							•
Telecommunications,office equip.							•
Meteorology, environment							•
Other *							

Main Activities

STS Sensor Technik Sirnach develops, produces and sells piezoresistive pressure sensors (integrated semiconductor strain gauge principle): Development, production, sales.

Specific Expertise

The sensors are based on a semiconductor strain gauge, integrated in monocrystalline, micromachined silicon chip. The most special feature of this chip is a proprietary diffusion process. The result is an extremely low thermal sensitivity shift of typically less than 1% from -40 to 125 degrees Celsius for both current and voltage excitation, with active compensation. This simplifies the excitation and compensation, and makes this chip an ideal choice for automotive and consumer product application. On the other hand, this same chip offers superior performance for high precision applications. Based on this chip, STS produces a complete line of pressure sensor products, namely cells, stainless steel encapsulated OEM transducers, analogue transmitters, and at the top of the line, intelligent digital transmitters.

Collaboration Sought

STS is open towards collaboration with other companies, and interested in licensing its chip to automotive and consumer product companies.

236 Trafag AG

Gufenhaldenweg 4
CH-8708 Männedorf

Tel: +41 1 9223 232
Fax: +41 1 9223 233
E-mail:

Affiliation:
Contact: Mr Herrmann, Sales Manager

Organisation information
No of employees >100
Turnover (ECU*)
Sensors technology activities information
No of employees
Turnover (ECU*)
*1ECU=DM 1.89,FFr 6.30,£0.78,Lira 1909,Peseta 159.1

Organisation activities
Research and Development
Manufacturer of sensor-based products
Manufacture of finished sensors

Application \ Signal Sensed	Radiation	Optical	Electrical	Magnetic	Thermal	Chemical	Mechanical
Automotive							
Consumer products							
Process industries, mining, energy					•		•
Manufacturing, machinery					•		•
Medicine, health care							
HVAC, fire control, security					•		•
Military, aerospace					•		•
Telecommunications,office equip.							
Meteorology, environment					•		•
Other *							

Main Activities

Manufacture of electro-mechanical and electronic appliances.

Specific Expertise

Thin film strain gauge pressure sensor technology.

237 Universitätsspital Zürich

Eye Department
Frauenklinikstr. 24
CH-8091 Zürich

Tel: +41 1 2551111
Fax: +41 1 2554438
E-mail:

Affiliation:
Contact: Prof Gloor, Director

Organisation information
No of employees
Turnover (ECU*)
Sensors technology activities information
No of employees
Turnover (ECU*)
*1ECU=DM 1.89,FFr 6.30,£0.78,Lira 1909,Peseta 159.1

Organisation activities
Research and Development
Manufacturer of sensor-based products
Testing, consulting, design

Application \ Signal Sensed	Radiation	Optical	Electrical	Magnetic	Thermal	Chemical	Mechanical
Automotive							
Consumer products							
Process industries, mining, energy							
Manufacturing, machinery							
Medicine, health care							•
HVAC, fire control, security							
Military, aerospace							
Telecommunications,office equip.							
Meteorology, environment							
Other *							

Main Activities

The Eye Department develops diagnostic instruments for ophthalmology.

Specific Expertise

The Department's main work comprises, the integration of miniaturised pressure sensors in diagnostic ophthalmic instrumentation.

238 Vibro-Meter SA

Moncor 4
CH-1701 Fribourg

Tel: +41 26 4071 111
Fax: +41 26 4071 301
E-mail:

Affiliation:
Contact: Dr R W Greaves, MDAerospace
Mr Di-Marco, Director R & D

Organisation information
No of employees >100
Turnover (ECU*) >5m

Sensors technology activities information
No of employees >100
Turnover (ECU*) >5m
*1ECU=DM 1.89,FFr 6.30,£0.78,Lira 1909,Peseta 159.1

Organisation activities
Research and Development
Manufacturer of sensor-based products
Testing, consulting, design
Manufacture of finished sensors

Application \ Signal Sensed	Radiation	Optical	Electrical	Magnetic	Thermal	Chemical	Mechanical
Automotive							
Consumer products							
Process industries, mining, energy					•		•
Manufacturing, machinery							•
Medicine, health care							
HVAC, fire control, security							
Military, aerospace					•		•
Telecommunications,office equip.							
Meteorology, environment							
Other *							

Main Activities

Vibro-Meter SA is primarily concerned with the design and manufacture of transducers and associated electronic equipment for the measurement of vibration and, in a more general way, the dynamic and static measurement of mechanical parameters. The activities and market areas include: Aviation; energy production; refineries; pipe lines and industrial instrumentation.

Specific Expertise

Vibro-Meter SA develops, manufactures and sells: piezoelectric accelerometers; piezoelectric pressure transducers; piezoelectric ice sensors; eddy current proximity transducers; eddy current test rig dynamometers and magnetic powder dynos; inductive/strain gauge torque measuring shafts; inductivie/eddy current displacement transducers; inductive speed probes; thin film/piezoresistive pressure transmitters; inductive/foil strain gauge load switching cells; mechanical deformation load switching cells.

Collaboration Offered

Build-up of sales representation.

Collaboration Sought

Market and application know-how.

239 Wintion AG

Analytical Instruments
CH-3115 Gerzensee

Tel: +41 31 7811 646
Fax: +41 31 7811 648
E-mail:

Affiliation:
Contact: W Willywalter, President

Organisation information
No of employees 1-10
Turnover (ECU*) <0.5m

Sensors technology activities information
No of employees 1-10
Turnover (ECU*) <0.5m
*1ECU=DM 1.89,FFr 6.30,£0.78,Lira 1909,Peseta 159.1

Organisation activities
Research and Development
Manufacturer of sensor-based products
Testing, consulting, design
Manufacture of finished sensors

Application \ Signal Sensed	Radiation	Optical	Electrical	Magnetic	Thermal	Chemical	Mechanical
Automotive							
Consumer products							
Process industries, mining, energy							
Manufacturing, machinery							
Medicine, health care							
HVAC, fire control, security							
Military, aerospace							
Telecommunications,office equip.							
Meteorology, environment						●	
Other *						●	

* Dairy Industry (Food)

Main Activities
Research, development, production and marketing service.

Specific Expertise
pH, °C, pION (pCa, pCl, pk, pNa, pNH_4, pNO_3, pF a.o.) systems.

Collaboration Offered
Portable analytical instrument systems.

240 Wyler AG

Elevation Systems and Measuring Equipment
Im Holderli
CH-8405 Winterthur

Tel: +41 52 2336 666
Fax: +41 52 2332 053
E-mail:

Affiliation:
Contact: Mr Morlet, President

Organisation information
No of employees 26-100
Turnover (ECU*) 1-5m
Sensors technology activities information
No of employees 1-10
Turnover (ECU*) 0.5-1m
*1ECU=DM 1.89,FFr 6.30,£0.78,Lira 1909,Peseta 159.1

Organisation activities
Research and Development
Manufacturer of sensor-based products
Manufacture of finished sensors

Application \ Signal Sensed	Radiation	Optical	Electrical	Magnetic	Thermal	Chemical	Mechanical
Automotive							•
Consumer products							•
Process industries, mining, energy							
Manufacturing, machinery							•
Medicine, health care							
HVAC, fire control, security							
Military, aerospace							•
Telecommunications,office equip.							
Meteorology, environment							
Other *							

Main Activities

Wyler AG has a very long experience in producing and marketing of spirit levels and setting measuring and clamping tools. The Electronic Department has a very long experience in producing and marketing angle measurement equipment, geometry measuring systems and angle sensors. These angle sensors and measuring equipment are supplied to machine manufacturers, aircraft industries, geophysics industries and many other markets.

Specific Expertise

Wyler angle (and acceleration under 1g) sensors base the measuring values on the pendulum properties of a friction-free supported disc of mass weighing less than one gram. Because of the construction and the capacitive measuring principle the sensors are very resistant against mechanical shocks and magnetic fields.

Collaboration Offered

Angle measuring equipment. Angle measuring sensors. Geometry measuring sensors.

United Kingdom

Entries 241 to 425

241 ABB Kent-Taylor Limited

Oldends Lane
Stonehouse
Gloucestershire GL10 3TA

Tel: +44 145 3826661
Fax: +44 145 3821478
E-mail:

Affiliation:

Contact: C Rushton
T Booth
A Bottom
J Bavester
F Coston

Organisation information

No of employees >100
Turnover (ECU*) >5m

Sensors technology activities information

No of employees 11-25
Turnover (ECU*)
*1ECU=DM 1.89,FFr 6.30,£0.78,Lira 1909,Peseta 159.1

Organisation activities

Research and Development
Manufacturer of sensor-based products
Testing, consulting, design
Manufacture of finished sensors

Application \ Signal Sensed	Radiation	Optical	Electrical	Magnetic	Thermal	Chemical	Mechanical
Automotive					•	•	
Consumer products							
Process industries, mining, energy			•		•	•	•
Manufacturing, machinery			•			•	
Medicine, health care					•	•	
HVAC, fire control, security							•
Military, aerospace					•		
Telecommunications,office equip.					•		
Meteorology, environment			•		•		•
Other *							

Main Activities

ABB Kent-Taylor Limited are known throughout the world as a manufacturing supplier of control, monitoring and analysis equipment for process industries. The company aims to be innovative in development and design, meticulous in quality control and competitive on cost and delivery. The company has over thirty years experience in manufacturing potentiometric sensors for pH, redox and concentration, to satisfy the diverse requirements of laboratory and plant use. The company's range is extremely comprehensive, including micro combination electrodes for measurements in small samples, toughened pH sensors for process use and ion-selective electrodes for use in beakers and on-line monitors.

Specific Expertise

Potentiometric sensor: pH ion selective and reference electrodes (for process and laboratory use). Associated systems and instrumentation including emperature compensation for relevant instrumentation. Dissolved gases e.g. NH_3 and SO_4. Amperometric sensors for dissolved oxygen and hydrazine

Collaboration Offered

Gas: thermal conductivity analysers, infrared analysers - non dispersing infrared solid state pyroelectric detection.

242 Ablewood Technology

Medina Croft
Winchester Road
Waltham Chase
Southampton SO32 2LG

Tel: +44 1489 891500
Fax: +44 1489 874513
E-mail:

Affiliation:
Contact: Dr S T Hughes, Principal Consultant
Miss W Lead, Personal Assistant

Organisation information
No of employees 1-25
Turnover (ECU*) <1m
Sensor technology activities information
No of employees 1-10
Turnover (ECU*) <0.5m
*1ECU=DM 1.89,FFr 6.30,£0.78,Lira 1909,Peseta 159.1

Organisation activities
Research and development on sensors sensor materials or applications
Testing, consulting, design

Application \ Signal Sensed	Radiation	Optical	Electrical	Magnetic	Thermal	Chemical	Mechanical
Automotive			•	•	•		•
Consumer products		•	•	•	•		
Process industries,mining energy		•	•	•	•		•
Manufacturing, machinery							
Medicine, health care							
HVAC, fire control, security		•					
Military, aerospace			•	•			•
Telecommunications,office equip.							
Meteorology environment							
Other (specify)							

Main Activities

Ablewood Technology offers applied research and consultancy expertise in a wide range of areas related to transducers (both sensors and actuators). Founded by a past director of the University of Southampton Institute of Transducer Technology (USITT), the Company continues to maintain strong links with the university and collaborates in many areas of mutual interest.

Specific Expertise

Ablewood carries out problem investigations, feasibility studies, product design and prototype manufacture in areas such as: pressure, level and flow instrumentation; electromagnetic and capacitive sensing devices; acoustics and ultrasonics; mixed signal electronics; precision analogue signal conditioning. In some cases, manufacturing can also be arranged through associate companies.

Collaboration Offered

Technical expertise in the above areas.

Collaboration Sought

Manufacturing and marketing.

243 Acam Instrumentation Limited

23 Thomas Street
Northampton NN1 3EN

Tel: +44 1604 28700
Fax: +44 1604 28700
E-mail:

Affiliation:
Contact: T C Cottam, Director

Organisation information
No of employees 1-10
Turnover (ECU*) <0.5m

Sensors technology activities information
No of employees 1-10
Turnover (ECU*) <0.5m
*1ECU=DM 1.89,FFr 6.30,£0.78,Lira 1909,Peseta 159.1

Organisation activities
Research and Development
Manufacturer of sensor-based products
Testing, consulting, design
Manufacture of finished sensors

Application \ Signal Sensed	Radiation	Optical	Electrical	Magnetic	Thermal	Chemical	Mechanical
Automotive							•
Consumer products							•
Process industries, mining, energy							•
Manufacturing, machinery							•
Medicine, health care							•
HVAC, fire control, security							
Military, aerospace							•
Telecommunications,office equip.							
Meteorology, environment							
Other *							

Main Activities

Acam's Main Activities are the production of pressure, load and strain transducers. Also offered are a strain gauging service and the design of new transducers to customers requirements.

Specific Expertise

Pressure transducers in the range 1 bar to 700 bar in any material using foil or semiconductor gauges. Temperatures to 300°C. Load cells in the range low grams to 300 tonnes in any material using foil or semiconductor gauges. Temperatures to 300°C. Strain transducers - any transducer using strain gauges as above and for example torque transducers. A strain gauging service for customers' own designed transducers or custom design of transducers.

Collaboration Offered

Amplifiers and power supplies for any transducer manufacturer.

Collaboration Sought

Anything to do with transducers.

244 AEA Technology Plc

Product Development Services Group
Building 552
Didcot
Oxfordshire OX11 0RA

Tel: +44 1235 434891
Fax: +44 1235 436322
E-mail:

Affiliation:
Contact: A Young

Organisation information
No of employees >100
Turnover (ECU*) >5m
Sensors technology activities information
No of employees 1-10
Turnover (ECU*) <0.5m
*1ECU=DM 1.89,FFr 6.30,£0.78,Lira 1909,Peseta 159.1

Organisation activities
Research and Development
Testing, consulting, design

Application \ Signal Sensed	Radiation	Optical	Electrical	Magnetic	Thermal	Chemical	Mechanical
Automotive						•	
Consumer products						•	
Process industries, mining, energy						•	
Manufacturing, machinery						•	
Medicine, health care						•	
HVAC, fire control, security						•	
Military, aerospace						•	
Telecommunications,office equip.							
Meteorology, environment						•	
Other *							

Main Activities

Research and development in gas and microelectrode sensors, and sensors consultancy.

Specific Expertise

Design and development of new materials for established types of sensors. Design and development of new types of sensors.

Collaboration Offered

Research and development in sensor and materials, and rod consultancy.

Collaboration Sought

Manufacturing capability and marketing.

245 AEA Technology

TH2R2 Marketing & Sales
Risley
Warrington WA3 6AT
UK

Tel: +44 1925 252000
Fax: +44 1925 254921
E-mail:

Affiliation:
Contact: Dr Davis, General Manager

Organisation information
No of employees >100
Turnover (ECU*) >5m
Sensors technology activities information
No of employees 1-10
Turnover (ECU*) 0.5-1m
*1ECU=DM 1.89,FFr 6.30,£0.78,Lira 1909,Peseta 159.1

Organisation activities
Research and Development
Manufacturer of sensor-based products
Testing, consulting, design
Manufacture of finished sensors

Application \ Signal Sensed	Radiation	Optical	Electrical	Magnetic	Thermal	Chemical	Mechanical
Automotive							
Consumer products							
Process industries, mining, energy	•		•	•	•	•	•
Manufacturing, machinery							
Medicine, health care							
HVAC, fire control, security							
Military, aerospace							
Telecommunications,office equip.							
Meteorology, environment							
Other *							

Main Activities

AEA Technology is one of the largest research and development organisations in Europe, having 30 years experience of being at the forefront of technology, developing equipment for the nuclear power industry. The special demands of this relatively new industry have created within AEA technology expertise in virtually all branches of science and technology.

Specific Expertise

Inductive sensors for process measurements in liquid metals;
inductive and capacitive displacement sensors for high temperature application;
ultrasonic sensors for high temperature including a solid coupled probe for corrosion monitoring;
radiation-tolerant electrodeless conductivity cell;
surface moisture sensor using laser;
special eddy current sensor to detect cracks in metals.

Collaboration Offered

Consultancy; Trouble shooting; Development of special sensors; Supply of bespoke complete sensors.

Collaboration Sought

Companies with measurement problems.

246 Aerosonic Limited

Offar's Business Park
Buttington
Welsh Pool
Powys SY21 8JF Wales

Tel: +44 1938 556644
Fax: +44 1938 556645
E-mail:

Affiliation:
Contact: J H Steynor, Director

Organisation information
No of employees 11-25
Turnover (ECU*) 1-5m
Sensors technology activities information
No of employees 11-25
Turnover (ECU*) 1-5m
*1ECU=DM 1.89,FFr 6.30,£0.78,Lira 1909,Peseta 159.1

Organisation activities
Research and Development
Manufacturer of sensor-based products
Testing, consulting, design

Application \ Signal Sensed	Radiation	Optical	Electrical	Magnetic	Thermal	Chemical	Mechanical
Automotive							
Consumer products							
Process industries, mining, energy	•			•			
Manufacturing, machinery	•			•			•
Medicine, health care							
HVAC, fire control, security							
Military, aerospace							
Telecommunications,office equip.							
Meteorology, environment							
Other *							

Main Activities

Aerosonic specialises in the design and manufacture of test equipment for the magnetic media industries and research establishments. Products include vibrating sample magnetometers (VSMs), signal loss evaluators, tape transports (testing, cleaning, etc.) and void detection equipment. All products are designed and built in the company's own premises on a modern technology park located in Mid Wales.

Specific Expertise

The VSM systems measure magnetic susceptibility of a sample. Sensitivity is high with moments measured down to 10^{-6} emu. The system measures the magnetic moment by vibrating the sample in a controlled magnetic field with pick-up coils sensing the sample output. The optical monitoring systems employ a high intensity IR light source with compatible sensing elements parallel processed for maximum response time. These are typically used for coated web products detecting coat loss along with the overall transmission properties.

Collaboration Offered

Vibrating sample magnetometers. Open reel video tape testers (helical scan mechanisms employed). High speed winders/cleaners with pin hole detection. Drop-out/signal loss detectors (electrical). Void processing equipment.

247 AFRC Silsoe Research Institute

Wrest Park
Silsoe
Bedford MK45 4HS

Tel: +44 1525 860000
Fax: +44 1525 860156
E-mail:

Affiliation:
Contact: Dr J V Stafford,
Dev/Applic of Sensors
Dr C R Bull,
Analysis Agric Products & Food
Barry Ambler, Image Analysis

Organisation information
No of employees
Turnover (ECU*)
Sensors technology activities information
No of employees
Turnover (ECU*)
*1ECU=DM 1.89,FFr 6.30,£0.78,Lira 1909,Peseta 159.1

Organisation activities
Research and Development
Testing, consulting, design

Application \ Signal Sensed	Radiation	Optical	Electrical	Magnetic	Thermal	Chemical	Mechanical
Automotive							
Consumer products							
Process industries, mining, energy	•		•				
Manufacturing, machinery							
Medicine, health care							
HVAC, fire control, security							
Military, aerospace							
Telecommunications,office equip.	•		•			•	•
Meteorology, environment							
Other *							

Main Activities

AFRC Silsoe Research Institute carries out research and development on a wide range of engineering topics associated with agriculture and, to a lesser but increasing extent, with the environment and with the food processing industry. The Sensors Group, within the Information Engineering Division, develops sensors and sensing techniques for routine application on the farm, including post-harvesting processes and on food processing lines.

Specific Expertise

Development of novel image capture techniques for agricultural applications.
Near infrared reflection and transmission for constituent analysis of agricultural products and part-processed food: water; protein; fats; gases. Use of image analysis techniques for dynamic determination of size and shape distributions of particles, flakes, aggregates and for characterisation of surface texture/topography. Development of non-intrusive sensors for mass flow and in-line bulk density of dry particulate materials. Development of sensors for monitoring crop and soil condition, including soil moisture content and nutrient status.

Collaboration Offered

Development of sensors and application of sensing techniques.

Collaboration Sought

Identification of sensing requirements and developments for agriculture and food processing.

248 Agema Infrared Systems

Arden House
West Street
Leighton Buzzard
Beds LU7 7DD

Tel: +44 1525 375660
Fax: +44 1525 579271
E-mail: 101473.3271@compuserve.com

Affiliation:
Contact: B McQueen Smith, Managing Director

Organisation information
No of employees 1-25
Turnover (ECU*) 5m-50m
Sensor technology activities information
No of employees 11-25
Turnover (ECU*) 1m-5m
*1ECU=DM 1.89,FFr 6.30,£0.78,Lira 1909,Peseta 159.1

Organisation activities
Manufacture of sensor-based products

Application \ Signal Sensed	Radiation	Optical	Electrical	Magnetic	Thermal	Chemical	Mechanical
Automotive					•		
Consumer products							
Process industries,mining energy					•		
Manufacturing, machinery					•		
Medicine, health care					•		
HVAC, fire control, security					•		
Military, aerospace					•		
Telecommunications,office equip.							
Meteorology environment					•		
Other							

Main Activities

Electro-optics/infrared thermal imaging.

Specific Expertise

FPA, PESI, SPRITE, CMT

Collaboration Sought

System support.

249 AG Electro-Optics Limited

29 Forest Road
Tarporley
Cheshire CW6 0HX

Tel: +44 829 733305
Fax: +44 829 733679
E-mail:

Affiliation:
Contact: Dr J A Gibson, Director

Organisation information
No of employees 1-25
Turnover (ECU*) 1-5m

Sensors technology activities information
No of employees 1-10
Turnover (ECU*) <0.5m
*1ECU=DM 1.89,FFr 6.30,£0.78,Lira 1909,Peseta 159.1

Organisation activities
Supply of raw materials or semi-finished products used to make sensors.

Application \ Signal Sensed	Radiation	Optical	Electrical	Magnetic	Thermal	Chemical	Mechanical
Automotive		•					
Consumer products							
Process industries, mining, energy		•					
Manufacturing, machinery		•					•
Medicine, health care							
HVAC, fire control, security							
Military, aerospace							
Telecommunications,office equip.		•					
Meteorology, environment		•					
Other *							

Main Activities

Using fibre optic components to develop the control of sensor systems. For example, fibre optic switches as a sensor location control.

Specific Expertise

Gas sensing using semiconductor laser diodes and LEDs, using fibre optics to control and duct the electromagnetic radiation.

Collaboration Offered

Fibre switches. Fibre attenuators.

Collaboration Sought

General.

250 Ai Cambridge

London Road
Pampisford
Cambridge CB2 2ND

Tel: +44 1223 834420
Fax: +44 1223 835050
E-mail: 100436.3674@compuserve.com

Affiliation: TGE
Contact: I Simpson, Tech Manager
M Ophield, MD

Organisation information
No of employees 26-100
Turnover (ECU*) 5m-50m
Sensor technology activities information
No of employees 1-10
Turnover (ECU*) <0.5m
*1ECU=DM 1.89,FFr 6.30,£0.78,Lira 1909,Peseta 159.1

Organisation activities
Manufacture of systems used in the manufacture, testing or development of sensors.
Manufacture of sensor-based products.

Application \ Signal Sensed	Radiation	Optical	Electrical	Magnetic	Thermal	Chemical	Mechanical
Automotive						•	•
Consumer products						•	•
Process industries,mining energy						•	•
Manufacturing, machinery						•	•
Medicine, health care						•	•
HVAC, fire control, security							
Military, aerospace						•	•
Telecommunications,office equip.							
Meteorology, environment							
Other *							

Main Activities

Manufacture of: pressure decay leak detection equipment; trace gas analysis equipment; chromatography equipment; differential pressure transducers.

Specific Expertise

Sensor instrumentation; measurement; applications; system; integration.

Collaboration Offered

Instrumentation design and build capability. Knowledge of trace gas and leak detection business.

Collaboration Sought

New sensor technology.

251 Airflow Developments Ltd

Lancaster Road
High Wycombe
Buckinghamshire HP12 3QP

Tel: +44 1494 525252
Fax: +44 1494 461073
E-mail:

Affiliation:
Contact: Dr P M Downing, Technical Dir

Organisation information
No of employees >100
Turnover (ECU*) >5m
Sensors technology activities information
No of employees >100
Turnover (ECU*) >5m
*1ECU=DM 1.89,FFr 6.30,£0.78,Lira 1909,Peseta 159.1

Organisation activities
Manufacturer of sensor-based products

Application \ Signal Sensed	Radiation	Optical	Electrical	Magnetic	Thermal	Chemical	Mechanical
Automotive							
Consumer products						•	
Process industries, mining, energy							
Manufacturing, machinery							
Medicine, health care							
HVAC, fire control, security					•	•	•
Military, aerospace							
Telecommunications,office equip.							
Meteorology, environment							
Other *							

Main Activities

Design, manufacture and sales of gas pressure, temperature, humidity and velocity measuring instruments for heating, ventilating and scientific use.

Specific Expertise

Manufacture of sensor based products including air velocity and pressure measuring instruments and a range of humidity sensors integral within domestic ventilation fans.

Collaboration Offered

Use of sensors within the above areas. “Packaging” of sensors into customer usable instruments.

252 Alan Cobham Engineering Limited

Holland Way
Blandford Forum
Dorset

Tel: +44 1258 451 481
Fax: +44 1258 451 414
E-mail:

Affiliation:
Contact: P R Wilbourne, Marketing
G Young, Sales

Organisation information
No of employees 26-100
Turnover (ECU*) 5m-50m
Sensor technology activities information
No of employees 1-10
Turnover (ECU*) .05m-1m
*1ECU=DM 1.89,FFr 6.30,£0.78,Lira 1909,Peseta 159.1

Organisation activities
Manufacture of sensor-based products

Application \ Signal Sensed	Radiation	Optical	Electrical	Magnetic	Thermal	Chemical	Mechanical
Automotive							
Consumer products							
Process industries,mining energy				•			•
Manufacturing, machinery				•			
Medicine, health care							
HVAC, fire control, security							
Military, aerospace				•			•
Telecommunications,office equip.				•			
Meteorology environment							
Other							

Main Activities

Alan Cobhanm Engineering provide M&C solutions in general for the liquids and hydrocaron markets. This involves product supply for level, pressure and flow together with packages as a systems integration.

Collaboration Offered

We have an extensive in-house test facility which includes nine storage tanks with total capacity of 100,000 litres. In addition we have a high capacity flow rig ($120m_3$/hr) with absolute calibration by weight.

Collaboration Sought

Exclusivity supplier for liquid hydrocarbon markets and full support to ACE in order to develop these markets.

253 Amot Controls

Western Way
Bury St Edmunds
Suffolk IP33 3SZ

Tel: +44 1284 762222
Fax: +44 1284 760256
E-mail:

Affiliation:
Contact: D Newton, Product Manager
S Spratte, Executive Vice President

Organisation information
No of employees 101-1000
Turnover (ECU*) 1m-5m

Sensor technology activities information
No of employees 26-100
Turnover (ECU*) 1m-5m
*1ECU=DM 1.89,FFr 6.30,£0.78,Lira 1909,Peseta 159.1

Organisation activities
Manufacture of finished sensors.
Manufacture of sensor-based products.

Application \ Signal Sensed	Radiation	Optical	Electrical	Magnetic	Thermal	Chemical	Mechanical
Automotive			●				
Consumer products							
Process industries,mining energy			●	●	●		●
Manufacturing, machinery			●	●	●		●
Medicine, health care							
HVAC, fire control, security			●		●		●
Military, aerospace							
Telecommunications,office equip.							
Meteorology, environment			●	●	●		●
Other *			●	●	●		●

* Marine

Main Activities

Manufacturer of pressure, temperature, flow, level and vibration switches and transmitters.

254 Amersham International Plc

Biosensors Group
Little Chalfont
Buckinghamshire HP8 4SP

Tel: +44 1494 544000
Fax: +44 1494 545271
E-mail:

Affiliation:
Contact: Dr Parnell, Business Manager

Organisation information
No of employees >100
Turnover (ECU*) >5m
Sensors technology activities information
No of employees 26-100
Turnover (ECU*) <0.5m
*1ECU=DM 1.89,FFr 6.30,£0.78,Lira 1909,Peseta 159.1

Organisation activities
Research and Development

Application \ Signal Sensed	Radiation	Optical	Electrical	Magnetic	Thermal	Chemical	Mechanical
Automotive							
Consumer products							
Process industries, mining, energy							
Manufacturing, machinery							
Medicine, health care	•		•			•	
HVAC, fire control, security							
Military, aerospace							
Telecommunications,office equip.							
Meteorology, environment	•		•			•	
Other *							

Main Activities

Products for clinical diagnostics, life science research and industrial measurement and control.
Biosensor Group - developing novel biosensor technology.

Specific Expertise

Optical biosensor based on surface plasmon resonance.
Real time measurement of biomolecular binding at a surface, including immunoassay, enzyme substrate, nucleic acids, ligand-receptor, etc.
Applications include: clinical testing; food and environmental testing; veterinary testing; defence.

Collaboration Sought

Co-operation in sales and marketing.

255 The Analytical Development Company Ltd

Pindar Road
Hoddesdon
Hertfordshire EN11 0AQ

Tel: +44 992 469638
Fax: +44 992 4444567
E-mail:

Affiliation:
Contact: Dr J Rutherford, Sales Director

Organisation information
No of employees 26-100
Turnover (ECU*) >5m
Sensors technology activities information
No of employees 26-100
Turnover (ECU*) >m
*1ECU=DM 1.89,FFr 6.30,£0.78,Lira 1909,Peseta 159.1

Organisation activities
Research and Development
Manufacturer of sensor-based products

Application \ Signal Sensed	Radiation	Optical	Electrical	Magnetic	Thermal	Chemical	Mechanical
Automotive						•	
Consumer products							
Process industries, mining, energy	•						
Manufacturing, machinery						•	
Medicine, health care						•	
HVAC, fire control, security						•	
Military, aerospace						•	
Telecommunications,office equip.							
Meteorology, environment						•	
Other *							

Main Activities

ADC Limited is a company that specialises in the design and manufacture of infrared gas analysis equipment.
These instruments have applications in a large number of diverse markets, including plant physiology, heat treatment, medical and food processing.

Specific Expertise

The company's instruments work on the principle of non-dispersive infrared gas analysis technique. This method has the advantages of being continuous, highly selective, leaving the sample unaffected and offers a wide range of sensitivity from parts per million (ppm) levels to 100% concentration.

The principle of analysis is such that gases whose molecules consist of dissimilar atoms absorb radiation over the wavelength range 2-15 micrometers in a unique characteristic manner. By passing infrared radiation through a sample of gas and measuring the absorption at the wavelength appropriate to the gas concerned, a quantitative determination of its concentration may be made.
Application areas: to measure photosynthesis and transpiration rates in plant samples; to measure hazardous gases in the work place eg. for use in land fill sites; to measure emissions of gases from smoke stacks; to control atmosphere packaged food products; heat treatment systems.

Collaboration Offered

Plant physiology infrared gas analysis equipment.
Infrared gas analysis equipment for a wide variety of diverse markets.

256 Applied Microengineering Limited

The Innovation Centre
68 Milton Park
Abingdon
Oxford OX14 4RX

Tel: +44 1235 833 934
Fax: +44 1235 833 935
E-mail: aml@aml.co.uk
WWW: http://www.aml.co.uk

Affiliation:
Contact: R Santilli, Director

Organisation information
No of employees 1-25
Turnover (ECU*) <1m

Sensor technology activities information
No of employees 1-10
Turnover (ECU*) <0.5m
*1ECU=DM 1.89,FFr 6.30,£0.78,Lira 1909,Peseta 159.1

Organisation activities
Research and development on sensors, sensor materials or applications
Supply of raw materials on semi-finished products used to make sensors.
Manufacture of finished sensors.
Manufacture of sensor-based products.

Application \ Signal Sensed	Radiation	Optical	Electrical	Magnetic	Thermal	Chemical	Mechanical
Automotive							
Consumer products	•					•	
Process industries,mining energy	•		•		•	•	
Manufacturing, machinery	•		•		•	•	•
Medicine, health care						•	•
HVAC, fire control, security	•				•	•	
Military, aerospace							
Telecommunications,office equip.							
Meteorology, environment						•	
Other *							

Main Activities

AML is an independent, industrial and commercial organisaiton, dedicated to microengineering. AML's mission is to help industry exploit microengineering predominantly via the development and manufacture of microengineered products (currently mainly sensors). In most cases the customer will be an OEM but in some cases the end user.

AML also provides inspection, test calibration and packaging services, hence a complete exploitation route is provided. AML will also distribute micro engineered products and other goods related to microengineering.

Specific Expertise

A large variety of work in micro engineering. Various silicon sensors and microstructures have been fabricated.

Non microengineering skills, products and services: sensor design and interfacing; electronics - analogue, digital; software and their integration i.e. instrumentation engineering; design, fabrication and supply of custom aniodic bonder machines for bonding silicon wafers to glass wafers.

Ultrasonic technology: design, manufacture and supply of ultrasonic PVdF transducers for the following applications - immersion transducers for NDT, medical skin scanning, flow measurement; also development work on underwater PVdF microphones.

Collaboration Offered

Microengineering (HST) expertise, including design and manufacturing services, targeted to realising micro sensors.

Collaboration Sought

Complimentary technical, marketing, and distribution capabilities.

257 Automatic Systems Laboratories Ltd

28 Blundells Road
Bradville
Milton Keynes MK13 7HF

Tel: +44 908 320666
Fax: +44 908 322564
E-mail:

Affiliation:
Contact: R Nash, Product Manager

Organisation information

No of employees 26-100
Turnover (ECU*) 1-5m

Sensors technology activities information

No of employees 26-100
Turnover (ECU*) 1-5m
*1ECU=DM 1.89,FFr 6.30,£0.78,Lira 1909,Peseta 159.1

Organisation activities

Research and Development
Manufacturer of sensor-based products
Testing, consulting, design
Manufacture of finished sensors

Application \ Signal Sensed	Radiation	Optical	Electrical	Magnetic	Thermal	Chemical	Mechanical
Automotive							•
Consumer products							•
Process industries, mining, energy					•		•
Manufacturing, machinery	•				•		•
Medicine, health care	•				•		•
HVAC, fire control, security					•		
Military, aerospace	•				•		•
Telecommunications,office equip.							
Meteorology, environment							
Other *							

Main Activities

Temperature measurement standard and calibration; linear and angular capacitive transducer; actuator-transducer servo systems; remote sensing with steered optical tracker; optical tracking systems; humidity sensors; calibration grade temperature controllers.

Specific Expertise

Expertise areas include: very high accuracy measurement (part per million or better), of temperature, angle, linear displacement; servo and control systems; combined mechanical-optical-electronic software; digital systems for tracking; sensing, measurement and control; EG robot calibration.

Collaboration Offered

Joint development of application. Specific projects.

Collaboration Sought

Application needs for the company's technology.

258 Barnes & Associates

12 Ulster Terrace
Regents Park
London NW1 4PJ

Tel: +44 171 935 3696
Fax: +44 171 935 3696
E-mail:

Affiliation:
Contact: P Lloyd, Tech Dir.
J Barnes, Tech Sales
Admiral E Martin, Consultant

Organisation information
No of employees 26-100
Turnover (ECU*)

Sensor technology activities information
No of employees 1-10
Turnover (ECU*) 0.5m-1m
*1ECU=DM 1.89,FFr 6.30,£0.78,Lira 1909,Peseta 159.1

Organisation activities
Research and development on sensors, sensor materials or applications.
Testing, consulting, design.
Manufacture of finished sensors.

Application \ Signal Sensed	Radiation	Optical	Electrical	Magnetic	Thermal	Chemical	Mechanical
Automotive							
Consumer products							
Process industries,mining energy			•				•
Manufacturing, machinery							
Medicine, health care							
HVAC, fire control, security							
Military, aerospace			•				•
Telecommunications,office equip.							
Meteorology, environment							
Other *			•				•

*Marine

Main Activities

Barnes and Associates are fluid power engineers established - as servotel controls - in 1959 for the development and manufacture of high tech hydraulic components. The Watchdog TM sensor has been developed to monitor hydraulic and lubrication oil systems.

Specific Expertise

The Watchdog TM sensor on operate on-line for continuous sensing, or portable for field use. The sensor gives a reading of the contamination level for NAS; SAE; ASTM; and AIA; on a "GO-NO GO" basis for contamination in lubrication and hydraulic oil systems. The sensor may be used with direct meter or computer readings.

Collaboration Offered

A fully patented and developed sensor named Watchdog TM for condition monitoring of hydraulic and lubricating oil systems using the blockage technique. The sensor was developed in conjunction with the Universities of Cranfield, Bath and Aston and the Nuclear Power Industry. We would now welcome collaboration with a major computer company to apply the sensor to condition monitoring programmes for hydraulic and lubrication systems.

Collaboration Sought

To further develop the Watchdog sensor for onboard aircraft installation with a major aircraft instrumentation company - to monitor hydraulic oil systems condition from its current ground equipment application.

259 Bartington Instruments Ltd

10 Thorey Leys Business Plant
Whitny
Oxford OX8 7GE

Tel: +44 1993 774813
Fax: +44 1993 774813
E-mail:

Affiliation:
Contact: C Chapman, Tech Sales Manager

Organisation information
No of employees 1-10
Turnover (ECU*) 0.5-1m
Sensors technology activities information
No of employees
Turnover (ECU*) 0.5-1
*1ECU=DM 1.89,FFr 6.30,£0.78,Lira 1909,Peseta 159.1

Organisation activities
Research and Development
Manufacturer of sensor-based products
Testing, consulting, design
Manufacture of finished sensors

Application \ Signal Sensed	Radiation	Optical	Electrical	Magnetic	Thermal	Chemical	Mechanical
Automotive							
Consumer products							
Process industries, mining, energy				•			
Manufacturing, machinery				•			
Medicine, health care				•			
HVAC, fire control, security							
Military, aerospace				•			
Telecommunications,office equip.				•			
Meteorology, environment				•			
Other *							

Main Activities

Bartington Instruments specialises in the design and manufacture of high performance magnetic measuring instruments.

Specific Expertise

Bartington Instruments manufactures and markets a wide range of magnetic instruments and sensors. This includes single and three axis fluxgate magnetometers with a sensitivity down to less than 1nT, and a fluxgate declinometer/inclinometer. The company also manufactures magnetic susceptibility equipment for laboratory, field and borehold measurements to 6000 metres. Bartington Instruments designs and develops specialised magnetic sensors to customer requirements. Equipment has been developed for both the Royal Australian Navy and the UK Ministry of Defence.

Collaboration Offered

Capability of developing specialised magnetic sensors for use in conjunction with other types of sensors.

260 Baumer Electric Limited

33-36 Shrivenham Hundred Business Park
Watchfield
Swindon SN6 8TZ

Tel: +44 1793 783839
Fax: +44 1793 783814
E-mail:

Affiliation:
Contact: Sharon Lear, Sales Admin

Organisation information

No of employees	1-25
Turnover (ECU*)	5m-50m

Sensor technology activities information

No of employees	11-25
Turnover (ECU*)	>5m

*1ECU=DM 1.89,FFr 6.30,£0.78,Lira 1909,Peseta 159.1

Organisation activities

Research and development on sensors sensor materials or applications
Testing, consulting, design
Supply of raw materials or semi-finished products used to make sensors
Manufacture of finished sensors
Manufacture of sensor-based products

Application \ Signal Sensed	Radiation	Optical	Electrical	Magnetic	Thermal	Chemical	Mechanical
Automotive		•	•	•			•
Consumer products							
Process industries,mining energy		•	•	•			•
Manufacturing, machinery		•	•	•			•
Medicine, health care		•	•	•			•
HVAC, fire control, security		•	•	•			•
Military, aerospace		•	•	•			•
Telecommunications,office equip.		•	•	•			•
Meteorology environment		•	•	•			•
Other							

Main Activities

Baumer Electronics Ltd develop and manufacture optical, electrical, magnetic and mechanical sensors.

261 Bedford Opto Limited

Lindsaylands
Biggar
Lanarkshire ML12 6NR

Tel: +44 1899 221221
Fax: +44 1899 221009
E-mail:

Affiliation:
Contact: B Stott, Design/Development
L Whitfield, Production Manager
A Stuthers, Sales/Progress Desk

Organisation information
No of employees
Turnover (ECU*)
Sensors technology activities information
No of employees
Turnover (ECU*)
*1ECU=DM 1.89,FFr 6.30,£0.78,Lira 1909,Peseta 159.1

Organisation activities
Research and Development
Manufacturer of sensor-based products
Testing, consulting, design
Manufacture of finished sensors

Application \ Signal Sensed	Radiation	Optical	Electrical	Magnetic	Thermal	Chemical	Mechanical
Automotive	●		●				●
Consumer products	●						●
Process industries, mining, energy	●		●				●
Manufacturing, machinery			●				●
Medicine, health care	●		●				●
HVAC, fire control, security	●		●				
Military, aerospace							
Telecommunications,office equip.			●				●
Meteorology, environment							
Other *							

Main Activities

Manufacturers of optoelectronic assemblies for the purpose of sensing and visual display. Sensing assemblies using light emitting diodes (visible or IR) in conjunction with detectors utilise manufacturers' standard products fitted to plastic mouldings or PCBs. Interfacing to control circuits may be direct or incorporate additional electronic components. Sensing may be in either a reflective or transmissive mode. The company's service includes design, development and manufacturing - a consultancy fee is charged when advice only is required.

Specific Expertise

Sensed signals: any visible or IR radiation mechanical movement; electrical signals. Applications: robots; office equipment; process control; security systems; medical electronics; communications. Sensor principles: semiconductor emitters; photo detectors; silicon photo transistors; photo diodes etc.

Collaboration Offered

Design and development. Production. Sales and progress service.

Collaboration Sought

Passive infrared detectors. UV solid state emitters.

262 BICC Thermoheat

Hedgeley Road
Hebburn
Tyne & Wear NE31 1XR

Tel: +44 191 428 3275
Fax: +44 191 483 4127
E-mail: bth.cmc@dial.pipex.com

Affiliation: BICC plc
Contact: C Corner, Business Manager
Optical Fibre Sensing Division
G Parker, Mineral Insulated Cable
Temperature Measurements
M Liddell, Gap Sensing

Organisation information

No of employees 101-1000
Turnover (ECU*) 5m-50m

Sensor technology activities information

No of employees 1-10
Turnover (ECU*) 1m-5m
*1ECU=DM 1.89,FFr 6.30,£0.78,Lira 1909,Peseta 159.1

Organisation activities

Research and development on sensors, sensor materials or applications.
Testing, consulting, design.
Supply of raw materials or semi-finished products used to make sensors.
Manufacture of systems used in the manufacture, testing or development of sensors.
Manufacture of finished sensors.
Manufacture of sensor-based products.

Application \ Signal Sensed	Radiation	Optical	Electrical	Magnetic	Thermal	Chemical	Mechanical
Automotive					•		
Consumer products							
Process industries,mining energy					•		
Manufacturing, machinery					•		
Medicine, health care					•		
HVAC, fire control, security					•		
Military, aerospace					•		
Telecommunications,office equip.							
Meteorology, environment					•		
Other *					•		

* Power Generation

Main Activities

BICC Thermoheat Ltd is a group company within BICC Plc. Our business is the manufacture and supply of mineral insulated cable systems, including temperature measurement, gap measurement. We are advancing into optical fibre sensing systems, firstly with temperature measurement.

Specific Expertise

Thermocouple technology; mineral insulated resistance temperature detectors; mineral insulated optical fibre temperature sensing capacitance; gap sensing.

263 Bolton Institute for Higher Education

Department of Electronic Engineering
Deane Road
Bolton BL3 5AB

Tel: +44 1204 528851
Fax: +44 1204 399074
E-mail:

Affiliation:
Contact: Dr R Jackson

Organisation information
No of employees
Turnover (ECU*)
Sensors technology activities information
No of employees
Turnover (ECU*)
*1ECU=DM 1.89,FFr 6.30,£0.78,Lira 1909,Peseta 159.1

Organisation activities
Research and Development
Testing, consulting, design

Application \ Signal Sensed	Radiation	Optical	Electrical	Magnetic	Thermal	Chemical	Mechanical
Automotive							
Consumer products			•				
Process industries, mining, energy							
Manufacturing, machinery							
Medicine, health care							
HVAC, fire control, security							
Military, aerospace							
Telecommunications,office equip.							
Meteorology, environment			•				
Other *							

Main Activities

Educational establishment with a strong departmental interest in the measurement of two component flow using capacitance techniques.
These methods have been applied to:
pneumatically conveyed solids;
gas liquid mixtures;
liquid/solid mixtures, to provide conveyed phase concentration and volume flow rates.

Specific Expertise

The capacitance sensors determine either: the average permittivity or, the fluctuations in permitivity of the sensing volume.
Typical applications: pneumatic conveying - shot peening; light phase conveying of dry cement powder; pulverised fuel for power stations; flows.
Hydraulic conveying - solids content in non-conducting slurries; gas bubbles concentration in oil/gas or water/gas mixtures; water content of wet steam; oil/vinegar ratio in salad cream. Sensor principle is parallel plate capacitor with variable dielectric.

Collaboration Offered

Reseach/consultancy on flow measurement.

Collaboration Sought

Research into measurement of two component flows in the process/food industry.

264 Bourdon Sedeme Limited

Horton Close
West Drayton
Middx UB7 8JA

Tel: +44 1895 432 832
Fax: +44 1985 432 834
E-mail:

Affiliation: Bourdon Sedeme SA
Contact: David Price, MD

Organisation information
No of employees 101-1000
Turnover (ECU*) 5m-50m
Sensor technology activities information
No of employees 26-100
Turnover (ECU*) 0.5m-1m
*1ECU=DM 1.89,FFr 6.30,£0.78,Lira 1909,Peseta 159.1

Organisation activities
Research and development on sensors sensor materials or applications
Manufacture of finished sensors
Manufacture of sensor-based products

Application \ Signal Sensed	Radiation	Optical	Electrical	Magnetic	Thermal	Chemical	Mechanical
Automotive							•
Consumer products							
Process industries,mining energy							•
Manufacturing, machinery							•
Medicine, health care							•
HVAC, fire control, security							
Military, aerospace							•
Telecommunications,office equip.							
Meteorology environment							•
Other							

Main Activities

Bourdon Sedeme manufacture a wide range of pressure and temperature instruments. We specialise in gauges, switches and sensors/transmitters. The sensors are manufactured to the Transbar(R) specification using the latest in Ceramic technology.

Specific Expertise

Ceramic sensors in areas of pressure and force. This technology is applied to areas in manufacturing, utilities and medical applications.

265 Bowers Internal Gauge Co Ltd

32 Leeds Old Road
Bradford
West Yorkshire BD3 8HU

Tel: +44 1895 430001
Fax: +44 1895 430002
E-mail:

Affiliation:
Contact: S White, Sales Director

Organisation information
No of employees 26-100
Turnover (ECU*) 1-5m
Sensors technology activities information
No of employees 26-100
Turnover (ECU*) 1-5m
*1ECU=DM 1.89,FFr 6.30,£0.78,Lira 1909,Peseta 159.1

Organisation activities
Research and Development
Manufacturer of sensor-based products

Application \ Signal Sensed	Radiation	Optical	Electrical	Magnetic	Thermal	Chemical	Mechanical
Automotive	●		●				●
Consumer products	●		●				●
Process industries, mining, energy	●		●				●
Manufacturing, machinery	●		●				●
Medicine, health care							
HVAC, fire control, security							
Military, aerospace	●		●				●
Telecommunications,office equip.							
Meteorology, environment							
Other *							

Main Activities

Design, development and manufacture of metrology equipment; specialists in internal bore measurements to high precision; manufacture of gauge specific to customers requirements; manufacture of a standard range of gauges including: standard bore, deep bore, blind holes, threaded holes, internal splines etc.

Specific Expertise

Mechanical transducers measuring to micron accuracy; electronic capacitive transducers to submicron accuracy; optical transducers measuring to submicron accuracy; all applied to metrology gauges.

Collaboration Offered

Potential reseach funding or joint development activity.

Collaboration Sought

Predevelopment research activity and/or development for optical or electronic sensors.

266 British Ceramic Research Limited

Queens Road
Stoke-on-Trent ST4 7LQ

Tel: +44 1782 845431
Fax: +44 1782 8412331
E-mail:

Affiliation:
Contact: N Oliver,
Chemical Analysis & Testing
J W Cotton,
Technical/Industrial Ceramics
R C Edgar,
General Research & Consultancy

Organisation information
No of employees >100
Turnover (ECU*) >5m
Sensors technology activities information
No of employees 1-10
Turnover (ECU*) <0.5m
*1ECU=DM 1.89,FFr 6.30,£0.78,Lira 1909,Peseta 159.1

Organisation activities
Research and Development
Testing, consulting, design

Application \ Signal Sensed	Radiation	Optical	Electrical	Magnetic	Thermal	Chemical	Mechanical
Automotive	•				•	•	
Consumer products	•					•	
Process industries, mining, energy					•	•	
Manufacturing, machinery					•	•	
Medicine, health care							
HVAC, fire control, security	•				•	•	
Military, aerospace	•					•	
Telecommunications,office equip.	•						
Meteorology, environment						•	
Other *							

Main Activities

British Ceramic Research Limited offers contract research, consultancy and testing facilities to all sectors of the ceramics industries and associated industries including suppliers and end-users. The Sensors Department is involved in research and development of sensors based on ceramics for use in infrared detection, thermal measurements and gaseous pollution monitoring.

Specific Expertise

IR detectors; thermistors; moisture detectors; gas detectors for monitoring the oxides of nitrogen, sulphur dioxide, carbon monoxide from combustion processes, both industrial and automotive.

Collaboration Offered

Chemical analysis. Microstructual characterisation. Physical testing of raw materials and ceramic products. Technical/industrial ceramics including sensors. General research and consultancy.

267 Brunel University

The Brunel Centre for Manufacturing Metrology
Uxbridge
Middlesex UB8 3PH

Tel: +44 1895 274000 x 2514
Fax: +44 1895 812556
E-mail: barry.jones@brunel.ac.uk

Affiliation:
Contact: Prof B E Jones, Director BCMM

Organisation information
No of employees 26-100
Turnover (ECU*) 1m-5m
Sensor technology activities information
No of employees 11-25
Turnover (ECU*) 0.5m-1m
*1ECU=DM 1.89,FFr 6.30,£0.78,Lira 1909,Peseta 159.1

Organisation activities
Research and development on sensors sensor materials or applications

Application \ Signal Sensed	Radiation	Optical	Electrical	Magnetic	Thermal	Chemical	Mechanical
Automotive		●			●		●
Consumer products							
Process industries,mining energy		●			●		●
Manufacturing, machinery		●			●		●
Medicine, health care							
HVAC, fire control, security		●			●		●
Military, aerospace							
Telecommunications,office equip.							
Meteorology environment							
Other (specify)							

Main Activities

The Brunel Centre for Manufacturing Metrology (BCMM) was inaugurated in 1985 to provide a teaching/training/R&D/consultancy base where skills in measurement science and technology, automatic inspection and quality engineering are enhanced for the benefit of educational establishments, industry and Government.
BCMM is recoginised as a Centre of Excellence in the field of Quality Education by the Institute of Quality Assurance. The Centre received a 1995 Metrology for World Class Manufacturing Award.

Specific Expertise

Areas of expertise are in the subject areas of automatic inspection, quality engineering and systems, conditions monitoring, sensors and actuators, intelligent and distributed instrumentation, optical metrology and testing, and surface measurement. There is a strong emphasis on the development of sensors and activators utilising resonant and optical technologies, and related sensory processing for in-process monitoring.

Collaboration Offered

The Centre has a wide national and international network of contacts with industrial companies, research institutes and universities.

Collaboration Sought

Most R&D projects are done in collaboration with industry and industrial partners are always welcome, particularly for applications led research.

268 Cambridge Consultants Limited

Science Park
Milton Road
Cambridge CB4 4DW

Tel: +44 1223 420024
Fax: +44 1223 423373
E-mail: gordon.oswald@comcon.co.uk

Affiliation: Arthur D Little
Contact: G K A Oswald, Radio Sensors Mgr
D Wilson, Bus Dev Mgr Sens & Instr
J Collier, Gp Ldr, Measurement Prod.

Organisation information
No of employees 101-1000
Turnover (ECU*) 5m-50m

Sensor technology activities information
No of employees 26-100
Turnover (ECU*) 1m-5m
*1ECU=DM 1.89,FFr 6.30,£0.78,Lira 1909,Peseta 159.1

Organisation activities
Research and development on sensors, sensor materials or applications.
Manufacture of finished sensors.
Manufacture of sensor-based products.

Application \ Signal Sensed	Radiation	Optical	Electrical	Magnetic	Thermal	Chemical	Mechanical
Automotive		•	•	•			
Consumer products							•
Process industries,mining energy		•	•	•			
Manufacturing, machinery		•	•	•			
Medicine, health care		•				•	•
HVAC, fire control, security							
Military, aerospace		•	•	•			
Telecommunications,office equip.		•	•	•			
Meteorology, environment		•	•	•			
Other *							

Main Activities
Contract design and development of products, processes and systems in aerospace, telecom, healthcare, general manufacturing and automotive industries.

Specific Expertise
Optics, electromagnetics, manufacturing processes.

Collaboration Offered
Expertise in innovation and engineering.

Collaboration Sought
Market and applications knowledge.

269 Campbell Scientific Limited

Campbell Park
Hathern Road
Shepshed
Leics LE12 9RP

Tel: +44 1509 601141
Fax: +44 1509 601091
E-mail: sales@campbellsci.co.uk

Affiliation: Campbell Scientific, Inc.,Utah, USA
Contact: Dick Saffell, MD
Andrew Sandford, Technical Manager
Mark Pimperton, Mkting Co-ordinator

Organisation information

No of employees	26-100
Turnover (ECU*)	1m-5m

Sensor technology activities information

No of employees	1-10
Turnover (ECU*)	<0.5m

*1ECU=DM 1.89,FFr 6.30,£0.78,Lira 1909,Peseta 159.1

Organisation activities
Manufacture of sensor-based products

Application \ Signal Sensed	Radiation	Optical	Electrical	Magnetic	Thermal	Chemical	Mechanical
Automotive			●		●		●
Consumer products							
Process industries,mining energy			●		●		●
Manufacturing, machinery			●		●		●
Medicine, health care							
HVAC, fire control, security			●		●		●
Military, aerospace							
Telecommunications,office equip.							
Meteorology environment	●		●		●		
Other (specify)							

Main Activities

We supply data acquisition systems based on a range of portable, battery-powered dataloggers. These sophisticated dataloggers are compatible with most common sensor types.
Our in-house and third party sensors complement the dataloggers, enabling us to offer monitoring solutions for a wide range of environmental and industrial applications.

Specific Expertise

Meteorology; micrometeorology; automotive testing; structural monitoring.

270 CaradonTrend Ltd

PO Box 34
Horsham
West Sussex RH12 2YF

Tel: +44 403 211888
Fax: +44 403 241608
E-mail:

Affiliation:
Contact: J Curran, Project Leader

Organisation information
No of employees >100
Turnover (ECU*) >5m

Advanced materials activities information
No of employees 1-10
Turnover (ECU*) 0.5m-1m
*1ECU=DM 1.89,FFr 6.30,£0.78,Lira 1909,Peseta 159.1

Organisation activities
Manufacture of finished sensors

Application \ Signal Sensed	Radiation	Optical	Electrical	Magnetic	Thermal	Chemical	Mechanical
Automotive							
Consumer products							
Process industries, mining, energy							
Manufacturing, machinery							
Medicine, health care							
HVAC, fire control, security			•		•	•	•
Military, aerospace							
Telecommunications,office equip.							
Meteorology, environment							
Other *							

Main Activities

Trend is the leading U.K. supplier of building management sensors (BMS), which involves the manufacture of microprocessor-based controllers and specialist software for the control of heating, ventilation and air conditioning systems. Trend manufactures sensors for these systems and also distributes other manufacturers' sensors.

Specific Expertise

Temperature sensors for air and liquid measurement; pressure, differential pressure and velocity measurement of air, especially at low levels <100 Pascals; humidity measurement; "intelligent" sensors.

271 Castle Group Limited

Salter Road
Scarborough YO11 3UZ

Tel: +44 1723 584250
Fax: +44 1723 583728
E-mail:

Affiliation:
Contact: S Bull, Sales Manager
P R Bull, Managing Director

Organisation information
No of employees 1-25
Turnover (ECU*) 1m-5m
Sensor technology activities information
No of employees 11-25
Turnover (ECU*) 0.5m-1m
*1ECU=DM 1.89,FFr 6.30,£0.78,Lira 1909,Peseta 159.1

Organisation activities
Research and development on sensors, sensor materials or applications.
Testing, consulting, design.
Manufacture of sensor-based products.

Application \ Signal Sensed	Radiation	Optical	Electrical	Magnetic	Thermal	Chemical	Mechanical
Automotive							
Consumer products							
Process industries,mining energy			•				•
Manufacturing, machinery			•				•
Medicine, health care							
HVAC, fire control, security			•				
Military, aerospace			•				•
Telecommunications,office equip.							
Meteorology, environment							
Other *							

Main Activities

Castle Associates specialises in the design and manufacture of sound level meters and vibration measuring devices utilising microphones and various vibration sensors.

Specific Expertise

The use of sensors for measuring levels of noise and vibration. We carry out research into new ways of using existing or new sensor technology for this field.

Collaboration Offered

Project management; expertise in noise and vibration; development capabilities; established organisation; marketing capabilities.

Collaboration Sought

Research capability; manpower; finance.

272 CEL Instruments Limited

35-37 Bury Mead Road
Hitchin
Herts SG5 1RT

Tel: +44 1462 422411
Fax: +44 1462 422511
E-mail:

Affiliation: Welch Allyn Inc
Contact: B Morgan, Marketing Manager
N Hill, General Sales Manager
T Myles, Managing Director
A Poole, Design Manager

Organisation information
No of employees 26-100
Turnover (ECU*) 5m-50m
Sensor technology activities information
No of employees 26-100
Turnover (ECU*) 1m-5m
*1ECU=DM 1.89,FFr 6.30,£0.78,Lira 1909,Peseta 159.1

Organisation activities
Manufacture of sensor-based products.

Application \ Signal Sensed	Radiation	Optical	Electrical	Magnetic	Thermal	Chemical	Mechanical
Automotive							•
Consumer products							•
Process industries,mining energy							•
Manufacturing, machinery							
Medicine, health care							
HVAC, fire control, security							
Military, aerospace							•
Telecommunications,office equip.							
Meteorology, environment							
Other *							

Main Activities

CEL specialise in the design and manufacture of noise measurement instruments and systems. These range from simple hand-held sound level meters for industrial noise measurements, dosimeters for personal noise exposure assessments and sound level analysers for the measurement of noise in the environment or for product noise reduction measures.

Specific Expertise

CEL have been designing and manufacturing acoustic instruments for 25 years and have obtained a reputation for the use of leading edge technology. Recently instruments have incorporated DSP technology together with memory functions which reduce the time previously given to setting an instrument and recording the results. A suite of software packages (the Soundtrack Software library) has also been developed to offer Windows capability in the processing of data.

Collaboration Offered

CEL has collaborated with a number of companies and can offer a worldwide marketing expertise to successfully introduce products to the relevant marketplace.

Collaboration Sought

Development skills for products allied to acoustic instrumentation for both the industrial and medical sectors.

273 Chemlab Instruments Limited

Unit 6, Zone B
Chelmford Road Industrial Estate
Great Dunmaw
Essex CM5 1HD

Tel: +44 1371 876210
Fax: +44 2371 874083
E-mail:

Affiliation:
Contact: T Champion, Sales Manager

Organisation information
No of employees
Turnover (ECU*)
Sensors technology activities information
No of employees
Turnover (ECU*)
*1ECU=DM 1.89,FFr 6.30,£0.78,Lira 1909,Peseta 159.1

Organisation activities
Research and Development
Manufacturer of sensor-based products
Testing, consulting, design
Manufacture of finished sensors

Application \ Signal Sensed	Radiation	Optical	Electrical	Magnetic	Thermal	Chemical	Mechanical
Automotive							
Consumer products							
Process industries, mining, energy							
Manufacturing, machinery							
Medicine, health care							
HVAC, fire control, security							
Military, aerospace							
Telecommunications,office equip.							
Meteorology, environment						•	
Other *							

Main Activities

Chemlab Instruments are working in co-operation with a number of other people on the development of both sensors, and analytical means of determining environmental pollution. The work is aimed at producing rapid laboratory and on-site analytical systems for the determination of the common water pollutants in rivers, potable waters, lakes, estuaries and the open sea. Current work includes using flow injection techniques, with electrochemical sensors to determine pH and conductivity on large numbers of sequentially presented samples. The work is being expanded to cover a wide range of determinands such as nitrate, fluoride and ammonia.

Specific Expertise

ChemLab Instruments sensor development work is currently concentrating on electrochemical, conductivity, and ion selective electrode technology.
The company is showing an active interest in other sensor types. In general the sensor will deliver a high impedance potential difference which can then be amplified and impedance converted in order to drive recording devices and computer data acquisition units.

Collaboration Offered

Marketing, Design and development, Direct sales.

Collaboration Sought

Pollution monitoring, Process control.

274 CIBA Corning Diagnostics

Northern Road
Chilton Industrial Estate
Sudbury
Suffolk CO10 6XD

Tel: +44 787 880022
Fax: +44 787 880033
E-mail:

Affiliation:
Contact: T J Allison, R&D manager

Organisation information
No of employees >100
Turnover (ECU*) >5m

Sensors technology activities information
No of employees 11-25
Turnover (ECU*) >5m
*1ECU=DM 1.89,FFr 6.30,£0.78,Lira 1909,Peseta 159.1

Organisation activities
Research and Development
Manufacturer of sensor-based products
Manufacture of finished sensors

Application \ Signal Sensed	Radiation	Optical	Electrical	Magnetic	Thermal	Chemical	Mechanical
Automotive							
Consumer products							
Process industries, mining, energy							
Manufacturing, machinery							
Medicine, health care						•	
HVAC, fire control, security							
Military, aerospace							
Telecommunications,office equip.							
Meteorology, environment						•	
Other *						•	

* General Laboratory

Main Activities

CIBA Corning Diagnostics develops, manufactures and sells a range of blood analysers and diagnostic kits. The product development group in Sudbury, UK develops sensor based analysers for manufacture, primarily by the adjoining plant. Present day sensors are electrochemical.

Specific Expertise

Electrochemical sensors, predominantly ion selective electrodes. The company's expertise is in refining and applying these sensors in clinical analysers, optimising chemistry/fluidic/electronic/ software/mechanical systems to give the best overall system performance. Parameters measured: Na^+, K^+, Ca^{++}, Cl^-,Li^+, pCO_2, pH, pO_2. Measurements are normally taken using fresh whole blood as the sample.

Collaboration Offered

Sensors capable of measuring blood constituents.

275 City Technology Ltd

City Technology Centre
Walton Road
Portsmouth PO6 1SZ

Tel: +44 705 325511
Fax: +44 705 386611
E-mail:

Affiliation:
Contact: S Hazell, Sales & Marketing

Organisation information
No of employees >100
Turnover (ECU*) >5m
Sensors technology activities information
No of employees >100
Turnover (ECU*) >5m
*1ECU=DM 1.89,FFr 6.30,£0.78,Lira 1909,Peseta 159.1

Organisation activities
Research and Development
Testing, consulting, design

Application \ Signal Sensed	Radiation	Optical	Electrical	Magnetic	Thermal	Chemical	Mechanical
Automotive						•	
Consumer products							
Process industries, mining, energy						•	
Manufacturing, machinery							
Medicine, health care						•	
HVAC, fire control, security						•	
Military, aerospace						•	
Telecommunications,office equip.							
Meteorology, environment						•	
Other *							

Main Activities

City Technology is the recognised world leader in the development and manufacture of high quality electrochemical gas sensors, all of which are originated from the company's advanced research and development facility. The Citicel (Registered Trade Mark), range of sensors, includes those for: oxygen, carbon monoxide, sulphur dioxide, hydrogen sulphide, nitric oxide, nitrogen dioxide, chlorine, hydrogen, hydrogen cyanide, hydrogen chloride, and ammonia

Specific Expertise

City Technology's main area of expertise is electrochemistry, with a more recent development in the catalytic oxidation field. The company works closely with its customers to develop appropriate products specific to their requirements.

276 City University

Dept Electrical, Electronic & Information Engineering
Northampton Square
London EC1V 0HB

Tel: +44 171 477 8120
Fax: +44 171 477 8121
E-mail: k.t.v.grattan@city.ac.uk

Affiliation:
Contact: K T V Grattan, Head of Dept

Organisation information
No of employees 101-1000
Turnover (ECU*) 1m-5m
Sensor technology activities information
No of employees 11-25
Turnover (ECU*) 0.5m-1m
*1ECU=DM 1.89,FFr 6.30,£0.78,Lira 1909,Peseta 159.1

Organisation activities
Research and development.

Application \ Signal Sensed	Radiation	Optical	Electrical	Magnetic	Thermal	Chemical	Mechanical
Automotive		•					
Consumer products		•					
Process industries,mining energy		•					
Manufacturing, machinery		•					
Medicine, health care		•					
HVAC, fire control, security							
Military, aerospace		•					
Telecommunications,office equip.		•	•	•			
Meteorology, environment		•				•	
Other *							

Main Activities

Expertise in research and development of sensors, especially optical, ultrasonic and imaging. Modelling expertise in various aspects of sensor development and understanding.

Specific Expertise

Research and development, education of undergraduate and postgraduate students in sensor technology.

Collaboration Offered

Research and development expertise, low overheads, advanced equipment trace, experience of senior staff.

Collaboration Sought

Industrial direction, certain specialised facilities to meet project needs.

277 Control Techniques Dynamics

South Way
Walworth Industrial Estate
Andover
Hampshire SP10 5AB

Tel: +44 264 324155
Fax: +44 264 356561
E-mail:

Affiliation:
Contact: C Newman, Industrial Sales Engineer
C Wells, Sales Engineer

Organisation information
No of employees >100
Turnover (ECU*) >5m
Sensors technology activities information
No of employees 11-25
Turnover (ECU*) 0.5-1m
*1ECU=DM 1.89,FFr 6.30,£0.78,Lira 1909,Peseta 159.1

Organisation activities
Research and Development
Manufacturer of sensor-based products
Testing, consulting, design
Manufacture of finished sensors

Application \ Signal Sensed	Radiation	Optical	Electrical	Magnetic	Thermal	Chemical	Mechanical
Automotive							•
Consumer products							
Process industries, mining, energy							•
Manufacturing, machinery							•
Medicine, health care							
HVAC, fire control, security							
Military, aerospace							•
Telecommunications,office equip.							
Meteorology, environment							
Other *							

Main Activities

Control Techniques Dynamics manufacture and supply electro-mechanical servo components. These include the following: DC permanent magnet motors; DC brushless motors; torque motors; AC motors; stepper motors; resolvers; encoders; brushless industrial servo motors; tachogenerators.

Specific Expertise

Control Techniques Dynamics manufacture and supply resolvers and encoders, for rotary position sensing, in military, aerospace and industrial applications. Typical performance details are as follows: Size range - 19mm - 90mm diameter; Voltage - 6-24V; Frequency - 400-10,000HZ; Resolution - 16 Bit maximum (65,536); Temperature range -45°C to +85°C; Electrical error - ± 3 mins maximum.

The company operates under full design and manufacturing approval of Def standard 05.21 (AQAP1) and CAA (Civil Aviation Authority). It also designs, manufactures and builds products to specific standards, such as Factory Mutual, BASEEFA and CENELEC; it is one of the first companies in the United Kingdom, to hold BASEEFA and Factory Mutual approvals.

278 Cropico Ltd

Hampton Road
Croydon CR9 2RU

Tel: +44 181 684 4025
Fax: +44 181 684 4094
E-mail:

Affiliation:
Contact: M Potter, Managing Director
B Bint, Sales Engineer

Organisation information
No of employees 1-25
Turnover (ECU*) <1m
Sensor technology activities information
No of employees 1-10
Turnover (ECU*) <0.5m
*1ECU=DM 1.89,FFr 6.30,£0.78,Lira 1909,Peseta 159.1

Organisation activities
Manufacture of systems used in the manufacture, testing or development of sensors.

Application \ Signal Sensed	Radiation	Optical	Electrical	Magnetic	Thermal	Chemical	Mechanical
Automotive					●		
Consumer products							
Process industries,mining energy					●		
Manufacturing, machinery							
Medicine, health care							
HVAC, fire control, security							
Military, aerospace					●		
Telecommunications,office equip.							
Meteorology, environment					●		
Other *							

Main Activities

Design and manufacture of temperature measuring and simulating instruments. For the precise measurement of temperature and the calibration/certification of probes.

Specific Expertise

High precision measurement of PRT's and thermocouples.

279 Crowcon Detection Instruments Limited

2 Blacklands Way
Abingdon Business Park
Abingdon
Oxon OX14 1DY

Tel: +44 1235 553057
Fax: +44 1235 553062
E-mail:

Affiliation: Part of Halma Group
Contact: L Greenham, Dev Director

Organisation information
No of employees 101-1000
Turnover (ECU*) 5m-50m

Sensor technology activities information
No of employees 1-10
Turnover (ECU*) <0.5m
*1ECU=DM 1.89,FFr 6.30,£0.78,Lira 1909,Peseta 159.1

Organisation activities
Manufacture of sensor-based products

Application \ Signal Sensed	Radiation	Optical	Electrical	Magnetic	Thermal	Chemical	Mechanical
Automotive							
Consumer products							
Process industries,mining energy						•	
Manufacturing, machinery						•	
Medicine, health care						•	
HVAC, fire control, security						•	
Military, aerospace						•	
Telecommunications,office equip.							
Meteorology, environment						•	
Other *							

Main Activities

Design and manufacture of gas detection instrumentation using various sensor technologies.

Specific Expertise

Use of electochemical gas sensors. Use of catalytic, thermal conductivity sensors for gas volume determination. Use and design of infrared benches.

Collaboration Offered

End user knowledge and evaluation facilities.

Collaboration Sought

Design of new sensors for new hazardous gases.

280 Cuel Ltd

Electrical & Optical Engineering
12 Tulip Tree Avenue
Kenilworth CV8 2BU

Tel: +44 1926 854293
Fax:
E-mail:

Affiliation:
Contact: D A Hill, Director

Organisation information
No of employees 1-10
Turnover (ECU*) 0.5-1m

Sensors technology activities information
No of employees 1-10
Turnover (ECU*) 0.5-1m
*1ECU=DM 1.89,FFr 6.30,£0.78,Lira 1909,Peseta 159.1

Organisation activities
Manufacturer of sensor-based products

Application \ Signal Sensed	Radiation	Optical	Electrical	Magnetic	Thermal	Chemical	Mechanical
Automotive							
Consumer products							
Process industries, mining, energy	•						
Manufacturing, machinery	•						
Medicine, health care							
HVAC, fire control, security							
Military, aerospace							
Telecommunications,office equip.							
Meteorology, environment							
Other *							

Main Activities

Since 1973 the company has applied light to awkward scientific and industrial problems. All wavelengths of light from the long wave UV to the near IR have been used. Lasers are not used. As creative engineers, activities range from designing and manufacturing complete systems and electro-optic rigs to separate components for customers to interface into their own systems and rigs.

Specific Expertise

The company is familiar with a wide range of components: optical, fibre optic, electro-optic, electronic and has extensive electrical, electronic and mechanical engineering facilities for development work.

Main areas of work: designing and building systems and rigs; coupling sensors into systems; signal processing; matching the geometry of sensors to that of targets; "image" acquisition; illumination. Sensors used: single elements eg. photodiodes, CCTV cameras.

281 DACO Scientific Ltd

Vulcan House
Calleva Industrial Park
Aldermaston
Berkshire RG7 8PB

Tel: +44 118 9817311
Fax: +44 118 9819963
E-mail:

Affiliation:
Contact: P Kerrison, MD
D Highton, Presssure Calibration

Organisation information
No of employees 11-25
Turnover (ECU*) 1-5m
Sensors technology activities information
No of employees 11-25
Turnover (ECU*) 1-5m
*1ECU=DM 1.89,FFr 6.30,£0.78,Lira 1909,Peseta 159.1

Organisation activities
Research and Development
Manufacturer of sensor-based products
Testing, consulting, design
Manufacture of finished sensors

Application \ Signal Sensed	Radiation	Optical	Electrical	Magnetic	Thermal	Chemical	Mechanical
Automotive							•
Consumer products							
Process industries, mining, energy							•
Manufacturing, machinery							•
Medicine, health care							
HVAC, fire control, security							
Military, aerospace							•
Telecommunications,office equip.							•
Meteorology, environment							
Other *							

Main Activities

Design and manufacture (ISO 9001 Registered) of joysticks, trackerballs and interface circuits. Strain gauge installation service. Pressure calibration lab (NAMAS approved). Sale of pressure calibration equipment. Sale of displacement sensors.

Specific Expertise

Force sensing, particularly in force sensitive joysticks; strain gauging - foil and semiconductor; pressure calibration, pressure transducer assessment.

Collaboration Offered

Design, development and manufacturing of special products. 'One-off' to medium volume production.

Collaboration Sought

Sales and marketing.

282 Delta Controls Limited

Island Farm Avenue
West Molesey
Surrey KT8 2UZ

Tel: +44 181 941 5166
Fax: +44 181 783 1163
E-mail: delcon@cix.compulink.co.uk

Affiliation: Concentric Group
Contact: W B Dent, Trans. Prod Mgr
D Grossmith, Development
J L Higginson, Overall Policy Mgr

Organisation information
No of employees 26-100
Turnover (ECU*) 5m-50m
Sensor technology activities information
No of employees 11-25
Turnover (ECU*) 0.5m-1m
*1ECU=DM 1.89,FFr 6.30,£0.78,Lira 1909,Peseta 159.1

Organisation activities
Manufacture of sensor-based products.

Application \ Signal Sensed	Radiation	Optical	Electrical	Magnetic	Thermal	Chemical	Mechanical
Automotive							
Consumer products							
Process industries,mining energy							•
Manufacturing, machinery							
Medicine, health care							
HVAC, fire control, security							•
Military, aerospace							•
Telecommunications,office equip.							
Meteorology, environment							•
Other *							

Main Activities

Our "traditional" business lies in electromechanical pressure sensors/controllers. Our expanding, new business lies in analogue/digital measurement of process variables, which gives a major interest in sensor technology. We have specialised in strain gauge based pressure measurement and our business areas are heavy industrial applications.

Specific Expertise

Strain gauge interface and advanced analogue and digital electronics. Specialist experience in materials for high-corrosion environments.

Collaboration Offered

Manufacturing and marketing infrastructure for high reliability sensors, especially pressure.

Collaboration Sought

Applied physics of actual sensor.

283 Delta-T Devices Ltd

128 Low Road
Burwell
Cambridge CB5 0EJ
2

Tel: +44 1638 742922
Fax: +44 1638 743155
E-mail:

Affiliation:
Contact: L Dodds, Mark. Co-ordinator
W B Dent, Div Mgr - Transducers

Organisation information

No of employees	26-100
Turnover (ECU*)	1m-5m

Sensor technology activities information

No of employees	1-10
Turnover (ECU*)	<0.5m

*1ECU=DM 1.89,FFr 6.30,£0.78,Lira 1909,Peseta 159.1

Organisation activities
Research or development on sensors, sensor materials or applications.
Manufacture of finsihed sensors.
Manufacture of sensor-based products.

Application \ Signal Sensed	Radiation	Optical	Electrical	Magnetic	Thermal	Chemical	Mechanical
Automotive					•		
Consumer products					•		
Process industries,mining energy							
Manufacturing, machinery							
Medicine, health care							
HVAC, fire control, security							
Military, aerospace							
Telecommunications,office equip.							
Meteorology, environment	•				•	•	•
Other *							

Main Activities

Instruments for crop science, data acquisition and meteorology. Some specialist agronomic sensors are manufactured.

Specific Expertise

Delta-T has designed tube solarimeters and tube net radiometers.
Tube solarimeters (tube pyranometers): two sizes (large and miniature); and two types (plain tube and filtered tube) are produced. Plain tubes respond to total solar radiation whilst the filtered version responds only to the infrared. The difference between the tube ouputs corresponds to the visible component of solar radiation.

Tube net radiometers: respond to the difference between the total radiation incident on a surface and the radiation reflected and emitted by that surface. They are designed for use in situations in which the distribution of radiant energy is non-uniform e.g. among the foliage of plants.
Sensors used in the Delta-T Weather Station are: raingauge (tipping bucket unit); energy sensor (silicon diode); wind vane (microtorque wirewound potentiometer); anemometer (reed switch); air and soil temperature probes (thermistors); relative humidity sensor (cracked chromium dioxide capacitance sensor).

284 De Montfort University

Serc Centre
Hawthorne Building
Santeway
Leicester LE1 9BH

Tel: +44 116 2551551
Fax: +44 116 2550307
E-mail:

Affiliation:
Contact: Prof G Chapman,
Head of Science & Engineering
Research Centre

Organisation information
No of employees
Turnover (ECU*)
Sensors technology activities information
No of employees
Turnover (ECU*)
*1ECU=DM 1.89,FFr 6.30,£0.78,Lira 1909,Peseta 159.1

Organisation activities
Research and Development
Testing, consulting, design

Application \ Signal Sensed	Radiation	Optical	Electrical	Magnetic	Thermal	Chemical	Mechanical
Automotive						•	
Consumer products							
Process industries, mining, energy							
Manufacturing, machinery							
Medicine, health care			•				•
HVAC, fire control, security							
Military, aerospace							
Telecommunications,office equip.							
Meteorology, environment							
Other *							

Main Activities

The Department of Applied Physics at the Polytechnic provides the focus for research into thin film and silicon-based sensors. It is currently developing humidity sensors and biomagnetic superconducting sensors on silicon.

Collaboration Offered

Custom sensor manufacture, using thin film and silicon technology.

285 Digitron Instrumentation Ltd

Technology House
Mead Lane
Hertfordshire SG13 7AW

Tel: +44 992 587441
Fax: +44 992 500028
E-mail:

Affiliation:
Contact: B Fox, Marketing Assistant

Organisation information
No of employees >100
Turnover (ECU*) >5m

Sensors technology activities information
No of employees
Turnover (ECU*)
*1ECU=DM 1.89,FFr 6.30,£0.78,Lira 1909,Peseta 159.1

Organisation activities
Research and Development
Manufacturer of sensor-based products
Manufacture of finished sensors

Application \ Signal Sensed	Radiation	Optical	Electrical	Magnetic	Thermal	Chemical	Mechanical
Automotive	•		•		•	•	
Consumer products					•		
Process industries, mining, energy	•		•		•	•	•
Manufacturing, machinery	•		•		•		•
Medicine, health care			•		•		
HVAC, fire control, security					•	•	•
Military, aerospace			•		•		
Telecommunications,office equip.			•		•		
Meteorology, environment					•	•	•
Other *							

Main Activities

Digitron Instrumentation has been a manufacturer and designer of test and measurement equipment for over fifteen years. It produces a range of handheld thermometers, dataprinters, dataloggers and panel mounting indicators, complete with a wide range of compatible probes and sensors.

Specific Expertise

Digitron Instrumentation has designed a specific range of products to meet the new food safety regulations. This range includes highly accurate and fast responding thermometers with a wide range of compatible probes including colour coded probes to avoid cross contamination of foodstuffs. Also included in the range are dataprinters and dataloggers to help prove due diligence. Other key markets as well as the food industry include heating, ventilating and air conditioning, processing, instrumentation and the electronics industry.

286 Douglas Randall Limited

7 Cobham Road
Ferndown Industrial Estate
Wimborne
Dorset BH21 7PE

Tel: +44 202 897969
Fax: +44 202 891918
E-mail:

Affiliation:
Contact: W Carlyle, Director
M Jamson, Product Marketing

Organisation information
No of employees >100
Turnover (ECU*) >5m

Sensors technology activities information
No of employees 26-100
Turnover (ECU*) 1-5m
*1ECU=DM 1.89,FFr 6.30,£0.78,Lira 1909,Peseta 159.1

Organisation activities
Research and Development
Manufacturer of sensor-based products
Testing, consulting, design
Manufacture of finished sensors

Application \ Signal Sensed	Radiation	Optical	Electrical	Magnetic	Thermal	Chemical	Mechanical
Automotive			•	•			•
Consumer products			•	•			•
Process industries, mining, energy				•			•
Manufacturing, machinery				•			•
Medicine, health care				•			•
HVAC, fire control, security				•			•
Military, aerospace				•			
Telecommunications,office equip.				•			•
Meteorology, environment				•			•
Other *							

Main Activities

Douglas Randall Limited designs, develops and manufactures switches, sensors, relays and power supplies. The company offers both standard and custom products and sub-assemblies. Quality is inherent and supported by BS 5750 Part 1, IEC 9000, AQAP 1, CAA and local product approvals to UL, CSA and VDE as appropriate.

Specific Expertise

Douglas Randall Limited designs, develops and manufactures fluid levels, proximity sensors and switches. Both standard and custom designed products are available, integrating detection sensing/switching with control or feedback circuitry.

Collaboration Offered

Advanced quality and manufacturing systems. Flexibility, know-how and international sales outlets.

Collaboration Sought

Additional products and manufacturing. New sales outlets.

287 Druck Limited

Fir Tree Lane
Groby
Leicester LE6 0FH

Tel: +44 116 2314314
Fax: +44 533 875022
E-mail:

Affiliation: Druck Holdings
Contact: N Richards, Sales Office Supervisor
R S Etherington,
Director & Marketing Manager

Organisation information
No of employees >100
Turnover (ECU*) >5m

Sensors technology activities information
No of employees >100
Turnover (ECU*) >5m
*1ECU=DM 1.89,FFr 6.30,£0.78,Lira 1909,Peseta 159.1

Organisation activities
Research and Development
Manufacturer of sensor-based products
Testing, consulting, design
Manufacture of finished sensors

Application \ Signal Sensed	Radiation	Optical	Electrical	Magnetic	Thermal	Chemical	Mechanical
Automotive							•
Consumer products							
Process industries, mining, energy							•
Manufacturing, machinery							•
Medicine, health care							•
HVAC, fire control, security							•
Military, aerospace							•
Telecommunications,office equip.							•
Meteorology, environment							•
Other *							

Main Activities

In 20 years Druck has grown into one of the leading manufacturers of pressure sensors and controllers; the emphasis has been on the total solution to measuring, displaying and controlling pressures.
To achieve this goal development, manufacturing and marketing activities have been established within the company.
These facilities include a silicon micromachining plant, numerically controlled machines for mechanical assemblies, with assembly processes such as ultrasonic thermal bonding, electrostatic (glass/silicon) bonding, electron beam welding, flow and vapour phase soldering with automatic pressure and temperature calibration.
These processes and technologies are operated within a total quality management system which includes formal approvals such as ISO 9001, CAA and MIL-STD-2000, whilst the pressure standards laboratory has been approved by NAMAS.

Collaboration Offered

Druck would be interested in collaborative ventures which include silicon micromachining and which utilise the manufacturing techniques and facilities described in Main Activities above.

Collaboration Sought

The company would seek a product which provides an equitable share of the final device based upon the technologies and capabilities of two or more companies.
Druck would be willing to supply a sensing element which the collaborator integrated within a system, or to consider alternative sensing techniques which could be built into a system using its digital and instrumentation experience.

288 Dynamic Logic Ltd

The Western Centre
Western Road
Bracknell
Berks RG12 1RW

Tel: +44 344 51915
Fax: +44 344 52253
E-mail:

Affiliation:
Contact: T D Heath,MD & Marketing

Organisation information
No of employees 26-100
Turnover (ECU*) 1-5m
Sensors technology activities information
No of employees 1-10
Turnover (ECU*) 0.5-1m
*1ECU=DM 1.89,FFr 6.30,£0.78,Lira 1909,Peseta 159.1

Organisation activities
Research and Development
Testing, consulting, design
Manufacture of finished sensors

Application \ Signal Sensed	Radiation	Optical	Electrical	Magnetic	Thermal	Chemical	Mechanical
Automotive							
Consumer products							
Process industries, mining, energy							•
Manufacturing, machinery							•
Medicine, health care							•
HVAC, fire control, security							•
Military, aerospace							
Telecommunications,office equip.							
Meteorology, environment							•
Other *							

Main Activities

Dynamic Logic Ltd is a leading manufacturer and supplier primarily of telemetry equipment. Part of the company, Shape Instruments, manufacture and supply level and pressure transducers and transmitters.

Specific Expertise

Design, manufacture and supply of level and pressure transducers and transmitters. Offering the latest in ceramic diaphragm technology, Shape Instruments are reliable and accurate. Various models are available with BASEEFA approvals.
Involved with the water industry, both in the U.K. and mainland Europe.

289 Edinburgh Sensors Limited

2 Barn Square
Kirkton Campus
Livingston EH54 7DQ

Tel: +44 1506 425300
Fax: +44 1506 425320
E-mail:

Affiliation: Edinburgh Instruments Group
Contact: P Bramley, Technical Manager

Organisation information
No of employees 1-10
Turnover (ECU*) <0.5m
Sensors technology activities information
No of employees 1-10
Turnover (ECU*) <0.5m
*1ECU=DM 1.89,FFr 6.30,£0.78,Lira 1909,Peseta 159.1

Organisation activities
Research and Development
Manufacturer of sensor-based products
Manufacture of finished sensors

Application \ Signal Sensed	Radiation	Optical	Electrical	Magnetic	Thermal	Chemical	Mechanical
Automotive							
Consumer products							
Process industries, mining, energy						•	
Manufacturing, machinery							
Medicine, health care						•	
HVAC, fire control, security						•	
Military, aerospace						•	
Telecommunications,office equip.							
Meteorology, environment						•	
Other *							

Main Activities

Edinburgh Sensors is based on the research park of Heriot-Watt University, and specialises in the design, development and manufacture of infrared gas sensors and gas products. The company is part of the Edinburgh Instruments Group which provides lasers and optical instruments to industry, research and government establishments.

Specific Expertise

Detection and monitoring of specific gases using a dual wavelength, infrared absorption principle. A range of gas sensor products for process control, personnel protection and environmental monitoring is provided, as well as basic sensors for inclusion in customers' systems. Applications include: safety in breweries and power stations; heating and ventilating systems; food packaging; horticulture; livestock protection; landfill site monitoring; process control; gas detection in mining/offshore applications.

Collaboration Offered

Research into new sensors and products. Customising/re-labelling of the company's products.

Collaboration Sought

Fibreoptic sensors. Sales and marketing of Edinbugh Sensors' products.

290 Electromotif Electronic Ltd

315 Staines Road
Twickenham TW2 5AU

Tel: +44 181 296 0650
Fax: +44 181 296 0649
E-mail:

Affiliation:
Contact: D Haydon, Director

Organisation information
No of employees 1-25
Turnover (ECU*) 1m-5m
Sensor technology activities information
No of employees 1-10
Turnover (ECU*) 1m-5m
*1ECU=DM 1.89,FFr 6.30,£0.78,Lira 1909,Peseta 159.1

Organisation activities
Supply of raw materials or semi-finished products used to make sensors
Manufacture of finished sensors.

Application \ Signal Sensed	Radiation	Optical	Electrical	Magnetic	Thermal	Chemical	Mechanical
Automotive				•			
Consumer products			•				•
Process industries,mining energy			•	•			•
Manufacturing, machinery			•	•			•
Medicine, health care							
HVAC, fire control, security			•	•			•
Military, aerospace			•	•			•
Telecommunications,office equip.							
Meteorology, environment							
Other *							

Main Activities

Electromotif Electronic Ltd designs, develops and produces electromagnetic speed, position and proximity sensors for a wide range of applications from low volume scientific, to high volume automotive.

Specific Expertise

Advanced materials for sensor design, namely: advanced adhesives, encapsulants, lubricants.

Collaboration Offered

Marketing ability.

291 ELE International

Eastman Way
Hemel Hempstead
Hertfordshire HP2 7HB

Tel: +44 1442 218355
Fax: +44 1442 252474
E-mail:

Affiliation:
Contact: G Miller, Divisional Manger

Organisation information
No of employees 26-100
Turnover (ECU*) >5m
Sensors technology activities information
No of employees 1-10
Turnover (ECU*) <0.5m
*1ECU=DM 1.89,FFr 6.30,£0.78,Lira 1909,Peseta 159.1

Organisation activities
Manufacturer of sensor-based products

Application \ Signal Sensed	Radiation	Optical	Electrical	Magnetic	Thermal	Chemical	Mechanical
Automotive							
Consumer products							
Process industries, mining, energy							
Manufacturing, machinery							
Medicine, health care							
HVAC, fire control, security							
Military, aerospace							
Telecommunications,office equip.							
Meteorology, environment					•	•	•
Other *							

Main Activities

ELE International manufactures and supplies materials testing and agronomics equipment to the UK market and worldwide. The agronomics and environmental divisions supply a comprehensive range of equipment including weather stations, mobile laboratories and various testing and monitoring instruments. The ELE environmental monitoring station has been developed as a versatile data acquisition device enabling a variety of sensors to be connected to a standard system.

Specific Expertise

Meteorological sensors; wind speed; wind direction; temperature; humidity; rainfall; surface wetness; barometric pressure; solarimeter; UV radiation; soil temperature. Gas detection sensors: carbon monoxide; sulphur dioxide; nitrogen dioxide. A variety of different types of sensors are used, those with a linear millivolt output being preferred.

Collaboration Offered

Environmental monitoring stations: weather, gas and water quality.

Collaboration Sought

Pollution monitoring.

292 EME (UK) Limited

6 Towerfield Close
Towerfield Industrial Estate
Shoeburyness Essex SS3 9QP

Tel: +44 702 295047
Fax: +44 702 297784
E-mail:

Affiliation:
Contact: C G Carver, Director
Allen Hines, Milled - Machined Parts

Organisation information
No of employees 26-100
Turnover (ECU*) 0.5-1m

Sensors technology activities information
No of employees 1-10
Turnover (ECU*) <0.5m
*1ECU=DM 1.89,FFr 6.30,£0.78,Lira 1909,Peseta 159.1

Organisation activities
Research and Development
Manufacturer of sensor-based products
Testing, consulting, design
Manufacture of finished sensors

Application \ Signal Sensed	Radiation	Optical	Electrical	Magnetic	Thermal	Chemical	Mechanical
Automotive				•	•		•
Consumer products							
Process industries, mining, energy				•	•		
Manufacturing, machinery				•	•		•
Medicine, health care					•		
HVAC, fire control, security				•	•	•	
Military, aerospace				•	•		•
Telecommunications,office equip.					•		
Meteorology, environment				•	•		
Other *				•	•	•	•

* Nuclear reactors

Main Activities

The Meclec Company was formed in 1968 to design and manufacture purpose-built transducers and measurement control systems. The scope of the Meclec Company is:
Transducers: development and manufacture of capacitive, piezoelectric, strain gauge and other types of measuring transducers.
Electronics: development, design and manufacture of signal conditioning equipment covering analogue amplifiers, digital equipment, indicators and controllers.
Control Systems: design and manufacture of control systems to operate various types of plant throughout the world. The control systems have, until recent years, been produced in electro-mechanical and solid state relays as specified by the end-user.
Distribution: due to expansion, it was found worthwhile to consider handling a number of inter-compatible products as a stockist distributor.

Specific Expertise

Development of devices for operating at 800°C in nuclear environments, and transducers for operating totally immersed in water or oil. Medical transducers and probes plus non-contacting dimension measuring equipment for use in NDT radioactive handling facilities. New eutectic metal bonding techniques have been developed for transducers used in liquid sodium for the UKAEA at Dounreay.

Collaboration Offered

Distribution Sales - box assemblies. Milled - machined parts. Transducers - instrumentation. Control systems - sub-assembly

Collaboration Sought

Sensors and transducers working on capacitive principles. Electronic equipment for use with capacitive transducers. Piezoelectric transducers and sensors.

293 Entran

19 Garston Park Parade
Garston
Herts WD2 6LQ

Tel: +44 1923 893999
Fax: +44 1923 893434
E-mail:

Affiliation:
Contact: Bernadette Dwyer, Mkting Manager

Organisation information
No of employees 101-1000
Turnover (ECU*)
Sensor technology activities information
No of employees >100
Turnover (ECU*)
*1ECU=DM 1.89,FFr 6.30,£0.78,Lira 1909,Peseta 159.1

Organisation activities
Manufacture of finished sensors

Application \ Signal Sensed	Radiation	Optical	Electrical	Magnetic	Thermal	Chemical	Mechanical
Automotive			●				●
Consumer products							
Process industries,mining energy			●			●	●
Manufacturing, machinery			●				●
Medicine, health care			●				●
HVAC, fire control, security							
Military, aerospace			●				●
Telecommunications,office equip.							
Meteorology environment			●				●
Other							

Main Activities

Manufacture ofminiature sensors employing techniques involving silicon, foil guage and thin film sensing technology.

294 Envin Scientific Products Limited

28 High Street
Aylburton, Nr Lydney
Gloucestershire GL15 5DA

Tel: +44 594 844707
Fax: +44 594 844722
E-mail:

Affiliation:
Contact: R A Yates, MD

Organisation information
No of employees 1-10
Turnover (ECU*) 0.5-1m
Sensors technology activities information
No of employees
Turnover (ECU*)
*1ECU=DM 1.89,FFr 6.30,£0.78,Lira 1909,Peseta 159.1

Organisation activities
Manufacturer of sensor-based products
Testing, consulting, design

Application \ Signal Sensed	Radiation	Optical	Electrical	Magnetic	Thermal	Chemical	Mechanical
Automotive							
Consumer products							
Process industries, mining, energy						•	
Manufacturing, machinery							
Medicine, health care						•	
HVAC, fire control, security						•	
Military, aerospace							
Telecommunications,office equip.							
Meteorology, environment							
Other *							

Main Activities

Distribution of sensors for toxic and inflammable gas, and oxygen, and for optical and security-related devices. Consulting services in these fields are also provided.

Specific Expertise

The sensing and monitoring of both gaseous atmospheric contaminants, and of oxygen deficiency is covered by both the distribution and consulting services offered. Optical equipment and infrared-based security devices are also within the orbit of current activities.

Collaboration Offered

Gas sensors. Optical devices. IR security products.

295 ERA Technology Ltd

Cleve Road
Leatherhead
Surrey KT22 7SA

Tel: +44 1372 367 000
Fax: +44 1372 367 099
E-mail: info@era.co.uk

Affiliation:
Contact: Dr V Skarda, Dept Manger
D Baynes, Information Services

Organisation information
No of employees >100
Turnover (ECU*) >5m
Sensors technology activities information
No of employees 11-25
Turnover (ECU*) 1-5m
*1ECU=DM 1.89,FFr 6.30,£0.78,Lira 1909,Peseta 159.1

Organisation activities
Research and Development
Testing, consulting, design

Application \ Signal Sensed	Radiation	Optical	Electrical	Magnetic	Thermal	Chemical	Mechanical
Automotive					•		•
Consumer products							
Process industries, mining, energy							•
Manufacturing, machinery					•		•
Medicine, health care					•		•
HVAC, fire control, security					•		
Military, aerospace					•		•
Telecommunications,office equip.							
Meteorology, environment					•	•	•
Other *							

Main Activities

ERA Technology is a research and development company. The Optoelectronics department is involved in the development of a range of optical fibre sensors.

Specific Expertise

Intrinsic and extrinsic optical fibre sensors to measure pressure, temperature, rotation rate, flow and displacement. Sensor multiplexing and distributed sensing are both being developed. Recent interest into optical evanescent field devices is directing research into chemical and biosensing areas. Application areas are diverse, the major activities are in the aerospace, automotive, medical and oil industries.

Collaboration Offered

Research, development and consultancy on optical fibre sensors.

296 ETC Ltd

25 North Street
Lewes
East Sussex

Tel: +44 273 480667
Fax: +44 273 480652
E-mail:

Affiliation:
Contact: T West, Sales Manager

Organisation information
No of employees 11-25
Turnover (ECU*) 0.5-1m
Sensors technology activities information
No of employees
Turnover (ECU*) <0.5
*1ECU=DM 1.89,FFr 6.30,£0.78,Lira 1909,Peseta 159.1

Organisation activities
Research and Development
Manufacturer of sensor-based products
Manufacture of finished sensors

Application \ Signal Sensed	Radiation	Optical	Electrical	Magnetic	Thermal	Chemical	Mechanical
Automotive							
Consumer products							
Process industries, mining, energy					•	•	•
Manufacturing, machinery							•
Medicine, health care							
HVAC, fire control, security					•	•	•
Military, aerospace							
Telecommunications,office equip.							
Meteorology, environment					•	•	•
Other *							

Main Activities

Manufacturer of sensors and electronic control equipment to control and monitor the combustion and environmental emissions related to industrial boilers and incinerators.

Specific Expertise

Boiler and incinerator (and similar combustion plant) flue gas sensor to measure oxygen concentration and temperature. (The former is an in situ device utilising a zirconia-based sensor). Control of this type of plant requiring the measurement of mechanical displacement using position sensors is based on Hall effect devices.

Collaboration Offered

Applications and control skills (patents).

Collaboration Sought

Infrared sensors applied to gas analysis.

297 ETI - Electronic Temperature Instruments

Dominion Way
Worthing
West Sussex BN14 8NW

Tel: +44 1903 202151
Fax: +44 1903 202445
E-mail:

Affiliation:
Contact: P Webb, MD
S Goodehill, Advertising

Organisation information

No of employees	26-100
Turnover (ECU*)	1-5m

Sensors technology activities information

No of employees	11-25
Turnover (ECU*)	0.5-1m

*1ECU=DM 1.89,FFr 6.30,£0.78,Lira 1909,Peseta 159.1

Organisation activities
Research and Development
Manufacturer of sensor-based products
Testing, consulting, design
Manufacture of finished sensors

Application \ Signal Sensed	Radiation	Optical	Electrical	Magnetic	Thermal	Chemical	Mechanical
Automotive					•		
Consumer products					•	•	
Process industries, mining, energy					•	•	
Manufacturing, machinery					•	•	
Medicine, health care					•		
HVAC, fire control, security					•	•	
Military, aerospace							
Telecommunications,office equip.					•		
Meteorology, environment					•	•	
Other *							

Specific Expertise

Manufacturer of:
digital instrumentation including thermometers, hygrometers, pH meters;
thermocouple and PRT, thermometer probes for use in association with ETI instrumentation.

298 Ferranti Technologies Limited

Cairo Mill
Waterlead
Oldham
Lancashire

Tel: +44 161 6240251
Fax: +44 161 6245244
E-mail:

Affiliation:
Contact: M Elwell, Sales Executive

Organisation information
No of employees >100
Turnover (ECU*) >5m
Sensors technology activities information
No of employees 11-25
Turnover (ECU*) 0.5-1m
*1ECU=DM 1.89,FFr 6.30,£0.78,Lira 1909,Peseta 159.1

Organisation activities
Research and Development
Manufacturer of sensor-based products
Testing, consulting, design
Manufacture of finished sensors

Application \ Signal Sensed	Radiation	Optical	Electrical	Magnetic	Thermal	Chemical	Mechanical
Automotive							
Consumer products							
Process industries, mining, energy							
Manufacturing, machinery							
Medicine, health care							
HVAC, fire control, security							
Military, aerospace	•		•	•	•	•	•
Telecommunications,office equip.							
Meteorology, environment							
Other *							

Main Activities

The Sensors group within the Ferranti Aerospace Components Limited Research Department is primarily concerned with the remote sensing of military targets. The technologies of acoustics, seismic, infrared, magnetic and radio frequency are used. The group is involved in pure research offering no design/development capability; this is done in other groups within Ferranti. The sensors research concentrates on converting commerically available 'dumb' sensor packages into smart sensor systems. Application areas: remote sensing of military targets on land, sea and air eg. tanks, helicopters, fixed wing aircraft.

Specific Expertise

Signal sensed - Acoustic: all 'in air' propagation properties and associated signal processing. Seismic: propagation of energy in the ground. Passive IR 3-5 micrometers and 8-14 micrometers. Active CO_2 laser. Magnetic - DC magnetometry. RF - passive radiated spurious RF detection; active MMW radar systems. Chemical - low vapour pressure explosive materials.

Collaboration Offered

Sensor research for remote target detection, location and identification, location and identification: acoustic, electro-optic, magnetic, RF, chemical. Commerical measurement and consultancy service for these technologies.

Collaboration Sought

'In air' acoustics, passive IR sensors, IR optics, DC and AC magnetometry seismic sensor, MMW components.

299 Ferraris Development & Engineering Co

26 Lea Valley Trading Estate
Angel Road
Edmonton N18 3JD

Tel: +44 181 8059055
Fax: +44 181 8074886
E-mail:

Affiliation: Stainless Metal Craft Plc & DST
Contact: R Linzell, Sales Manager

Organisation information

No of employees 26-100
Turnover (ECU*) 1-5m

Sensors technology activities information

No of employees 26-100
Turnover (ECU*) 1-5m

*1ECU=DM 1.89,FFr 6.30,£0.78,Lira 1909,Peseta 159.1

Organisation activities

Research and Development
Manufacturer of sensor-based products
Testing, consulting, design
Manufacture of finished sensors

Application \ Signal Sensed	Radiation	Optical	Electrical	Magnetic	Thermal	Chemical	Mechanical
Automotive							
Consumer products							
Process industries, mining, energy							
Manufacturing, machinery							
Medicine, health care							•
HVAC, fire control, security							
Military, aerospace							
Telecommunications,office equip.							
Meteorology, environment							
Other *							

Main Activities

Manufacturer of medical instruments.

Specific Expertise

Monitoring patients' tidal and minute volume and breath per minute in anaesthesia and intensive care.
Measurements based on rotation of a turbine wheel.

300 GEC Marconi Materials Technology Ltd

Caswell, Towcester
Northants

Tel: +44 1327 350581
Fax: +44 1327 353410
E-mail:

Affiliation:
Contact: W G Stewart

Organisation information
No of employees >100
Turnover (ECU*)
Sensors technology activities information
No of employees
Turnover (ECU*)
*1ECU=DM 1.89,FFr 6.30,£0.78,Lira 1909,Peseta 159.1

Organisation activities
Research and Development
Testing, consulting, design

Application \ Signal Sensed	Radiation	Optical	Electrical	Magnetic	Thermal	Chemical	Mechanical
Automotive							
Consumer products							
Process industries, mining, energy					●	●	
Manufacturing, machinery							
Medicine, health care							
HVAC, fire control, security					●		
Military, aerospace			●		●		●
Telecommunications,office equip.							
Meteorology, environment						●	
Other *							

Main Activities

The Optoelectronics Department is involved in the research and development from feasibility study to engineered laboratory prototype of fibre optic sensors and components for use within sensors including optical sensors, detectors, modulators, switches and integrated optics.

Specific Expertise

Intrinsic: hydrophone (marine); temperature distribution sensors (the profile of the temperature existing along a long fibre optic cable, fire detection, process control etc.).

Extrinsic: pressure (aerospace, marine); electric field (comms.); linear/rotary position (process control, aerospace); liquid level (process control); methane sensing (mining); magnetic fields; temperature; fluorimetry.

Collaboration Offered

Research and development of fibre optic and integrated optic sensors and optical/ optoelectronic components for sensors.

301 Gentech International Ltd

Grangestone Industrial Estate
Girvan
Ayrshire
KA26 9PS

Tel: +44 1465 713581
Fax: +44 1465 714974
E-mail:

Affiliation:
Contact: J J Curry, MD

Organisation information
No of employees 101-1000
Turnover (ECU*) 5m-50m
Sensor technology activities information
No of employees >100
Turnover (ECU*) >5m
*1ECU=DM 1.89,FFr 6.30,£0.78,Lira 1909,Peseta 159.1

Organisation activities
Research and development on sensors, sensor materials or applications.
Testing, consulting, design.
Supply of raw materials or semi-finished products used to make sensors.
Manufacture of finished sensors.

Application \ Signal Sensed	Radiation	Optical	Electrical	Magnetic	Thermal	Chemical	Mechanical
Automotive				●			●
Consumer products				●			●
Process industries,mining energy				●			●
Manufacturing, machinery							●
Medicine, health care							
HVAC, fire control, security					●		●
Military, aerospace							
Telecommunications,office equip.				●			●
Meteorology, environment				●			●
Other *							

Main Activities

Gentech International has over 30 years' experience in the design, development, manufacture and marketing of sensing and switching devices. The company's long-established expertise in magnetic reed switch/reed relay technology is employed to create a wide variety of sensor types. Fluid management, position, motion, inertia and impact sensors meet application specifications across a wide spectrum of industries including automotive, food, petro-chemicals, process control and security. The Gentech design engineering team is fully equipped to provide a prompt and comprehensive service to customers for the most demanding specialised sensor designs.

Specific Expertise

Flexible contacting sensor material; magnetic reed switch sensors; hall effect sensor technology; optical sensors; electromechanical sensors.

Collaboration Offered

Technology transfer, training support market knowledge, continuous product improvement.

Collaboration Sought

Technology transfer, training support market knowledge, continuous product improvement.

302 Gestra (UK) Ltd

Bancroft Court
Hitchin
Herts
SG5 1PH

Tel: +44 1462 431681
Fax: +44 1462 420396
E-mail:

Affiliation: Siebe plc
Contact: S Samad, Technical Director

Organisation information
No of employees 26-100
Turnover (ECU*) 5m-50m

Sensor technology activities information
No of employees 11-25
Turnover (ECU*) 0.5m-1m
*1ECU=DM 1.89,FFr 6.30,£0.78,Lira 1909,Peseta 159.1

Organisation activities
Research and development on sensors, sensor materials or applications.

Application \ Signal Sensed	Radiation	Optical	Electrical	Magnetic	Thermal	Chemical	Mechanical
Automotive							
Consumer products							
Process industries,mining energy		•	•		•		
Manufacturing, machinery							
Medicine, health care							
HVAC, fire control, security							
Military, aerospace							
Telecommunications,office equip.							
Meteorology, environment		•	•				
Other *							

Main Activities
Manufacturer and selling of sensors and controllers for process medium.

Specific Expertise
Sensors for control of pressure, temperature, level and flow of process fluids.

Collaboration Offered
Expertise in steam and process controls.

Collaboration Sought
In-depth electronics and fuzzy logic control expertise.

303 Graticules Ltd

Morley Road
Tonbridge TN9 1RN

Tel: +44 1732 359061
Fax: +44 1732 770217
E-mail:

Affiliation:
Contact: W Hogben, Director

Organisation information
No of employees 26-100
Turnover (ECU*) 1m-5m

Sensor technology activities information
No of employees 1-10
Turnover (ECU*) <0.5m
*1ECU=DM 1.89,FFr 6.30,£0.78,Lira 1909,Peseta 159.1

Organisation activities
Manufacture of systems used in the manufacture, testing or development of sensors.
Manufacture of finished sensors.
Manufacture of sensor-based products.

Application \ Signal Sensed	Radiation	Optical	Electrical	Magnetic	Thermal	Chemical	Mechanical
Automotive							
Consumer products							
Process industries,mining energy							
Manufacturing, machinery		●					
Medicine, health care		●					
HVAC, fire control, security							
Military, aerospace		●					
Telecommunications,office equip.							
Meteorology, environment							
Other *		●					

* Metrology

Main Activities

Moire Fringe Sensors for position sensing in measurement.

304 Hadland Photonics Limited

Newhouse Laboratories
New House Road
Bovingdon
Herts HP3 0EL

Tel: +44 1442 832525
Fax: +44 1442 833733
E-mail:

Affiliation:
Contact: A Hilton, Sales Mgr
A Needham, Engineering Mgr

Organisation information
No of employees 26-100
Turnover (ECU*) 5m-50m
Sensor technology activities information
No of employees 26-100
Turnover (ECU*) 1m-5m
*1ECU=DM 1.89,FFr 6.30,£0.78,Lira 1909,Peseta 159.1

Organisation activities
Manufacture of sensor-based products

Application \ Signal Sensed	Radiation	Optical	Electrical	Magnetic	Thermal	Chemical	Mechanical
Automotive		•					
Consumer products							
Process industries,mining energy		•					
Manufacturing, machinery							
Medicine, health care		•					
HVAC, fire control, security							
Military, aerospace		•					
Telecommunications,office equip.		•					
Meteorology, environment		•					
Other *							

Main Activities

Design, manufacture and sales of high speed image recording and analysis systems for scientific, military and research applications.

Specific Expertise

High speed image capture, recording and analysis systems to 100 million images per second, analysis of short duration luminous events to 300 femto second resolution.

305 Hampton Electronics Ltd

Wrekin House, Market Street
Wellington
Telford TF1 1DT

Tel: +44 952 242939
Fax:
E-mail:

Affiliation:
Contact: B A James, MD

Organisation information
No of employees
Turnover (ECU*)
Sensors technology activities information
No of employees
Turnover (ECU*)
*1ECU=DM 1.89,FFr 6.30,£0.78,Lira 1909,Peseta 159.1

Organisation activities
Manufacture of finished sensors

Application \ Signal Sensed	Radiation	Optical	Electrical	Magnetic	Thermal	Chemical	Mechanical
Automotive							•
Consumer products							
Process industries, mining, energy							•
Manufacturing, machinery							
Medicine, health care							
HVAC, fire control, security							
Military, aerospace							
Telecommunications,office equip.							
Meteorology, environment							
Other *							

Specific Expertise

Manufacture a standard range of magnetic transducers.

306 Hawco Limited

Cathedral Hill Industrial Estate
Guildford
Surrey GU2 5YB

Tel: +44 1483 560606
Fax: +44 1483 575978
E-mail:

Affiliation:
Contact: J Wainwright, Managing Director
G Westwood, Product Manager
T Berry, Sales Director
G Sandeman, Marketing Manager

Organisation information
No of employees 26-100
Turnover (ECU*) 5m-50m
Sensor technology activities information
No of employees 11-25
Turnover (ECU*) 0.5m-1m
*1ECU=DM 1.89,FFr 6.30,£0.78,Lira 1909,Peseta 159.1

Organisation activities
Manufacture of systems used in the manufacture, testing or development of sensors.
Manufacture of finished sensors.
Manufacture of sensor-based products.

Application \ Signal Sensed	Radiation	Optical	Electrical	Magnetic	Thermal	Chemical	Mechanical
Automotive					•		
Consumer products					•		
Process industries,mining energy					•		
Manufacturing, machinery					•		
Medicine, health care							
HVAC, fire control, security					•		
Military, aerospace							
Telecommunications,office equip.					•		
Meteorology, environment					•		
Other *							

Main Activities

Specialist in temperature sensing and control.

Specific Expertise

Temperature sensors generally, including thermocouples, resistance thermometers, thermistors (NTC and PTC), IC's, infrared. Also fibre optics.

307 Hawker Electronics Ltd

43 Melchett Road
Kings Norton Business Centre
Kings Norton
Birmingham B30 3HP

Tel: +44 121 459 8911
Fax: +44 121 433 3041
E-mail:

Affiliation:
Contact: J Slevin, Director
C Babington, Sales & Mark. Mgr

Organisation information
No of employees 26-100
Turnover (ECU*) 1m-5m
Sensor technology activities information
No of employees 1-10
Turnover (ECU*) 0.5m-1m
*1ECU=DM 1.89,FFr 6.30,£0.78,Lira 1909,Peseta 159.1

Organisation activities
Manufacture of sensor-based products

Application \ Signal Sensed	Radiation	Optical	Electrical	Magnetic	Thermal	Chemical	Mechanical
Automotive							
Consumer products							
Process industries,mining energy			•				
Manufacturing, machinery							
Medicine, health care			•				
HVAC, fire control, security			•				
Military, aerospace							
Telecommunications,office equip.							
Meteorology, environment			•				
Other *							

Main Activities

Manufacturer of level control systems for liquids and free flowing powders.

Specific Expertise

Ultrasonic sensors; capacitance sensors.

Collaboration Offered

Ultrasonic technology.

Collaboration Sought

Ultrasonic technology, intelligence.

308 Herga Electric Limited

Northern Way
Bury St Edmunds
Suffolk IP32 6NN

Tel: +44 1284 701422
Fax: +44 1284 753112
E-mail: herga.electric@dial.pipex.com

Affiliation:
Contact: R C Chatham, Director
R G Oscroft, Technical Director

Organisation information
No of employees 101-1000
Turnover (ECU*) 5m-50m
Sensor technology activities information
No of employees 11-25
Turnover (ECU*) 0.5m-1m
*1ECU=DM 1.89,FFr 6.30,£0.78,Lira 1909,Peseta 159.1

Organisation activities
Manufacture of finished sensors.
Manufacture of sensor-based products.

Application \ Signal Sensed	Radiation	Optical	Electrical	Magnetic	Thermal	Chemical	Mechanical
Automotive			●				
Consumer products		●	●				●
Process industries,mining energy			●				
Manufacturing, machinery		●	●				●
Medicine, health care		●	●				●
HVAC, fire control, security		●					
Military, aerospace							
Telecommunications,office equip.							
Meteorology, environment							
Other *							

Main Activities

Herga Electric manufactures fibre optic sensors, built into the following products: pressure sensitive safety mats; pressure sinsitive edge skips; pressure sensitive bumpers.

Specific Expertise

Herga Electric excels in four areas of sensing: fibre optic sensors; footswitches; air switching systems; pressure and vacuum switches. ISO 9002 accredited.

Collaboration Offered

Twelve years of experience of fibre optic sensors and applications.

309 HSE - Health & Safety Executive

Research & Laboratory Services Division
Broad Lane
Sheffield S3 7HQ

Tel: +44 144 2892000
Fax: +44 144 2892500
E-mail:

Affiliation:
Contact: Dr Peter T Walsh,
Section Head, Sensor Applications
Dr S J Gentry,
Group Head, Chemistry and Gases
Dr S C Thorpe,
Sensor Development

Organisation information
No of employees >100
Turnover (ECU*)

Sensors technology activities information
No of employees 1-10
Turnover (ECU*)
*1ECU=DM 1.89,FFr 6.30,£0.78,Lira 1909,Peseta 159.1

Organisation activities
Research and Development

Application \ Signal Sensed	Radiation	Optical	Electrical	Magnetic	Thermal	Chemical	Mechanical
Automotive							
Consumer products							
Process industries, mining, energy						•	
Manufacturing, machinery							
Medicine, health care						•	
HVAC, fire control, security							
Military, aerospace							
Telecommunications,office equip.							
Meteorology, environment						•	
Other *						•	

* Occupational hygiene

Main Activities

HSE's objectives are to promote health and safety at work through development of standards and codes of practice, provision of guidance and enforcement of legislation. The Occupational Medicine and Hygiene Laboratory provides scientific support through research and development and provision of analytical services.

Specific Expertise

Applications of commercially available sensors for specialised workplace air monitoring tasks.
Development of sensors and instruments for:
detection of flammable and toxic gases at concentrations around LELs and OELs, eg. calorimetric sensors for flammables, inorganic semiconductors (SnO_2, ZnO);
for general organic vapours and fire detection, organic semiconductors;
for nitrogen dioxide detection, multi-sensor systems for atmospheric monitoring.

Collaboration Offered

A wide range of experience of hazardous gas monitoring.
Part of policy and legislating organisation for health and safety.
Access to sophisticated gas testing rigs and materials production techniques.

Collaboration Sought

Provision of novel materials, techniques for gas sensing.
Manufacturing/commercial base to exploit developments.

310 H Tinsley & Co Ltd

275 King Henry's Drive
New Addington
Croydon CR0 0AE

Tel: +44 689 800799
Fax: +44 689 800405
E-mail:

Affiliation:
Contact: C Carter, Chief Engineer

Organisation information
No of employees 26-100
Turnover (ECU*) 1-5m
Sensors technology activities information
No of employees 1-10
Turnover (ECU*) <0.5m
*1ECU=DM 1.89,FFr 6.30,£0.78,Lira 1909,Peseta 159.1

Organisation activities
Manufacturer of sensor-based products
Testing, consulting, design
Manufacture of finished sensors

Application \ Signal Sensed	Radiation	Optical	Electrical	Magnetic	Thermal	Chemical	Mechanical
Automotive			●		●		●
Consumer products							
Process industries, mining, energy							
Manufacturing, machinery							
Medicine, health care	●						
HVAC, fire control, security							
Military, aerospace			●		●		
Telecommunications,office equip.			●		●		
Meteorology, environment			●		●		
Other *			●		●		

* Standards Laboratory

Main Activities

H Tinsley and Company Ltd manufacture a wide range of electrical, optical and thermal measuring instruments and sensors.

Specific Expertise

The company manufactures a range of precision temperature sensors (0.01°C to 0.001°C) together with a range of thermometer bridges. A full range of resistance strain gauges are also manufactured. Transducers for overload detection in lifts are produced alongside a range of strain measurement bridges. Other products include sensors for detecting the number of turns, resistance or shortcircuits in coil windings, and optical sensors produced include flour colour graders and glaucoma detectors for the human eye.

311 H Warner & Son Ltd

Arclion House
Hadleigh Road
Ipswich
Suffolk IP2 0EQ

Tel: +44 1473 253702
Fax: +44 1473 233330
E-mail:

Affiliation:
Contact: M H Warner, Chairman & Joint MD

Organisation information
No of employees 26-100
Turnover (ECU*) 1m-5m
Sensor technology activities information
No of employees 1-10
Turnover (ECU*) <0.5m
*1ECU=DM 1.89,FFr 6.30,£0.78,Lira 1909,Peseta 159.1

Organisation activities
Manufacture of sensor-based products.

Application \ Signal Sensed	Radiation	Optical	Electrical	Magnetic	Thermal	Chemical	Mechanical
Automotive							
Consumer products					•		•
Process industries,mining energy		•	•	•	•		•
Manufacturing, machinery				•			
Medicine, health care							
HVAC, fire control, security			•	•	•		•
Military, aerospace							
Telecommunications,office equip.							
Meteorology, environment			•	•			•
Other *							

Main Activities
Electrical control panel designers, manufacturers and installation and also control systems. Mechanical and electrical services engineers - heating, ventilation, air conditioning, piped services, and associated electricals (commercial, industrial and large domestic). Manufacture of ARCHION delayed Action Ball valves. Water engineering services - boreholes, wells, pumps, and water treatment. Domestic Service Engineers - Aga and Raeburn cookers.

Specific Expertise
As above.
Level sensor applications within tanks (water). Level sensor applications within boreholes/wells.

Collaboration Offered
Experience and continuing applications for "problem solving" techniques needed.

312 IMO Industries Limited

T/A GEMS/Transinstruments
Lennox Road
Basingstoke
Hampshire RG22 4AW

Tel: +44 1256 20244
Fax: +44 1256 473680
E-mail:

Affiliation:
Contact: P C Tack, Chief Engineer
T Thexton, European Sales Dir
J G Goodson, European Mkting dir

Organisation information
No of employees 101-1000
Turnover (ECU*) 5m-50m
Sensor technology activities information
No of employees 11-25
Turnover (ECU*) 1m-5m
*1ECU=DM 1.89,FFr 6.30,£0.78,Lira 1909,Peseta 159.1

Organisation activities
Manufacture of finished sensors.

Application \ Signal Sensed	Radiation	Optical	Electrical	Magnetic	Thermal	Chemical	Mechanical
Automotive							
Consumer products							
Process industries,mining energy							•
Manufacturing, machinery							•
Medicine, health care							
HVAC, fire control, security							
Military, aerospace							•
Telecommunications,office equip.							
Meteorology, environment							•
Other *							

Main Activities
Research and development, manufacture and sales of pressure, level and flow transducers and switches.

Specific Expertise
Key area of expertise is thin film depositions using sputtering and chemical vapour deposition. In addition to this we are well versed with electronics packaging for minimum space requirements.

Collaboration Offered
As part of a $450m corporation with a good size European and North American sales team we offer a good "route to market" for smaller organisations.

Collaboration Sought
Complimentary product lines or technology to improve product coverage in target markets.

313 International Instruments Ltd

Cross Lances Road
Hounslow TW3 2AF
Middlesex

Tel: +44 81 5695253
Fax: +44 81 5702682
E-mail:

Affiliation:
Contact: Mr Orchard-Lisle, Technical Support

Organisation information
No of employees
Turnover (ECU*) 1-5m
Sensors technology activities information
No of employees
Turnover (ECU*) 0.5-1m
*1ECU=DM 1.89,FFr 6.30,£0.78,Lira 1909,Peseta 159.1

Organisation activities
Research and Development
Manufacturer of sensor-based products
Testing, consulting, design

Application \ Signal Sensed	Radiation	Optical	Electrical	Magnetic	Thermal	Chemical	Mechanical
Automotive						•	
Consumer products						•	•
Process industries, mining, energy							
Manufacturing, machinery						•	•
Medicine, health care							
HVAC, fire control, security						•	•
Military, aerospace							
Telecommunications,office equip.							•
Meteorology, environment							
Other *							

Main Activities

Wide range of instrumentation for industry covering relative humidity sensors and transmitters, low pressure measurement and control, chart recorders, flowmeters and shock test machines.

Specific Expertise

All applications involving low pressure sensing and relative humidity sensing (ie. HVAC, pharmaceutical etc).

314 International Mining Consultants

Bretly Business Park
Ashby Road
Burton-on-Trent
Staffordshire DE15 0QD

Tel: +44 283 550500
Fax: +44 283 553060
E-mail:

Affiliation:
Contact: Mr S C Benett,
Head of Environmental Sciences
Mr W E Hindmarsh, Director OBE

Organisation information
No of employees >100
Turnover (ECU*) >5m
Sensors technology activities information
No of employees
Turnover (ECU*)
*1ECU=DM 1.89,FFr 6.30,£0.78,Lira 1909,Peseta 159.1

Organisation activities
Research and Development
Testing, consulting, design

Application \ Signal Sensed	Radiation	Optical	Electrical	Magnetic	Thermal	Chemical	Mechanical
Automotive							
Consumer products							
Process industries, mining, energy	•		•	•	•	•	•
Manufacturing, machinery							
Medicine, health care							
HVAC, fire control, security							
Military, aerospace							
Telecommunications,office equip.							
Meteorology, environment							
Other *							

Main Activities

International Mining Consultants, internationally one of the largest coal companies, has its major mining technical support activities centred at Burton-on-Trent in the Midlands. Sensor related work is undertaken mainly within Research Division. Work is primarily undertaken to meet coal industry needs. A wide range of sensors are developed and evaluated. These must generally be intrinsically safe and suitable for underground coal mine application.

Specific Expertise

Product quality monitors on-line determination of ash, moisture, sulphur and calorific value mainly using ionising radiation based sensors.
Environmental motors for gases, fire sensing airborne dust and flow.
Machine guidance, coak thickness, strain orientation.
Drill probe sensing, orientation, discontinuity.
Condition monitoring, wear, water in oil, liquid level, strain.

Collaboration Offered

Research and development in sensor technology.
Sensor maintenance.

Collaboration Sought

Reasearch and development in sensor technology.
Sensors being marketed.
Service facilities.

315 Integrated Photomatrix Limited

IPL
The Grove Trading Estate
Dorchester
Dorset DT1 1SY

Tel: +44 1305 263673
Fax: +44 1305 263679
E-mail:

Affiliation: B Elliot Group of Companies
Contact: Mr C Deuchars, MD
Mr I Whiting, Product Mgr

Organisation information
No of employees 26-100
Turnover (ECU*) 5m-50m

Sensor technology activities information
No of employees 26-100
Turnover (ECU*) 1m-5m
*1ECU=DM 1.89,FFr 6.30,£0.78,Lira 1909,Peseta 159.1

Organisation activities
Research and development on sensors sensor materials or applications
Testing, consulting, design
Manufacture of finished sensors

Application \ Signal Sensed	Radiation	Optical	Electrical	Magnetic	Thermal	Chemical	Mechanical
Automotive		•					
Consumer products							
Process industries,mining energy	•	•					
Manufacturing, machinery		•					
Medicine, health care		•					
HVAC, fire control, security		•					
Military, aerospace		•					
Telecommunications,office equip.		•					
Meteorology environment		•					
Other							

Main Activities

Design, development and manufacture of opto-electronic sensors and sensor sub-assemblies. ISO9001 registered company.

Specific Expertise

Specialist silicon photodetectors, development and manufacture of photodetectors with custom response characteristics, special geometries; design and packaging of sensor elements with optics, optical filters, gratings, emitters; manufacture of sensor assemblies to full opto-electronic instruments.

Collaboration Offered

Opto-electronic sensor and instrument design capability/expertise; production engineering skills; 'whole-project' package, sensor/instrument design, test, production; sensor manufacture capability, prototype through to full scale - manufacture, including wafer processing chip on board, SMD, hybrids, standard and custom packaging.

Collaboration Sought

Market knowledge for their perceived product; part funding of developments.

316 I/O Sensor Operations

49 Hurrican Way
Airport Industrial Estate
Norwich

Tel: +44 1603 400516
Fax: +44 1603 415143
E-mail:

Affiliation:
Contact: Brett Goodyear, Regional Sales Manager

Organisation information
No of employees 101-1000
Turnover (ECU*) 5m-50m

Sensor technology activities information
No of employees 26-100
Turnover (ECU*) >5m
*1ECU=DM 1.89,FFr 6.30,£0.78,Lira 1909,Peseta 159.1

Organisation activities
Research and development on sensors sensor materials or applications
Testing, consulting, design
Manufacture of finished sensors
Manufacture of sensor-based products

Application \ Signal Sensed	Radiation	Optical	Electrical	Magnetic	Thermal	Chemical	Mechanical
Automotive							
Consumer products							
Process industries,mining energy							
Manufacturing, machinery				•			•
Medicine, health care							
HVAC, fire control, security							
Military, aerospace				•			•
Telecommunications,office equip.							
Meteorology environment							
Other				•			•

Main Activities
We are a manufacturer of velocity sensors targeted specifically at the Seismic Industry. We provide these sensors ("Geophones") and related products to field areas engaged in the geophysical exploration for oil and gas. We are the largest of only three companies in the world that produce geophones and operate with 90% vertical integration.

Specific Expertise
We produce our own velocity sensor from the basic components of two precision engineered springs, magnet, coil and case to provide a mass spring system that works on the theory of electro-magnetic energy. When coupled to a surface, physical motion will be transferred into an analogous electrical signal, able to be interpreted by a system with an in-built amplifier.

Collaboration Offered
We provide a ruggedised high quality response vibration/velocity transducer type sensor, that can be used in any number of applications require discriminatory data or a moving/vibrating source, e.g. vibration monitoring, security, structural evaluation.

Collaboration Sought
To collaborate with a system-based company involved in any of the above described industries, who have a system capable of acquiring and analysing the data output from our sensor.

317 IRD Condition Monitoring

Hithercroft Road
Wallingford
Oxford OX10 9DG

Tel: +44 1491 825845
Fax: +44 1491 825855
E-mail:

Affiliation: IRD Mechanalysis, Columbus, Ohio, USA
Contact: J G Drinkwater, Sales/Mark Mgr

Organisation information
No of employees 26-100
Turnover (ECU*) 1m-5m
Sensor technology activities information
No of employees 1-10
Turnover (ECU*) 0.5m-1m
*1ECU=DM 1.89,FFr 6.30,£0.78,Lira 1909,Peseta 159.1

Organisation activities
Manufacture of finished sensors.

Application \ Signal Sensed	Radiation	Optical	Electrical	Magnetic	Thermal	Chemical	Mechanical
Automotive							
Consumer products							
Process industries,mining energy			•	•			
Manufacturing, machinery			•	•			
Medicine, health care							
HVAC, fire control, security							
Military, aerospace							
Telecommunications,office equip.							
Meteorology, environment							
Other *							

Main Activities

Design, manufacture and sales of on-line machinery protection systems in the oil, gas, petro-chemical and power industries.

Specific Expertise

On going development of Eddy Current Probes and Accelerometers for use with our on line monitoring products.

318 ISA Controls Ltd

Hackworth Industrial Park
Shildon
Co Durham DL4 1LH

Tel: +44 388 773065
Fax: +44 388 774888
E-mail:

Affiliation:

Contact: L Gibson, Engineering Director
R Cullen, Marketing & Sales

Organisation information
No of employees 26-100
Turnover (ECU*) >5m

Sensors technology activities information
No of employees 26-100
Turnover (ECU*) 1-5m
*1ECU=DM 1.89,FFr 6.30,£0.78,Lira 1909,Peseta 159.1

Organisation activities
Research and Development
Manufacturer of sensor-based products
Testing, consulting, design
Manufacture of finished sensors

Application \ Signal Sensed	Radiation	Optical	Electrical	Magnetic	Thermal	Chemical	Mechanical
Automotive							
Consumer products							
Process industries, mining, energy					•		•
Manufacturing, machinery							
Medicine, health care							
HVAC, fire control, security					•		•
Military, aerospace					•		•
Telecommunications,office equip.							
Meteorology, environment					•		•
Other *							

Main Activities

Design and manufacture of skid-mounted modular systems packages for process industries, together with level sensing and temperature measurement products accredited to BS5750 Part 1/ ISO 9001.

Specific Expertise

Design and manufacture of differential pressure flow sensors, which can be incorporated into full systems for metering. Manufacture of thermowells, thermocouples and resistance bulbs for temperature sensing within vessels and piping. Level sensing within vessels for use both onshore and offshore.

Collaboration Offered

Full mechanical design engineering and manufacture to internationally recognised quality assurance standards (ISO 9001).

Collaboration Sought

Electronic/computerisation for signal transfer, handling and control from the mechanical sensors.

319 ICI Engineering

PO Box 7
Bruner House
Winnington
Northwich, Cheshire CW8 4DJ

Tel: +44 606 704444
Fax:
E-mail:

Affiliation:
Contact: Prof J P Clarke, Engineering

Organisation information
No of employees >100
Turnover (ECU*) >5m
Sensors technology activities information
No of employees 11-25
Turnover (ECU*) <0.5m
*1ECU=DM 1.89,FFr 6.30,£0.78,Lira 1909,Peseta 159.1

Organisation activities
Research and Development
Manufacturer of sensor-based products
Testing, consulting, design
Manufacture of finished sensors

Application \ Signal Sensed	Radiation	Optical	Electrical	Magnetic	Thermal	Chemical	Mechanical
Automotive							
Consumer products							
Process industries, mining, energy	•		•	•	•	•	•
Manufacturing, machinery							
Medicine, health care							
HVAC, fire control, security							
Military, aerospace							
Telecommunications,office equip.							
Meteorology, environment	•		•		•	•	•
Other *							

Main Activities

Development, research, consulting, educational, system design, construction, commissioning, sampling systems, transducers and signal processing for process chemicals.

Specific Expertise

Fibre optic; electrochemical; chemical transducers. Applied to chemical industry including: safety; hygiene; environmental monitoring. Sampling systems; signal processing; sensor validation.

Collaboration Offered

Consultancy. Market opportunities. Collaboration.

Collaboration Sought

Chemical measurement systems. Sampling. Transduction. Signal processing.

320 ICI Tracerco

PO Box 1
Billingham
Cleveland TS23 1LB

Tel: +44 642 370620
Fax: +44 642 370704
E-mail:

Affiliation:
Contact: Dr K James, Development Manger

Organisation information
No of employees >100
Turnover (ECU*) >5m

Sensors technology activities information
No of employees 11-25
Turnover (ECU*) 0.5-1m
*1ECU=DM 1.89,FFr 6.30,£0.78,Lira 1909,Peseta 159.1

Organisation activities
Research and Development
Manufacturer of sensor-based products
Testing, consulting, design
Manufacture of finished sensors

Application \ Signal Sensed	Radiation	Optical	Electrical	Magnetic	Thermal	Chemical	Mechanical
Automotive							●
Consumer products							●
Process industries, mining, energy							●
Manufacturing, machinery							●
Medicine, health care							●
HVAC, fire control, security							●
Military, aerospace							●
Telecommunications,office equip.							
Meteorology, environment							●
Other *							

Main Activities

Tracerco offers services based on industrial uses of radioisotopes.
Activities cover analysis, process investigation and plant instrumentation.

Specific Expertise

A significant part of Tracerco's work involves the development of specialised nucleonic instrument systems to solve customer's industrial measurement problems. Such projects involve a variety of different nuclear radiation sensors.
Research and development work is not aimed at developing sensors, but is more concerned with choosing sensors to optimise system performance.

321 ITT Barton UK

3 Steyning Way
Southern Cross Trading Estate
Bognor Regis
West Sussex

Tel: +44 1243 826741
Fax: +44 1243860263
E-mail:

Affiliation: ITT Barton
Contact: L Orchard, Mark/Docum. Co-ord

Organisation information
No of employees 26-100
Turnover (ECU*) 5m-50m
Sensor technology activities information
No of employees 1-10
Turnover (ECU*) 1m-5m
*1ECU=DM 1.89,FFr 6.30,£0.78,Lira 1909,Peseta 159.1

Organisation activities
Manufacture of sensor-based products.

Application \ Signal Sensed	Radiation	Optical	Electrical	Magnetic	Thermal	Chemical	Mechanical
Automotive							
Consumer products							
Process industries,mining energy			•	•	•		•
Manufacturing, machinery							
Medicine, health care							
HVAC, fire control, security							
Military, aerospace			•				
Telecommunications,office equip.							
Meteorology, environment							
Other *							

Main Activities

ITT Baron UK develops and manufactures a wide range of sensors for various special technical purposes.

322 Jackson Instruments & Controls Co Ltd

101 Knighton Fields Road West
Leicester LE2 6LH

Tel: +44 1162 832014
Fax: +441162 832790
E-mail:

Affiliation:
Contact: J C Jackson

Organisation information
No of employees 1-10
Turnover (ECU*) <0.5m

Sensors technology activities information
No of employees 1-10
Turnover (ECU*) <0.5m
*1ECU=DM 1.89,FFr 6.30,£0.78,Lira 1909,Peseta 159.1

Organisation activities
Research and Development
Manufacturer of sensor-based products
Testing, consulting, design
Manufacture of finished sensors

Application \ Signal Sensed	Radiation	Optical	Electrical	Magnetic	Thermal	Chemical	Mechanical
Automotive							
Consumer products							
Process industries, mining, energy	•		•	•	•		
Manufacturing, machinery	•		•	•	•		
Medicine, health care							
HVAC, fire control, security	•		•	•	•		
Military, aerospace							
Telecommunications,office equip.							
Meteorology, environment	•		•	•	•		
Other *							

Main Activities

Design and manufacturer of electrical controls and instrumentation, in particular: rotation switches and speed monitoring equipment; airflow switches and controls; electronic thermostats; fan speed controls; fan and pump changeover controls.

Specific Expertise

Development of sensors usually with bought-in elements, manufacture of complete sensors with their control electronics and external interface circuits.

Collaboration Offered

Expertise in design, development, application and manufacturing, together with overall system design and manufacture.

Collaboration Sought

After initial design with application work.

323 Jaltek Ltd

Eaton House
Eaton Green Road
Luton LU2 9LD

Tel: +44 1582 457255
Fax: +44 1582 480895
E-mail:

Affiliation:
Contact: R Howard, Marketing Mgr

Organisation information
No of employees 1-25
Turnover (ECU*) 1m-5m
Sensor technology activities information
No of employees 11-25
Turnover (ECU*) 1m-5m
*1ECU=DM 1.89,FFr 6.30,£0.78,Lira 1909,Peseta 159.1

Organisation activities
Research and development on sensors, sensor materials or applications.
Manufacture of sensor-based products.

Application \ Signal Sensed	Radiation	Optical	Electrical	Magnetic	Thermal	Chemical	Mechanical
Automotive			•				
Consumer products			•				
Process industries,mining energy							
Manufacturing, machinery			•				
Medicine, health care			•				
HVAC, fire control, security			•				
Military, aerospace							
Telecommunications,office equip.			•				
Meteorology, environment							
Other *							

Main Activities

Electronic design-development of own products and sub contract services. Design and manufacturing engineers.

Specific Expertise

Manufacture products using sensors for application in security and protection equipment.

Collaboration Offered

A knowledge of electronic applications. Design and manufacture of expertise to provide market exploitation.

324 Jandel Engineering Ltd

1 Redwood Close
Wing, Leighton Buzzard
Beds LU7 0TP

Tel: +44 525 378554
Fax: +44 525 381945
E-mail:

Affiliation:
Contact: J Clark, MD

Organisation information
No of employees 1-10
Turnover (ECU*) <0.5m

Sensors technology activities information
No of employees 1-10
Turnover (ECU*) <0.5m
*1ECU=DM 1.89,FFr 6.30,£0.78,Lira 1909,Peseta 159.1

Organisation activities
Research and Development
Manufacturer of sensor-based products
Manufacture of finished sensors

Application \ Signal Sensed	Radiation	Optical	Electrical	Magnetic	Thermal	Chemical	Mechanical
Automotive							
Consumer products							
Process industries, mining, energy						•	
Manufacturing, machinery							
Medicine, health care							
HVAC, fire control, security							
Military, aerospace							
Telecommunications,office equip.							
Meteorology, environment							
Other *			•				

* Characterisation of semiconducter materials

Main Activities

Scientific instrument designers and manufacturers. Manufacture of electrolytic moisture cells and elements for two UK manufacturers of moisture measuring systems. Manufacture of four-point probe heads used to evaluate semiconductor material resistivity together with the associated measuring electronics.

Specific Expertise

Moisture cells/elements comprise a bi-filar winding of platinum wire on a fused silica former. The winding is coated with phosphoric acid. Gas flows over the coating and when water is absorbed, the coating becomes conducting - a current flows proportionally to the quantity of water contained in the gas. Water content of a sample gas from 0.05 vpm @ 1 bar upwards to 1000 vpm. The principle is an absolute method based on Faraday's Laws of Electrolysis.

325 JMS Consultants

Ravenscliffe
Lands Lane
Knaresborough
N Yorkshire HG5 9DE

Tel: +44 1423 869593
Fax: +44 1423 860404
E-mail: john@jmsknars.demon.co.uk

Affiliation:
Contact: J Sonley, Partner

Organisation information
No of employees 1-25
Turnover (ECU*) <1m
Sensor technology activities information
No of employees 1-10
Turnover (ECU*) 0.5m-1m
*1ECU=DM 1.89,FFr 6.30,£0.78,Lira 1909,Peseta 159.1

Organisation activities
Research and development of sensors, sensor materials or applications.
Testing, consulting, design.

Application \ Signal Sensed	Radiation	Optical	Electrical	Magnetic	Thermal	Chemical	Mechanical
Automotive							
Consumer products							
Process industries,mining energy		•	•			•	
Manufacturing, machinery							
Medicine, health care							
HVAC, fire control, security							
Military, aerospace							
Telecommunications,office equip.							
Meteorology, environment			•			•	
Other *							

Main Activities

JMS is a specialist consultancy involved specifically with flammable and toxic gas detection. Activities include sensor consultancy; training and surveys frequently within the oil industry. John Sonley is Chairman of the British Standards National Committee on gas detectors - EEL/29

Specific Expertise

Catalytic and semiconductor, flammable and toxic gas detectors. Infra-red flammable gas detection.

Collaboration Offered

Specialist consultancy for gas detection.

326 John Caunt Scientific Limited

Oakfield Industrial Estate
Eynsham
Oxford OX8 1JA

Tel: +44 865 880479
Fax: +44 865 881935
E-mail:

Affiliation:
Contact: J F Caunt, MD

Organisation information
No of employees 1-10
Turnover (ECU*)
Sensors technology activities information
No of employees 1-10
Turnover (ECU*)
*1ECU=DM 1.89,FFr 6.30,£0.78,Lira 1909,Peseta 159.1

Organisation activities
Research and Development
Manufacturer of sensor-based products
Testing, consulting, design
Manufacture of finished sensors

Application \ Signal Sensed	Radiation	Optical	Electrical	Magnetic	Thermal	Chemical	Mechanical
Automotive	•						
Consumer products	•						
Process industries, mining, energy	•						
Manufacturing, machinery	•						
Medicine, health care	•						
HVAC, fire control, security	•						
Military, aerospace	•						
Telecommunications,office equip.	•						
Meteorology, environment	•						
Other *							

Specific Expertise

Production of scintillation crystals mounted on photodiodes complete with preamplifiers. Also signal processing electronics mostly PC based. The sensors measure gamma and X-rays of 100keV upwards. Designed originally for medical applications but could be used wherever radioisotopes are used. The company is interested in looking at new ideas and developments of all types of radiation monitors.

Collaboration Offered

Miniature detectors for cardiac monitoring, human dynamic organ function studies and industrial plant monitoring.

Collaboration Sought

All forms of radiation monitor development.

327 Kobold Instruments Limited

Hobold House
8/9 Brunts Business Centre
Samuel Brunts Ways
Masfield
Notts NG18 2AH

Tel: +44 1623 427701
Fax: +44 1623 427702
E-mail:

Affiliation:
Contact: P Wharmby, General Manager

Organisation information
No of employees 26-100
Turnover (ECU*) >5m

Sensors technology activities information
No of employees 26-100
Turnover (ECU*) >5m
*1ECU=DM 1.89,FFr 6.30,£0.78,Lira 1909,Peseta 159.1

Organisation activities
Research and Development
Manufacturer of sensor-based products
Testing, consulting, design
Manufacture of finished sensors

Application \ Signal Sensed	Radiation	Optical	Electrical	Magnetic	Thermal	Chemical	Mechanical
Automotive					●		●
Consumer products							
Process industries, mining, energy					●		●
Manufacturing, machinery					●		●
Medicine, health care							
HVAC, fire control, security					●		●
Military, aerospace							
Telecommunications,office equip.					●		●
Meteorology, environment					●		●
Other *							

Main Activities

Manufacture, sales and distribution of instrumentation for the measurement, monitoring and control of flow, pressure, level and temperature.

Specific Expertise

Flow sensors use magnetic floats to trigger inductive and reed contacts. Others use the Hall effect, transforming magnetically-induced pulses into digital frequency signals. Level sensors use magnetically operated reed switches, conductive electrodes, piezoelectric exciters, ultrasonics, thyristors, capacitance.

Collaboration Offered

Flow, pressure, level, temperature sensing, measurement, monitoring and control.

Collaboration Sought

Agents required to sell and distribute Kobold range of instrumentation.

328 Kollmorgen Motion Technologies Group

Suite 34
Wyvols Court
Swallowfield
Nr Reading RG7 1PY

Tel: +44 118 9880 253
Fax: +44 1189 886 404
E-mail:

Affiliation:
Contact: R J Lacey, VP Bussiness Dev.
F Dubby, Marketing Director

Organisation information
No of employees >1000
Turnover (ECU*) >50m

Sensor technology activities information
No of employees 26-100
Turnover (ECU*) >5m
*1ECU=DM 1.89,FFr 6.30,£0.78,Lira 1909,Peseta 159.1

Organisation activities
Manufacture of finished sensors.

Application \ Signal Sensed	Radiation	Optical	Electrical	Magnetic	Thermal	Chemical	Mechanical
Automotive							
Consumer products							
Process industries,mining energy							
Manufacturing, machinery		•	•				•
Medicine, health care							
HVAC, fire control, security							
Military, aerospace		•	•				•
Telecommunications,office equip.							
Meteorology, environment							
Other *							

Main Activities

Suppliers of drives and controls (inc position and velocity sensors)

Specific Expertise

Position feedback rotary position sensors: resolvers (brushless); RVDT; encoders.

329 LabFacility Ltd

99 Waldegrave Road
Teddington
Middlesex TW11 8LR

Tel: +44 181943 5331
Fax: +44 181943 4351
E-mail:

Affiliation:
Contact: N J Riddett, Chairman
S Frost, Sales Director

Organisation information
No of employees 26-100
Turnover (ECU*) 1m-5m
Sensor technology activities information
No of employees 26-100
Turnover (ECU*) 1m-5m
*1ECU=DM 1.89,FFr 6.30,£0.78,Lira 1909,Peseta 159.1

Organisation activities
Testing, consulting, design.
Supply of raw materials or semi-finished products used to make sensors.
Manufacture of systems used in the manufacture, testing or development of sensors.
Manufacture of finished sensors.

Application \ Signal Sensed	Radiation	Optical	Electrical	Magnetic	Thermal	Chemical	Mechanical
Automotive					•		
Consumer products					•		
Process industries,mining energy					•		
Manufacturing, machinery					•		
Medicine, health care							
HVAC, fire control, security					•		
Military, aerospace					•		
Telecommunications,office equip.							
Meteorology, environment							
Other *							

Main Activities

Manufactureres of temperature and electrochemical sensors, and suppliers of associated accessories: measuring, controlling and calibrating instruments.

Specific Expertise

Thermocouples - fabricated and mineral insulated platinum resistance thermometers, thermistor sensors.

Collaboration Offered

Temperature sensor know-how.

330 Lambda Photometrics Limited

Lambda House
Batford Mill
Harpenden
Herts AL2 1SR

Tel: +44 1582 764334
Fax: +44 1582 712084
E-mail:

Affiliation: Polytec-PI GmbH, Wadbronn
Contact: R Traynor, Prod Engineer
P Harrison, Joint MD

Organisation information
No of employees 1-25
Turnover (ECU*) 5m-50m
Sensor technology activities information
No of employees 1-10
Turnover (ECU*) 0.5m-1m
*1ECU=DM 1.89,FFr 6.30,£0.78,Lira 1909,Peseta 159.1

Organisation activities
Supply of raw materials or semi-finished products used to make sensors.
Manufacture of systems used in the manufacture, testing or development of sensors.
Manufacture of finished sensors.

Application \ Signal Sensed	Radiation	Optical	Electrical	Magnetic	Thermal	Chemical	Mechanical
Automotive		●		●	●		●
Consumer products		●			●		●
Process industries,mining energy		●			●		●
Manufacturing, machinery		●			●		●
Medicine, health care							
HVAC, fire control, security							
Military, aerospace		●		●	●		●
Telecommunications,office equip.							
Meteorology, environment		●					
Other *				●			

* EMI/EMC

Main Activities
Our activities include the sale and technical support of lasers and light based instruments, support equipment (optical tables, hardware, optical components, fibre optics, detectors and signal processing electronics) and piezo-electric materials and finished components. We sell and provide technical and service support for remote laser-based measurements of vibration, fluid velocity and surface speed. Lambda also manufacture and supply/support medical lasers for skin blemish treatment and tattoo removal.

Specific Expertise
PZT ceramics and sensor components. Fibre optic sensors for temperature, pressure, RI, current measurement. Fibre and airpath vibrometers for sensor development.

Collaboration Offered
Working with university/R&D groups with Lambda-supplied equipment for sensor optimisation (measurement).

Collaboration Sought
New applications to develop instrumentation.

331 Land Combustion

Stubley Lane
Dronfield
Sheffield S18 6NQ

Tel: +44 246 417691
Fax: +44 246 290274
E-mail:

Affiliation:
Contact: D Stuart, Research Physicist

Organisation information
No of employees 26-100
Turnover (ECU*) >5m
Sensors technology activities information
No of employees 26-100
Turnover (ECU*) >5m
*1ECU=DM 1.89,FFr 6.30,£0.78,Lira 1909,Peseta 159.1

Organisation activities
Research and Development
Manufacturer of sensor-based products
Manufacture of finished sensors

Application \ Signal Sensed	Radiation	Optical	Electrical	Magnetic	Thermal	Chemical	Mechanical
Automotive							
Consumer products							
Process industries, mining, energy	•				•	•	
Manufacturing, machinery							
Medicine, health care							
HVAC, fire control, security							
Military, aerospace							
Telecommunications,office equip.							
Meteorology, environment							
Other *							

Main Activities

Land Combustion manufactures, markets and distributes instrumentation for the combustion industry.
The range includes: gas monitoring, flame monitoring, heat flux and research probes.
The Reseach Department is active in developing new instruments for all of these sectors.

Specific Expertise

Land's interest and activity in sensors is currently limited to source monitoring of gases from combustion plant.
The primary interest is for gases used in process control or subject to legislation omission limits.
The range of gases is being expanded and covers a variety of techniques including optical, electrochemical and infrared.

332 Lee-Dickens Ltd

Rushton Road
Desborough
Kettering
Northants NN14 2QW

Tel: +44 536 760156
Fax: +44 536 762552
E-mail:

Affiliation:
Contact: G Dickens, Marketing & Export Sales
T Meadows, Chief Engineer

Organisation information
No of employees 26-100
Turnover (ECU*) 1-5m
Sensors technology activities information
No of employees 11-25
Turnover (ECU*) 1-5m
*1ECU=DM 1.89,FFr 6.30,£0.78,Lira 1909,Peseta 159.1

Organisation activities
Manufacturer of sensor-based products

Application \ Signal Sensed	Radiation	Optical	Electrical	Magnetic	Thermal	Chemical	Mechanical
Automotive							
Consumer products					•	•	
Process industries, mining, energy					•	•	
Manufacturing, machinery							
Medicine, health care					•	•	
HVAC, fire control, security					•	•	
Military, aerospace					•	•	
Telecommunications,office equip.					•	•	
Meteorology, environment					•	•	
Other *							

Main Activities

Designers and manufacturers of electronic process control instrumentation and telemetry and SCADA systems. The process control instrumentation covers signal conditioning equipment, relative humidity/temperature monitoring equipment and digital process indicators. The company is assessed to AQAP1/AQAP13 for the design and manufacture of hardware/ software based products for the Ministry of Defence (MOD).

Specific Expertise

Lee-Dickens manufactures instrumentation and digital process indicators associated with the majority of process variables, but most specifically with relative humidity and temperature. The company manufactures 3 comprehensive ranges of signal conditioning equipment, comprising DC and mains powered, surface, DIN rail and Eurocard rack mounting instruments. The instruments cover a range of functions such as trip amplifiers, signal convertors, signal isolators, signal linearisers, etc. Each type can in general accept sensor input directly such as thermocouples and RTDs. The company has also designed and now manufactures a range of duct and wall mounting relative humidity and temperature probes/transmitters. For these it buys in a relative humidity sensor, normally capacative, and a temperature sensor and manufacture the conditioning circuitry.

333 Lee-Integer Ltd

1-3 Bowling Green Road
Kettering
Northants NN15 7QW

Tel: +44 536 511010
Fax: +44 536 513653
E-mail:

Affiliation:
Contact: Mr Bailey, General Manager

Organisation information
No of employees 26-100
Turnover (ECU*) >5m
Sensors technology activities information
No of employees 11-25
Turnover (ECU*) <0.5m
*1ECU=DM 1.89,FFr 6.30,£0.78,Lira 1909,Peseta 159.1

Organisation activities
Manufacturer of sensor-based products
Testing, consulting, design

Application \ Signal Sensed	Radiation	Optical	Electrical	Magnetic	Thermal	Chemical	Mechanical
Automotive							
Consumer products							
Process industries, mining, energy						•	
Manufacturing, machinery						•	
Medicine, health care						•	
HVAC, fire control, security						•	
Military, aerospace						•	
Telecommunications,office equip.						•	
Meteorology, environment						•	
Other *							

Main Activities

Development of relative humidity sensors in association with research subsidiary Dutch Sensor Technologie Ltd.
Manufacture of humidity/temperature probes/transmitters and dewpoint converters.

Specific Expertise

Supply of relative humidity probes/transmitters for all applications in industry and research in range 0-100% RH, -40°C to 100°C dewpoint and -50°C to 200°C temperature.

Collaboration Offered

Design and manufacturing licensing of temperature/humidity probes.

Collaboration Sought

Expansion of sales into other geographical areas.

334 Littlemore Scientific Engineering Company

Railway Lane
Littlemore
Oxford OX4 4PZ

Tel: +44 865 747437
Fax: +44 865 747780
E-mail:

Affiliation:
Contact: W D Hall, Metal Detectors

Organisation information
No of employees
Turnover (ECU*)
Sensors technology activities information
No of employees
Turnover (ECU*)
*1ECU=DM 1.89,FFr 6.30,£0.78,Lira 1909,Peseta 159.1

Organisation activities
Research and Development
Manufacturer of sensor-based products
Testing, consulting, design
Manufacture of finished sensors

Application \ Signal Sensed	Radiation	Optical	Electrical	Magnetic	Thermal	Chemical	Mechanical
Automotive							
Consumer products				•			•
Process industries, mining, energy				•			
Manufacturing, machinery				•			
Medicine, health care							
HVAC, fire control, security							
Military, aerospace				•			•
Telecommunications,office equip.							
Meteorology, environment				•			
Other *							

Main Activities

Manufacturer of electronic instruments.

Specific Expertise

Magnetic Sensors: proton magnetometers - sense earth's magnetic field.
Used in geology and finding ferrous objects; magnetic observatories; fluxgate magnetometers.
Metal Detectors: pulse induction; underwater diver held; personal search.

Collaboration Offered

Magnetometers. Metal detectors.

335 Litre Meter

Rabans Lane Industrial Estate
50/53 Rabans Close
Aylesbury HP19 3RS

Tel: +44 1296 20341
Fax: +44 1296 436446
E-mail:

Affiliation:
Contact: John Collins, MD
Charles Wemyss, Eng, Director

Organisation information

No of employees	1-25
Turnover (ECU*)	1m-5m

Sensor technology activities information

No of employees	11-25
Turnover (ECU*)	0.5m-1m

*1ECU=DM 1.89,FFr 6.30,£0.78,Lira 1909,Peseta 159.1

Organisation activities
Manufacture of systems used in the manufacture, testing or development of sensors
Manufacture of finished sensors
Manufacture of sensor-based products

Application \ Signal Sensed	Radiation	Optical	Electrical	Magnetic	Thermal	Chemical	Mechanical
Automotive							
Consumer products							
Process industries,mining energy				•	•		•
Manufacturing, machinery				•	•		
Medicine, health care					•		
HVAC, fire control, security							•
Military, aerospace				•			
Telecommunications,office equip.							
Meteorology environment				•	•		•
Other (specify)							

Main Activities

Manufacture of flow meter systems including water and other liquid flow; air and other gas flow; etc.

Specific Expertise

Manufacture of a sensor for detecting motion of a rotor with the ability to measure up to 1,000Hz at elevated pressures up to 795 bar. Temperatures to 135°C.

336 Loughborough Projects Ltd

Swingbridge Road
Loughborough
Leicestershire LE11 0JB

Tel: +44 509 262042
Fax: +44 509 262517
E-mail:

Affiliation:
Contact: M Meehn, Marketing Dir

Organisation information
No of employees 11-25
Turnover (ECU*) 1-5m
Sensors technology activities information
No of employees 11-25
Turnover (ECU*) 0.5-1m
*1ECU=DM 1.89,FFr 6.30,£0.78,Lira 1909,Peseta 159.1

Organisation activities
Manufacturer of sensor-based products
Testing, consulting, design
Manufacture of finished sensors

Application \ Signal Sensed	Radiation	Optical	Electrical	Magnetic	Thermal	Chemical	Mechanical
Automotive							•
Consumer products							
Process industries, mining, energy							•
Manufacturing, machinery							•
Medicine, health care							
HVAC, fire control, security							
Military, aerospace							
Telecommunications,office equip.							
Meteorology, environment							•
Other *							

Main Activities

Design, development and manufacture of specialised instrumentation systems and electro-mechanical equipment.

Specific Expertise

Safe load indicator; torque measurement systems; data logging.

Collaboration Offered

Mechanical to electronic engineering expertise and experience in the company's fields.

Collaboration Sought

New sensor concepts - other markets.

337 Loughborough Consultants Limited

Loughborough University
Leicestershire LE11 3TF

Tel: +44 509 222597
Fax: +44 509 231983
E-mail:

Affiliation:
Contact: W D Hall, Partner

Organisation information
No of employees 1-10
Turnover (ECU*) 1-5m
Sensors technology activities information
No of employees 1-10
Turnover (ECU*) <0.5m
*1ECU=DM 1.89,FFr 6.30,£0.78,Lira 1909,Peseta 159.1

Organisation activities
Research and Development
Manufacturer of sensor-based products
Testing, consulting, design
Manufacture of finished sensors

Application \ Signal Sensed	Radiation	Optical	Electrical	Magnetic	Thermal	Chemical	Mechanical
Automotive	•		•				•
Consumer products							
Process industries, mining, energy							
Manufacturing, machinery							
Medicine, health care							
HVAC, fire control, security	•					•	•
Military, aerospace			•	•			
Telecommunications,office equip.							
Meteorology, environment	•					•	
Other *							

Main Activities

Loughborough Consultants Limited was established in 1969, by Loughborough University of Technology to market the expertise of the academic departments to industry and commerce. Thus it has access to over 450 academic staff over a wide range of disciplines.
The University has very strong departments and sections in all areas of technology, pure and applied science, computer and business studies, sport science and recreation, creative and engineering design.

Specific Expertise

Current research interests reflect the following:
hydrodynamic amperometric sensors for the determination of nitrate ions in hydroponic systems;
multi-analyte immunoassay techniques with fibre optic probes;
the use of pulsed amperometry for the qualitative determination of organic vapours.
Work with microelectrodes is being extended to include the investigation of sub-micron diameter devices for the voltametric determination of gas phase species.
Infrared, acoustic and magnetic sensors are being developed for security purposes and the automotive industry.
The use of visual sensoring with image processing is also being investigated.

Collaboration Offered

Fully confidential consultancy service for long and short term research and development, measurement tasks and prototype development.

Collaboration Sought

Information on both services offered by other organisations and offers of co-operation.

338 Loughborough University of Technology

Department of Chemistry
Loughborough
Leicestershire LE11 3TU

Tel: +44 509 222570
Fax: +44 509 223925
E-mail:

Affiliation:
Contact: Prof J N Miller
Dr T Edmonds
Dr A G Fogg

Organisation information
No of employees
Turnover (ECU*)
Sensors technology activities information
No of employees
Turnover (ECU*)
*1ECU=DM 1.89,FFr 6.30,£0.78,Lira 1909,Peseta 159.1

Organisation activities
Research and Development
Testing, consulting, design

Application \ Signal Sensed	Radiation	Optical	Electrical	Magnetic	Thermal	Chemical	Mechanical
Automotive							
Consumer products							
Process industries, mining, energy						•	
Manufacturing, machinery							
Medicine, health care						•	
HVAC, fire control, security							
Military, aerospace							
Telecommunications,office equip.							
Meteorology, environment						•	
Other *							

Main Activities

The Analytical Research Group of the Department of Chemistry is involved in research projects on chemical sensing. Fibre optic, electrochemical (amperometric, coulometric and voltametric) and piezoelectric transducers are used to probe immunoassay and classical analytical chemistries.

Specific Expertise

The Loughborough University of Technology is currently involved in the following projects: dual analyte fluorescence immunoassays for clinical species are being developed in conjunction with fibre-optic probes; amperometric sensors for nitrate and phosphate based on capilliary fill devices are being developed; a research programme into the sensing of Theophylline by an immunoassay method coupled with electrochemical detection is in progress; research is being carried out into the use of piezoelectric crystals as sensors for gaseous species.

Collaboration Offered

Basic research into a range of chemical sensor technologies.

339 LTH Electronics Ltd

Chaul End Lane
Luton LU4 8EZ

Tel: +44 1582 593693
Fax: +44 1582 598036
E-mail:

Affiliation:
Contact: A Hayward, Sales Manager
K Higgs, Technical Manager
R Earl, Technical Director
G Williams, Managing Director

Organisation information

No of employees	26-100
Turnover (ECU*)	1m-5m

Sensor technology activities information

No of employees	11-25
Turnover (ECU*)	0.5m-1m

*1ECU=DM 1.89,FFr 6.30,£0.78,Lira 1909,Peseta 159.1

Organisation activities
Research and development on sensors, sensor materials or applications.
Testing, consulting, design.
Manufacture of finished sensors.

Application \ Signal Sensed	Radiation	Optical	Electrical	Magnetic	Thermal	Chemical	Mechanical
Automotive							
Consumer products							
Process industries,mining energy						•	
Manufacturing, machinery							
Medicine, health care							
HVAC, fire control, security							
Military, aerospace							
Telecommunications,office equip.							
Meteorology, environment						•	
Other *							

Main Activities

Design and manufacture of sensors and instrumentation for liquid analysis, in industrial "on-line" applications.

Specific Expertise

Measurement of electrolytic conductivity, resistivity for determining levels of dissolved salts in solution using, electrode and electrodeless (inductive) sensors. Measurement of dissolved oxygen in solution using a membrane galvanic cell.

340 Lucas Automotive Ltd

Advanced Engineering Centre
Dog Kennel Lane
Shirley
Solihull
West Midlands B90 4JJ

Tel: +44 21 6274141
Fax: +44 21 6273584
E-mail:

Affiliation:
Contact: D Hodgson,
Manager of Technology Dept

Organisation information
No of employees >100
Turnover (ECU*) >5m
Sensors technology activities information
No of employees 26-100
Turnover (ECU*) 1-5m
*1ECU=DM 1.89,FFr 6.30,£0.78,Lira 1909,Peseta 159.1

Organisation activities
Research and Development
Testing, consulting, design

Application \ Signal Sensed	Radiation	Optical	Electrical	Magnetic	Thermal	Chemical	Mechanical
Automotive			●		●	●	●
Consumer products							
Process industries, mining, energy			●		●		●
Manufacturing, machinery							
Medicine, health care							
HVAC, fire control, security							
Military, aerospace			●		●		●
Telecommunications,office equip.							
Meteorology, environment							
Other *							

Main Activities
Lucas Industries is a major international supplier of aerospace, automotive and industrial systems and components. The Advanced Engineering Centre is responsible for advanced technology development and evaluation. Lucas has to be selective in its technologies and development programmes, and concentrate on those which are judged to be most critical for its future business. Sensors have been clearly identified as one of those critical technologies. The Sensors and Microelectronics Department comprises multi-disciplinary engineers whose expertise ranges from electronics to physics, computing, silicon processing, chemistry, microjoining and mechanical engineering.

Specific Expertise
Sensors research and development carried out within the Advanced Engineering Centre is focused on meeting the Company's needs of robust, highly accurate and cost effective sensors to work over wide dynamic ranges in hostile environments. Types of signal sensed include air mass flow, pressure, acceleration, speed, position and combustion. Application areas include: advanced control systems for petrol and diesel engines; vehicle braking and suspension systems; industrial instrumentation; aerospace engine and flight control systems. Sensor technologies currently under application development include; micro-engineered silicon sensors, thin film and thick film sensors, magnetic sensors and optical sensors. In addition, work is also concentrated on modelling techniques, electronic interface circuitry and smart/intelligent sensor concepts.

Collaboration Offered
Technology development.
Automotive/aerospace applications.

341 Lucas-Varity

Lucas Automotive Electronics
Phoenix Way
Cirencester
Gloucestershire GL7 1QG

Tel: +44 285 657981
Fax: +44 285 658258
E-mail:

Affiliation:
Contact: I Hervey, Engineering Manager

Organisation information
No of employees >100
Turnover (ECU*) >5m
Sensors technology activities information
No of employees 26-100
Turnover (ECU*) 0.5-1m
*1ECU=DM 1.89,FFr 6.30,£0.78,Lira 1909,Peseta 159.1

Organisation activities
Research and Development
Manufacture of finished sensors

Application \ Signal Sensed	Radiation	Optical	Electrical	Magnetic	Thermal	Chemical	Mechanical
Automotive	•		•	•	•	•	•
Consumer products							
Process industries, mining, energy							
Manufacturing, machinery							
Medicine, health care							
HVAC, fire control, security							
Military, aerospace							
Telecommunications,office equip.							
Meteorology, environment							
Other *							

Main Activities

Lucas Industries Plc is a major worldwide supplier of components and systems to the automotove, aerospace and industrial sectors. Lucas Automotive Sensors is a new business unit formed in August 1989 as a part of Lucas Automotive Limited. Its role is to develop and supply advanced technology sensors to the vehicle industry on a worldwide basis.

Specific Expertise

Design and development of a range of low cost, advanced sensors suited to the emerging needs of advanced vehicle control systems including: non-contact position sensors based on eddy current techniques; linear 5mm to >300mm range; rotary <30° to >150°; high frequency response; temperature, -40/+125°C. Accelerometers: AC or DC types; 1g to 50g; silicon based; speed/position; variable reluctance; magnetoresistors. Optical speed-over-ground. Steering angle: Hall effect; optical; etc.

Collaboration Offered

Non-contact position sensors. Accelerometers. Speed/position sensors.

Collaboration Sought

Any sensor technologies capable of meeting automotive costs and environmental requirements.

342 Macam Photometrics Limited

10 Kelvin Square
Livingstone
Scotland EH54 5PF

Tel: +44 1506 437391
Fax: +44 1506 438543
E-mail:

Affiliation:
Contact: C Maclellen, MD

Organisation information
No of employees 1-10
Turnover (ECU*) 1-5m
Sensors technology activities information
No of employees 1-10
Turnover (ECU*)
*1ECU=DM 1.89,FFr 6.30,£0.78,Lira 1909,Peseta 159.1

Organisation activities
Manufacturer of sensor-based products
Manufacture of finished sensors

Application \ Signal Sensed	Radiation	Optical	Electrical	Magnetic	Thermal	Chemical	Mechanical
Automotive	•						
Consumer products							
Process industries, mining, energy	•						
Manufacturing, machinery							
Medicine, health care	•						
HVAC, fire control, security							
Military, aerospace							
Telecommunications,office equip.							
Meteorology, environment	•						
Other *							

Main Activities

Macam designs and manufactures high quality, high performance optical instruments and systems for radiometry and photometry of natural and artificial light sources, ranging from the ultra violet through to the invisible infrared region of the spectrum.
The product range includes: photometers, radiometers, spectroradiometers, spot photometers, safety monitors, electrometers, OEM contracts, optical detectors, quantum sensors, filters, monochromators, integrating spheres, fibre optic probes and stable light sources.
Macam also provides services in optical calibration, design of special systems and consultation and technical advice on optical measuring problems.

Specific Expertise

Sensors for:
ultra violet radiation detection in the UVA, UVB and UVC regions;
photometric sensors;
radiometric sensors in the visible and near IR spectral regions;
quantum sensors for plant science measurements.

All use solid state photodiodes and optical filters and are available for laboratory or outside applications.

Collaboration Offered

Optical radiation sensors.

343 Marchbanks Measurement Systems Ltd

Bryns Bett House
St Matins Road
Lymington
Hampshire SO41 9HA

Tel: +44 590 673007
Fax: +44 590 670443
E-mail:

Affiliation:
Contact: R J Marchbanks, MD

Organisation information
No of employees 1-10
Turnover (ECU*) <0.5m

Sensors technology activities information
No of employees 1-10
Turnover (ECU*) <0.5m
*1ECU=DM 1.89,FFr 6.30,£0.78,Lira 1909,Peseta 159.1

Organisation activities
Research and Development
Manufacturer of sensor-based products
Manufacture of finished sensors

Application \ Signal Sensed	Radiation	Optical	Electrical	Magnetic	Thermal	Chemical	Mechanical
Automotive							
Consumer products							
Process industries, mining, energy							•
Manufacturing, machinery							
Medicine, health care							•
HVAC, fire control, security							
Military, aerospace							•
Telecommunications,office equip.							
Meteorology, environment							
Other *							•

* Environment noise monitoring

Main Activities

MMS Limited is an independent research and development organisation which specialises in technology transfer from UK Universities. The company also undertakes small scale production of specialised sensors and measurement systems for customers in industry, government departments, and the health service.

The company was the 1988 winner of the TIE Innovators competition and in 1989 won a Department of Trade and Industry Small Firm's Merit Award for Research and Technology (SMART).

Specific Expertise

Pressure, flow, volume displacement, altitude and acoustic sensors. Applied to industrial, environmental health and biomedical area.
Specific sensors: highly sensitive wideband microphones/pressure transducers (DC-20kHZ) for detection of infrasonics, normal acoustic signals, barometric or altitude related pressure changes. Patented volume displacement measurement sensors (<1 nanolitre). Patented ultra-low flow measurement sensors (<10 nanolitres per second).

Collaboration Offered

Ultra-sensitive pressure sensors. Infrasonic microphones. Low flow and displacement sensors. Consultancy, licencing opportunities. Research for new sensors (pressure, acoustic, flow displacement).

Collaboration Sought

Highly sensitive pressure and flow sensors.

344 Maywood Instruments Limited

Rankine Road
Daneshill Industrial Estate
Basingstoke RG24 0PP

Tel: +44 256 57572
Fax: +44 256 840937
E-mail:

Affiliation:
Contact: M Baker, Director of Engineering

Organisation information
No of employees 26-100
Turnover (ECU*) 1-5m

Sensors technology activities information
No of employees 26-100
Turnover (ECU*) 1-5m
*1ECU=DM 1.89,FFr 6.30,£0.78,Lira 1909,Peseta 159.1

Organisation activities
Research and Development
Manufacturer of sensor-based products
Testing, consulting, design
Manufacture of finished sensors

Application \ Signal Sensed	Radiation	Optical	Electrical	Magnetic	Thermal	Chemical	Mechanical
Automotive							•
Consumer products							•
Process industries, mining, energy							•
Manufacturing, machinery							•
Medicine, health care							•
HVAC, fire control, security							•
Military, aerospace							•
Telecommunications,office equip.							•
Meteorology, environment							•
Other *							

Main Activities

Design and manufacture of force and load transducers, pressure transducers and associated electronic instrumentation, including digital readout and amplifying equipment. Ability and willingness to either design complete transducer systems to meet specific requirements where ready made solutions do not exist, or alternatively, a willingness to adapt company's standard transducer products to meet specific requirements.

Collaboration Offered

As per summary.

Collaboration Sought

Any strain gauge based transducers.

345 Measurements Group UK Limited

Stroudley Road
Basingstoke
Hants RG24 8FW

Tel: +44 1256 462131
Fax: +44 1256 471441
E-mail:

Affiliation:
Contact: A L Window, Managing Director
J J Stagg, Marketing Director
G C Mordan, Comm Dir/MD
K Scott, Tech Liaison Mgr

Organisation information
No of employees 26-100
Turnover (ECU*) 1m-5m
Sensor technology activities information
No of employees 26-100
Turnover (ECU*) 1m-5m
*1ECU=DM 1.89,FFr 6.30,£0.78,Lira 1909,Peseta 159.1

Organisation activities
Testing, consulting, design.
Manufacture of finished sensors.
Manufacture of sensor-based products.

Application \ Signal Sensed	Radiation	Optical	Electrical	Magnetic	Thermal	Chemical	Mechanical
Automotive							•
Consumer products							•
Process industries,mining energy							•
Manufacturing, machinery							•
Medicine, health care							•
HVAC, fire control, security							
Military, aerospace							•
Telecommunications,office equip.							•
Meteorology, environment							•
Other *							

Main Activities

Manufacturers and suppliers of resistive strain gauges and associated products and instrumentation. Strain gauge installation services for both prototypes and high volume production of transducers and sensors incorporating foil strain gauges. Hands-on training in strain gauge technology.

Specific Expertise

Collaboration with transducer designers/manufacturers to provide strain gauges which are the best "match" to the design of the transducer and the spring element material. There are over 50,000 catalogue designs plus custom designs as necessary. Gauges can be supplied to compensate for creep in the transducer spring material or for change of spring modulus with temperature. Careful choice of gauge can maximise performance and minimise cost.

Collaboration Offered

Free consultancy on the choice of gauge to meet the user's requirements, and the cost constraints, designing new gauge patterns if necessary.

Collaboration Sought

The opportunity to assist at the early design stage to maximise the performance/manufacturing cost ratio.

346 Measuretech Ltd

Foxdene House
Ripplesmore Close
Sandhurst Camberley
Surrey GU17 8PE

Tel: +44 344 771114
Fax: +44 344 771114
E-mail:

Affiliation:
Contact: E G Baker, Sales Manager

Organisation information
No of employees 1-10
Turnover (ECU*) <0.5

Sensors technology activities information
No of employees 1-10
Turnover (ECU*) <0.5
*1ECU=DM 1.89,FFr 6.30,£0.78,Lira 1909,Peseta 159.1

Organisation activities
Manufacturer of sensor-based products

Application \ Signal Sensed	Radiation	Optical	Electrical	Magnetic	Thermal	Chemical	Mechanical
Automotive							
Consumer products						•	
Process industries, mining, energy							
Manufacturing, machinery							
Medicine, health care							
HVAC, fire control, security							
Military, aerospace							
Telecommunications,office equip.							
Meteorology, environment						•	
Other *							

Main Activities

Measuretech specialises in on-line and automated analytical systems and services, providing cost effective solutions to gas and liquid analysis and analyser systems to the processing and allied industries. Scope of supply includes on-line chromatography, electrochemical sensor systems and calibration systems applicable to all gas analysis devices.

Specific Expertise

Measurement of pH, dissolved oxygen, redox, conductivity/resistivity, heavy metals, selective ion measurement.

Collaboration Offered

Application expertise in use of sensor technology in control and instrumentation.

347 Mechtric Engineering Ltd

Armstrong Road
Benfleet
Essex SS7 4PW

Tel: +44 268 759131
Fax: +44 268 565578
E-mail:

Affiliation:

Contact: T J Deadman, Marketing Manager
R H Petheram
D J Runchman

Organisation information
No of employees 26-100
Turnover (ECU*) 1-5m

Sensors technology activities information
No of employees 11-25
Turnover (ECU*) 0.5-1m
*1ECU=DM 1.89,FFr 6.30,£0.78,Lira 1909,Peseta 159.1

Organisation activities
Manufacturer of sensor-based products

Application \ Signal Sensed	Radiation	Optical	Electrical	Magnetic	Thermal	Chemical	Mechanical
Automotive							•
Consumer products							
Process industries, mining, energy							•
Manufacturing, machinery							•
Medicine, health care							•
HVAC, fire control, security							
Military, aerospace							•
Telecommunications,office equip.							•
Meteorology, environment							
Other *							

Main Activities

Manufacturers of test instruments in the dynamic force measuring field. Two main products: dynamometers for torque, speed and power measurement up to 25kW and 40,000 rpm; materials testing machines for light compression and tensile testing up to 10,000N force. Both products are precision electronic instruments with full computer interface.

Specific Expertise

Application of force and distance measuring sensors to test equipment for industry.

348 Menvier Hybrids Ltd

Southam Road
Banbury
Oxon OX16 7RX

Tel: +44 295 273939
Fax: +44 295 255325
E-mail:

Affiliation:
Contact: C R Westwood, MD

Organisation information
No of employees 26-100
Turnover (ECU*) 1-5m

Sensors technology activities information
No of employees 26-100
Turnover (ECU*) <0.5m
*1ECU=DM 1.89,FFr 6.30,£0.78,Lira 1909,Peseta 159.1

Organisation activities
Research and Development
Manufacture of finished sensors

Application \ Signal Sensed	Radiation	Optical	Electrical	Magnetic	Thermal	Chemical	Mechanical
Automotive							•
Consumer products							•
Process industries, mining, energy					•		•
Manufacturing, machinery					•		•
Medicine, health care						•	
HVAC, fire control, security							
Military, aerospace							
Telecommunications,office equip.							
Meteorology, environment							•
Other *							

Main Activities

Manufacturers of thick film resistors, hybrid and surface mount assemblies, electronic level sensors, temperature sensors, pressure and chemical sensors.

Specific Expertise

Menvier Hybrids Ltd are specialists in: level sensors (resistive and capacitive fluid level sensing); temperature sensors (thermistor and semiconductor temperature sensing); pressure sensors (resistor on ceramic strain gauge pressure sensing); chemical sensors (chemical sensing using platinum electrode structures).

Collaboration Offered

Low development and tooling costs. Manufacturing, sales and marketing capability.

Collaboration Sought

Research and development input.

349 MKS Instruments

2 St Georges Court
Hanover Business Park
Altringham
Cheshire, WA14

Tel: +44 161 929 5500
Fax: +44 161 929 5511
E-mail:

Affiliation: MKS Inc, USA
Contact: Dr A Dixon, Marketing Mgr

Organisation information
No of employees 1-25
Turnover (ECU*) 1m-5m
Sensor technology activities information
No of employees 11-25
Turnover (ECU*) 1m-5m
*1ECU=DM 1.89,FFr 6.30,£0.78,Lira 1909,Peseta 159.1

Organisation activities
Testing, consulting, design.
Manufacture of finished sensors.
Manufacture of sensor-based products.

Application \ Signal Sensed	Radiation	Optical	Electrical	Magnetic	Thermal	Chemical	Mechanical
Automotive							•
Consumer products							•
Process industries,mining energy							•
Manufacturing, machinery							•
Medicine, health care							•
HVAC, fire control, security							•
Military, aerospace							•
Telecommunications,office equip.							
Meteorology, environment							
Other *							•

* Semi conductor manufacturing

Main Activities

The supply, service and calibration of products for process control. Pressure sensors, gas analysis sensors, gas flow sensors. PID control systems. Vacuum sensors.

Specific Expertise

Manufacture the Baratron(R) capacitance manometer. Expertise in process control. Primarily in semiconductor manufacturing.

350 Moisture Control and Measurement Ltd

Thorp Arch Trading Estate
Wetherby
West Yorkshire LS23 7BJ

Tel: +44 937 588389
Fax: +44 937 587024
E-mail:

Affiliation:
Contact: R Berha, Sales Manager

Organisation information
No of employees 26-100
Turnover (ECU*) 1-5m
Sensors technology activities information
No of employees 26-100
Turnover (ECU*) 1-5m
*1ECU=DM 1.89,FFr 6.30,£0.78,Lira 1909,Peseta 159.1

Organisation activities
Research and Development
Manufacturer of sensor-based products
Manufacture of finished sensors

Application \ Signal Sensed	Radiation	Optical	Electrical	Magnetic	Thermal	Chemical	Mechanical
Automotive						•	
Consumer products							
Process industries, mining, energy						•	
Manufacturing, machinery						•	
Medicine, health care						•	
HVAC, fire control, security						•	
Military, aerospace						•	
Telecommunications,office equip.						•	
Meteorology, environment						•	
Other *							

Main Activities

Manufacture of hygrometers, gas dryers, moisture generators and moisture related control or analytical systems, moisture calibration service.

Specific Expertise

The company manufactures unique hygrometers based on a silicon chip moisture sensor, under the name of Si-Grometer (Registered Trade Mark). The Dewluxe (Registered Trade Mark) model has integral temperature control to the sensor to provide thermal independence. A fully automatic self-calibrating moisture analyser with built-in moisture generator is available, known as the Auto Veri-Si (Registered). All models have the exclusive Push Purge (Registered) sensor stripping system for rapid dry down.

351 Monode Ltd

Celtic House
44 Ballmoor
Buckingham Industrial Park
Buckingham MK18 1RQ

Tel: +44 1280 814171
Fax: +44 1280 814169
E-mail:

Affiliation:
Contact: J Vince, MD

Organisation information
No of employees 1-10
Turnover (ECU*) 0.5-1m

Sensors technology activities information
No of employees 1-10
Turnover (ECU*) <0.5m
*1ECU=DM 1.89,FFr 6.30,£0.78,Lira 1909,Peseta 159.1

Organisation activities
Manufacturer of sensor-based products

Application \ Signal Sensed	Radiation	Optical	Electrical	Magnetic	Thermal	Chemical	Mechanical
Automotive							
Consumer products							
Process industries, mining, energy						•	
Manufacturing, machinery							
Medicine, health care							
HVAC, fire control, security					•	•	
Military, aerospace							
Telecommunications,office equip.			•				
Meteorology, environment							
Other *							

Main Activities

Sub-contract design and manufacture of microprocessor-based measurement and control systems.

Specific Expertise

Design and manufacture of voltage, current and derived parameter measuring equipment.
Design and manufacture of temperature and humidity instrumentation.

352 Monitran Limited

Monitor House
Hazlemere Road
Penn
Bucks HP10 8AD

Tel: +44 1494 816 569
Fax: +44 1494 812 256
E-mail:

Affiliation:
Contact: Mr C Hansford, MD

Organisation information
No of employees 26-100
Turnover (ECU*) 1m-5m
Sensor technology activities information
No of employees 26-100
Turnover (ECU*)
*1ECU=DM 1.89,FFr 6.30,£0.78,Lira 1909,Peseta 159.1

Organisation activities
Research and development on sensors sensor materials or applications
Testing, consulting, design
Manufacture of systems used in the manufacture, testing or development of sensors
Manufacture of finished sensors
Manufacture of sensor-based products

Application \ Signal Sensed	Radiation	Optical	Electrical	Magnetic	Thermal	Chemical	Mechanical
Automotive			●	●			●
Consumer products			●	●			●
Process industries,mining energy			●	●			●
Manufacturing, machinery			●	●			●
Medicine, health care			●	●			●
HVAC, fire control, security			●	●			●
Military, aerospace			●	●			●
Telecommunications,office equip.							
Meteorology environment			●	●			●
Other							

Main Activities

Designers and manufacturers of transducers/sensors for measurement of vibration, acceleration, position, proximity and speed. Instrumentation equipment.

353 MR Sensors Limited

8-10 Clos Menter
Excelsior Business Park
Cardiff CF4 3AT

Tel: +44 222 520022
Fax: +44 222 520178
E-mail:

Affiliation:
Contact: Dr A J Collins, MD
Dr Selvaratnam, Tech & Mark Director

Organisation information
No of employees 26-100
Turnover (ECU*) 1m-5m
Sensor technology activities information
No of employees 1-10
Turnover (ECU*) <0.5m
*1ECU=DM 1.89,FFr 6.30,£0.78,Lira 1909,Peseta 159.1

Organisation activities
Research and development on sensors, sensor materials or applications
Manufacture of sensor-based products.

Application \ Signal Sensed	Radiation	Optical	Electrical	Magnetic	Thermal	Chemical	Mechanical
Automotive							
Consumer products							
Process industries,mining energy							
Manufacturing, machinery							
Medicine, health care							
HVAC, fire control, security				•			
Military, aerospace							
Telecommunications,office equip.				•			
Meteorology, environment							
Other *							

Main Activities

Design, develop, manufacture and sell thin film ferromagnetic magneto resistive sensors and associated components and electronic systems. Developing an interest in other sensors and their applications.

Specific Expertise

Main expertise lies in thin film ferromagnetic magneto resistive sensors. Undertaking basic research in universities on multi layer magnetoresistive sensors.

Collaboration Offered

Interested in collaborating in the application of existing sensors and new sensors and associated systems.

354 NAIS - Matsushita Automation Controls Ltd

Sunrise Parkway
Linford Wood
Milton Keynes MK14 6LF

Tel: +44 908 231555
Fax: +44 908 231599
E-mail:

Affiliation:
Contact: B Narfield, Technical Manager

Organisation information
No of employees >100
Turnover (ECU*) >5m
Sensors technology activities information
No of employees >100
Turnover (ECU*) .5m
*1ECU=DM 1.89,FFr 6.30,£0.78,Lira 1909,Peseta 159.1

Organisation activities
Manufacturer of sensor-based products
Manufacture of finished sensors

Application \ Signal Sensed	Radiation	Optical	Electrical	Magnetic	Thermal	Chemical	Mechanical
Automotive	•						
Consumer products	•						
Process industries, mining, energy							
Manufacturing, machinery	•						
Medicine, health care							
HVAC, fire control, security	•						
Military, aerospace							
Telecommunications,office equip.	•						
Meteorology, environment							
Other *							

Main Activities

Technical sales of photosensors, vision systems, PLCs and automation components.

Specific Expertise

Specification and design of vision inspection system for 100% production and quality visual inspection, including software development of vision systems and concept development for photosensors.

355 Neotronics Ltd

Parsonage Road
Takeley
Bishop's Stortford
Herts CM22 6PU

Tel: +44 1279 870182
Fax: +44 1279 870377
E-mail: jsaffell@aol.com

Affiliation: Neotronics Technology plc
Contact: J Saffell, Group Research Dir
G Escott, Sensor Manuf Mgr
S Collins, Netronics Scientific Dir
H Feldman, Group Mark. Mgr

Organisation information

No of employees 101-1000
Turnover (ECU*) 5m-50m

Sensor technology activities information

No of employees 26-100
Turnover (ECU*) 1m-5m

*1ECU=DM 1.89,FFr 6.30,£0.78,Lira 1909,Peseta 159.1

Organisation activities

Research and development on sensors, sensor materials or applications.
Manufacture of finished sensors.
Manufacture of sensor-based products.

Application \ Signal Sensed	Radiation	Optical	Electrical	Magnetic	Thermal	Chemical	Mechanical
Automotive							
Consumer products							
Process industries,mining energy		●			●	●	
Manufacturing, machinery		●			●	●	
Medicine, health care						●	
HVAC, fire control, security		●			●	●	
Military, aerospace						●	
Telecommunications,office equip.							
Meteorology, environment		●	●		●	●	
Other *							

Main Activities

Neotronics manufacture/design electrochemical cells and pellistors (catalytic sensors) for use in our own gas detection equipment. Another company in the group manufactures 1.8-4.4 μm LEDs and a third company manufactures/develops sensor arrays in our electronic nose equipment. A fourth company, Solonat, uses sensor technology (in-company and external) for indoor air and water quality monitoring.

Specific Expertise

Electrochemical sensors; catalytic sensors; infrared (2-4 μm) LEDs; polymer sensor arrays.

Collaboration Offered

Electrochemical sensor technology; catalysts; polypyrroles and additives; III-V semiconductors.

Collaboration Sought

New catalysts, new electrochemical sensor processes, materials; microheater sensors, polymer film technology.

356 NPL - National Physical Laboratory

Queens Road
Teddington
Middlesex TW11 0LW

Tel: +44 81 9437095
Fax: +44 81 9436755
E-mail:

Affiliation: Govt agency with the Dept of Trade & Industry
Contact: Dr P Woods, Head Environ Stds
Dr R Partridge, Quantum Metrology

Organisation information
No of employees >100
Turnover (ECU*) >5m
Sensors technology activities information
No of employees 26-100
Turnover (ECU*) >5m
*1ECU=DM 1.89,FFr 6.30,£0.78,Lira 1909,Peseta 159.1

Organisation activities
Research and Development

Application \ Signal Sensed	Radiation	Optical	Electrical	Magnetic	Thermal	Chemical	Mechanical
Automotive						•	
Consumer products							
Process industries, mining, energy						•	
Manufacturing, machinery						•	
Medicine, health care						•	
HVAC, fire control, security						•	
Military, aerospace						•	
Telecommunications,office equip.						•	
Meteorology, environment	•					•	
Other *							

Main Activities

The National Physical Laboratory (NPL) is the UK National Standards Laboratory, providing the basis for the National System of Measurements.
The Enviromental Standards Section develops optical remote sensing for atmospheric gas measurements, and reference gas standards which provide the basis for nationally-traceable and internationally-acceptable air pollution and gas emission measurements.

Specific Expertise

NPL has developed systems for remote and long-path measurements of gas concentrations at ground level, in the troposphere and in the stratosphere. These range from portable monitors measuring average gas concentrations to a sophisticated lidar system measuring 3-dimensional gas concentrations over areas up to 3km square. Stratosphere measurements are made with ground-based laser heterodyne spectrometer, using the sun as radiation source.
NPL has proved these systems in extensive field trials at oil refineries, chemical plants, industrial process areas, coke ovens, landfill sites and agricultural sites. The portable monitor is to be offered for sale by a commercial partner, and similar commercialisation is planned for the lidar system.

Collaboration Offered

Expertise in the optical measurement of gases, and in the preparation of gas mixtures for calibration purposes.

Collaboration Sought

Commerical exploitation of NPL developed systems.
Collaboration in the development of more advanced techniques.

357 Nulectrohms

Meppershall
Shefford
Bedfordshire SG17 5LX

Tel: +44 462 813000
Fax: +44 462 816807
E-mail:

Affiliation:
Contact: T Shipton, Sales Manager

Organisation information
No of employees 26-100
Turnover (ECU*) 1-5m
Sensors technology activities information
No of employees 26-100
Turnover (ECU*) 1-5m
*1ECU=DM 1.89,FFr 6.30,£0.78,Lira 1909,Peseta 159.1

Organisation activities
Research and Development
Manufacturer of sensor-based products
Testing, consulting, design
Manufacture of finished sensors

Application \ Signal Sensed	Radiation	Optical	Electrical	Magnetic	Thermal	Chemical	Mechanical
Automotive					•		
Consumer products					•		
Process industries, mining, energy	•				•		
Manufacturing, machinery					•		
Medicine, health care					•		
HVAC, fire control, security					•		
Military, aerospace					•		
Telecommunications,office equip.					•		
Meteorology, environment					•		
Other *							

Main Activities

The company manufactures a wide range of temperature sensors. Electrical instrumentation can be provided to suit by factoring.

Specific Expertise

Temperature sensors: resistance thermometer; thermocouple; thermistor; optical infrared; spot sensors; averaging sensors.

Collaboration Offered

Consultancy. Research. Development of new and prototype sensors. Custom building sensors with copper, nickel and platium wires.

Collaboration Sought

Any temperature or temperature associated product.

358 Oliver Valves Ltd

Parkgate Industrial Estate
Knutsford
Cheshire WA16 8DX

Tel: +44 565 632636
Fax: +44 565 654089
E-mail:

Affiliation:
Contact: R Oliver, MD

Organisation information
No of employees 26-100
Turnover (ECU*) 1-5m
Sensors technology activities information
No of employees 1-10
Turnover (ECU*) <0.5m
*1ECU=DM 1.89,FFr 6.30,£0.78,Lira 1909,Peseta 159.1

Organisation activities
Research and Development

Application \ Signal Sensed	Radiation	Optical	Electrical	Magnetic	Thermal	Chemical	Mechanical
Automotive							
Consumer products							
Process industries, mining, energy							•
Manufacturing, machinery							
Medicine, health care							
HVAC, fire control, security							
Military, aerospace							
Telecommunications,office equip.							
Meteorology, environment							
Other *							

Main Activities

The main activity of Oliver Valves Ltd is in the design and manufacture of instrument valves and manifolds for high pressure applications. The range includes needle, ball, rising plug and bellows sealed patterns and also a range of accessories.

Specific Expertise

Oliver Valves Ltd has been involved in the design of optical fibre sensors for intrinsically safe applications. Several projects have been followed to allow remote sensing of relief valves and gauge protectors, also remote sensing of shaft rotation speed using an optical fibre sensing head. Work has also been carried out on an optical microswitch, again allowing remote sensing of the state of the microswitch. A method for non-contact height measurement has also been achieved. Working models of these designs have been produced and can be seen at the Knutsford factory.

Collaboration Offered

The above designs of optical fibre sensors.

Collaboration Sought

Applications for and help in implementing the above designs.

359 Ometron Limited

Kelvin House
Worsley Bridge Road
London SE26 5BX

Tel: +44 181 461 5555
Fax: +44 818 461 4628
E-mail: 100735,556@compuserve.comm

Affiliation:
Contact: D J Berry, MD
R H Owens, Sales Manager

Organisation information
No of employees 1-25
Turnover (ECU*) 1m-5m
Sensor technology activities information
No of employees 1-10
Turnover (ECU*) 0.5m-1m
*1ECU=DM 1.89,FFr 6.30,£0.78,Lira 1909,Peseta 159.1

Organisation activities
Manufacture of finished sensors
Manufacture of senso-based products

Application \ Signal Sensed	Radiation	Optical	Electrical	Magnetic	Thermal	Chemical	Mechanical
Automotive							•
Consumer products							•
Process industries,mining energy							•
Manufacturing, machinery							•
Medicine, health care							•
HVAC, fire control, security							•
Military, aerospace							•
Telecommunications,office equip.							•
Meteorology environment							
Other (specify)							

Main Activities

Ometron has developed and manufactured for more than 10 years, optical measuring systems for fatigue stress and mechanical vibration.

Specific Expertise

Optical scanning control and software signal processing. Integration of data within the computer aided engineering environment.

Collaboration Offered

Industrial knowledge and experience.
Marketing and development experience
Systems integration.

Collaboration Sought

New types of scanning optical sensors.

360 Oxford Brookes University

Biomedical Instrumentation Group
School of Engineering
Oxford OX3 0BP

Tel: +44 1865 741111
Fax: +44 1865 483637
E-mail:

Affiliation:
Contact: Dr C Mcleod

Organisation information
No of employees 26-100
Turnover (ECU*)
Sensors technology activities information
No of employees 1-10
Turnover (ECU*)
*1ECU=DM 1.89,FFr 6.30,£0.78,Lira 1909,Peseta 159.1

Organisation activities
Research and Development

Application \ Signal Sensed	Radiation	Optical	Electrical	Magnetic	Thermal	Chemical	Mechanical
Automotive							
Consumer products							
Process industries, mining, energy							
Manufacturing, machinery							
Medicine, health care						•	•
HVAC, fire control, security							
Military, aerospace							
Telecommunications,office equip.							
Meteorology, environment							
Other *							

Main Activities

Teaching, research, development and consultancy.

Specific Expertise

Optical, optoelectronic and fibre optic sensors for biomedical applications.
Specific sensors:
immunosensors based on Brewster angle reflectometry;
immunosensors based on surface plasmon resonance;
laser Doppler blood flow meters.

Collaboration Offered

Consultancy: Brewster angle reflectometry, surface plasmon resonance, laser Doppler flow measurement, pulse oximetry.

Collaboration Sought

Brewster angle reflectometry. Ellipsometry. Surface plasmon resonance. Laser Doppler flow measurement. Pulse oximetry.

361 Oxford Instruments Limited

Old Station Way
Eynsham
Oxford OX8 1TL

Tel: +44 865 882855
Fax: +44 865 881567
E-mail:

Affiliation:
Contact: C Rowes, Customer Support Enginer
S Bailey

Organisation information
No of employees >100
Turnover (ECU*) 1-5m
Sensors technology activities information
No of employees 1-10
Turnover (ECU*) <0.5m
*1ECU=DM 1.89,FFr 6.30,£0.78,Lira 1909,Peseta 159.1

Organisation activities
Manufacturer of sensor-based products
Manufacture of finished sensors

Application \ Signal Sensed	Radiation	Optical	Electrical	Magnetic	Thermal	Chemical	Mechanical
Automotive							
Consumer products							
Process industries, mining, energy							
Manufacturing, machinery							
Medicine, health care							
HVAC, fire control, security				●			
Military, aerospace							
Telecommunications,office equip.							
Meteorology, environment							
Other *			●	●	●		

* Scientific research

Main Activities

Activities are the manufacture and supply of a wide range of superconducting magnet and low temperature systems using liquid nitrogen and helium cryogens. The products are used principally in scientific research and medicines and are sold worldwide to universities and government and corporate research establishments.

Specific Expertise

Thermal: a wide range of sensors for applications from above room temperature down to millikelvin temperature for application with all types of cryogenic equipment. Sensor types include various thermocouples, carbon and carbon glass, germanium, silicon diode, platinum, CLTS and CMN thermometry. There is also a sensor calibration service.
Magnetic: Hall effect magnetometers for measurement of magnetic fields 0-20 Tesla with a resolution of 0.01 mT. The probe may be specified for axial or transverse magnetic fields with probe at room temperature or 4.2K.
Electrical: a) Liquid helium level meter using the immersed superconducting wire principle. Analogue and digital readout versions are available.
b) A liquid nitrogen level indicator which operates using the coaxial tube capacitor transducer. Both above level meters reusable with liquid helium and nitrogen cryostats and storage vessels.

Collaboration Offered

Cryospares service - worldwide supply of off-the-shelf spare parts and accessories for cryogenics and superconducting magnet application.

362 PA Consulting Group

Cambridge Laboratory
Melbourn, Royston
Hertfordshire SG8 6DP

Tel: +44 763 261222
Fax: +44 763 260023
E-mail:

Affiliation:
Contact:

Organisation information
No of employees
Turnover (ECU*)
Sensors technology activities information
No of employees
Turnover (ECU*)
*1ECU=DM 1.89,FFr 6.30,£0.78,Lira 1909,Peseta 159.1

Organisation activities
Research and Development
Testing, consulting, design

Application \ Signal Sensed	Radiation	Optical	Electrical	Magnetic	Thermal	Chemical	Mechanical
Automotive						•	•
Consumer products			•	•	•	•	•
Process industries, mining, energy	•		•	•	•	•	•
Manufacturing, machinery	•		•	•	•	•	•
Medicine, health care	•				•	•	•
HVAC, fire control, security			•	•	•		
Military, aerospace	•		•	•	•	•	•
Telecommunications,office equip.	•		•	•			
Meteorology, environment							
Other *							

Main Activities

The technology division of PA Consulting Group undertakes new product and process developments under contract for industrial and government clients.

Specific Expertise

The Applied Sciences Group contains experts in the areas of: chemical sensing; optoelectronics; applied optics and fibre optics; magnetic field/eddy current; spectroscopy. In addition, the Biotechnology Group has expertise in the field of biosensors.

Collaboration Offered

Consultancy. Contract development. Technical audit. Market opportunities surveys.

Collaboration Sought

Optoelectronic/fibre optic sensors. Chemical sensors.

363 Ped Limited

Exning Road
Newmarket
Suffolk CB8 0AX

Tel: +44 638 665161
Fax: +44 638 660718
E-mail:

Affiliation:
Contact: B Gibbs, Dir of Engineering
R Wood, MD

Organisation information
No of employees
Turnover (ECU*)
Sensors technology activities information
No of employees
Turnover (ECU*)
*1ECU=DM 1.89,FFr 6.30,£0.78,Lira 1909,Peseta 159.1

Organisation activities
Research and Development
Manufacturer of sensor-based products
Testing, consulting, design
Manufacture of finished sensors

Application \ Signal Sensed	Radiation	Optical	Electrical	Magnetic	Thermal	Chemical	Mechanical
Automotive					•		
Consumer products							
Process industries, mining, energy							
Manufacturing, machinery					•		
Medicine, health care							
HVAC, fire control, security							
Military, aerospace							
Telecommunications,office equip.							
Meteorology, environment					•		
Other *							

Main Activities

Manufacture of electro-mechanical and electronic components for the white, brown, automotive and professional electronics industry.

Specific Expertise

Heat sensing thermostat for professional applications. Available in tubular form for tank and pipe insertion or in flat form for surface mounting. Stainless steel bodies available for dairy and food processing.

Collaboration Offered

Design and custom build.

Collaboration Sought

Offers of co-operation.

364 Penny & Giles Position Sensors Ltd

15 Airfield Road
Christchurch
Dorset BH23 3TJ

Tel: +44 1202 409409
Fax: +44 1202 409410
E-mail:

Affiliation:
Contact: D Gardener, Sales/Mkting Director

Organisation information
No of employees >100
Turnover (ECU*) 1-5m
Sensors technology activities information
No of employees >100
Turnover (ECU*) 1-5m
*1ECU=DM 1.89,FFr 6.30,£0.78,Lira 1909,Peseta 159.1

Organisation activities
Research and Development
Testing, consulting, design
Manufacture of finished sensors

Application \ Signal Sensed	Radiation	Optical	Electrical	Magnetic	Thermal	Chemical	Mechanical
Automotive							•
Consumer products							
Process industries, mining, energy							•
Manufacturing, machinery							•
Medicine, health care							
HVAC, fire control, security							•
Military, aerospace							•
Telecommunications,office equip.							•
Meteorology, environment							•
Other *							

Main Activities

Specialists in the design and manufacture of high reliability long life sensors and transducers for control and indication of displacement.
Custom design facilities for those applications that cannot be met from the company's own extensive range of stock items of linear, rotary and tilt sensors and joystick controllers.

Specific Expertise

Design and manufacture of displacement sensors and controllers using potentiometric LVDT, inductor or encoding principles.
Large stock of standard items held. Others designed to suit particular applications. Sensor movement includes linear, rotary, tilt, joysticks and tracker balls.

Collaboration Offered

Custom design of specialised displacement sensors and controllers.

365 Pressure Sensors (Marketing) Ltd

19 Cachfield Road
Hurst Pier Point
Sussex BN6 9RP

Tel: +44 1273 835599
Fax: +44 1273 835951
E-mail:

Affiliation:
Contact: G Taylor, Managing Director

Organisation information
No of employees
Turnover (ECU*) 1m-5m

Advanced materials activities information
No of employees 26-100
Turnover (ECU*) 1m-5m
*1ECU=DM 1.89,FFr 6.30,£0.78,Lira 1909,Peseta 159.1

Organisation activities
Research or development on sensors, sensor materials or applications
Testing, consulting, design
Supply of raw materials or semi-finished products used to make sensors
Manufacture of finished sensors

Application \ Signal Sensed	Radiation	Optical	Electrical	Magnetic	Thermal	Chemical	Mechanical
Automotive							●
Consumer products							●
Process industries, mining, energy							●
Manufacturing, machinery							●
Medicine, health care							●
HVAC, fire control, security							●
Military, aerospace							●
Telecommunications,office equip.							●
Meteorology, environment							●
Other *							

Main Activities

Manufacturers of level and pressure transmitters for the full spectrum of industrial application.

Specific Expertise

Liquid level and pressure transmitters for most applications in industry, including marine, chemical, processing food and beverage, oil, water distribution and treatment and heavy industries.

Collaboration Offered

Exclusive distributorship plus local manufacture/assembly.

366 Processor Scientific Instruments Ltd

Lady Lane Industrial Estate
Hadleigh
Ipswich IP7 6BQ

Tel: +44 473 823005
Fax: +44 473 824095
E-mail:

Affiliation:
Contact: F Prosser, Managing Director
D C Jessop, Engineering Manager

Organisation information
No of employees 26-100
Turnover (ECU*) 1m-5m

Advanced materials activities information
No of employees 1-10
Turnover (ECU*) <0.5m
*1ECU=DM 1.89,FFr 6.30,£0.78,Lira 1909,Peseta 159.1

Organisation activities
Research or development on sensors, sensor materials or applications
Testing, consulting, design
Manufacture of finished sensors
Manufacture of sensor-based products

Application \ Signal Sensed	Radiation	Optical	Electrical	Magnetic	Thermal	Chemical	Mechanical
Automotive							•
Consumer products							
Process industries, mining, energy							
Manufacturing, machinery							
Medicine, health care							•
HVAC, fire control, security							•
Military, aerospace			•				•
Telecommunications,office equip.							
Meteorology, environment							•
Other *							•

* Research

Main Activities

Design and manufacture of: marine instruments (including barometers); air flow instruments (including hot wire and rotating cup sensor types); test and measurement instruments; phase meters and function generators. Sub-contract design and manufacture of electronic equipment and assemblies. Approved to 150 9002 and AQAP4. Supply of thermal convertors for AC/DC transfer measurements.

Specific Expertise

Development of sensors using 'hot wire' technique; thermal converters for AC/DC transfer measurements; signal conditioning and equipment, incorporating pressure sensors, rotating cup anemometers and hot wire sensors.

Collaboration Offered

Expertise in flow measurement using the 'hot wire' technique.

367 Porvair Technology Limited

Clywedog Road South
Wrexham Industrial Estate
Wrexham
Clwyd LL13 9XS

Tel: +44 1978 661144
Fax: +44 1978 664554
E-mail:

Affiliation: Part of the Porvair Group (plc)
Contact: P Brennan, Prod Group Mgr
A Hughes, Tech Sales Engineer

Organisation information
No of employees 101-1000
Turnover (ECU*) 5m-50m

Sensor technology activities information
No of employees 11-25
Turnover (ECU*) <0.5m
*1ECU=DM 1.89,FFr 6.30,£0.78,Lira 1909,Peseta 159.1

Organisation activities
Research and development on sensors, sensor materials or applications.
Testing, consulting, design.
Manufacture of sensor-based products.

Application \ Signal Sensed	Radiation	Optical	Electrical	Magnetic	Thermal	Chemical	Mechanical
Automotive							
Consumer products		●	●			●	
Process industries,mining energy		●	●			●	
Manufacturing, machinery		●	●			●	
Medicine, health care		●	●			●	
HVAC, fire control, security		●	●			●	
Military, aerospace		●	●			●	
Telecommunications,office equip.						●	
Meteorology, environment							
Other *							

Main Activities

Porvair Technology's core competency is sintering metals and polymers to produce highly controlled porous media (controlled pore size, porosity, density, strength and permeability) and providing testing services to ensure compliance of these products to the relevant standards e.g. EECS for stainless steel flame arrestors (BS and ISO standard tests).

Specific Expertise

Production of porous stainless steel flame arrestor/gas detection protectors (disc, sheet and mouldings). Manufacture of "total package" design and qualification testing of gas sensor flame sinters, pellistor/body. Sintered polymer moisture barriers for "environmental" detectors, e.g. geological probes.

Collaboration Offered

Novel metal/polymer medias e.g. PEEK or Hastalloy. Design engineering of sinters shape/form and production (for volume manufacture). Prototyping and a wide variety of testing/qualification. Tooling/moulding skills (polymer and metal).

Collaboration Sought

Application/performance details. Open feedback.

368 Pulsar Developments Limited

Spracklen House
Dukes Place
Marlow
Bucks SL7 2QH

Tel: +44 1628 473555/474324
Fax: +44 1628 474325
E-mail:

Affiliation:
Contact: A Hogg, Director
B Murphy, Director

Organisation information
No of employees 1-25
Turnover (ECU*) <1m
Sensor technology activities information
No of employees 1-10
Turnover (ECU*) <0.5m
*1ECU=DM 1.89,FFr 6.30,£0.78,Lira 1909,Peseta 159.1

Organisation activities
Supply of raw materials or semi-finished products used to make sensors.

Application \ Signal Sensed	Radiation	Optical	Electrical	Magnetic	Thermal	Chemical	Mechanical
Automotive			●				
Consumer products			●				
Process industries,mining energy							
Manufacturing, machinery							
Medicine, health care			●				
HVAC, fire control, security			●				
Military, aerospace			●				
Telecommunications,office equip.			●				
Meteorology, environment			●				
Other *							

Main Activities

Supply of finished piezo ceramic bimorphs (as elements, loudspeakers and driver units). Low frequency through to ultrasonic frequencies.

Specific Expertise

Material - lead zirconate titanate wafers mounted on corrugated brass foil or brass wire mesh.

369 Radio Frequency Investigation

Ewhurst Park
Ramsdell
Basingstoke
Hants RG26 5RQ

Tel: +44 1256 855400
Fax: +44 1256 861192
E-mail: sales@rfi.co.uk

Affiliation: RFI, Dunlop, Ayshire
Contact: J J Heath-Caldwell, Comm Mgr

Organisation information
No of employees 26-100
Turnover (ECU*) 5m-50m
Sensor technology activities information
No of employees 1-10
Turnover (ECU*) <0.5m
*1ECU=DM 1.89,FFr 6.30,£0.78,Lira 1909,Peseta 159.1

Organisation activities
Testing, consulting, design.

Application \ Signal Sensed	Radiation	Optical	Electrical	Magnetic	Thermal	Chemical	Mechanical
Automotive	•		•	•			
Consumer products	•		•	•			
Process industries,mining energy	•		•	•			
Manufacturing, machinery	•		•	•			
Medicine, health care	•		•	•			
HVAC, fire control, security	•		•	•			
Military, aerospace							
Telecommunications,office equip.	•		•	•			
Meteorology, environment	•		•	•			
Other *							

Main Activities

Electromagnetic Compatibility Testing Service (EMC)

Specific Expertise

Electromagnetic Compatibility Testing Service (EMC)

370 Rank Taylor Hobson

2 New Star Road
Thurmaston
Leicester

Tel: +44 116 276 3771
Fax: +44 116 246 0579
E-mail:

Affiliation: The Rank Organisation
Contact: G Anderson, Marketing Mgr

Organisation information
No of employees 101-1000
Turnover (ECU*) 5m-50m
Sensor technology activities information
No of employees 26-100
Turnover (ECU*) 1m-5m
*1ECU=DM 1.89,FFr 6.30,£0.78,Lira 1909,Peseta 159.1

Organisation activities
Research and development on sensors, sensor materials or applications.
Testing, consulting, design.
Manufacture of systems used in the manufacture, testing or development of sensors.
Manufacture of finished sensors.
Manufacture of sensor-based products.

Application \ Signal Sensed	Radiation	Optical	Electrical	Magnetic	Thermal	Chemical	Mechanical
Automotive		•	•				•
Consumer products		•	•				•
Process industries,mining energy		•	•				•
Manufacturing, machinery		•	•				•
Medicine, health care		•	•				•
HVAC, fire control, security							
Military, aerospace		•	•				•
Telecommunications,office equip.							
Meteorology, environment							
Other *							

Main Activities

Manufacturer of precision metrology and ultra precision machine tools.

Specific Expertise

Contact stylus gauging for surface and roundness measuring systems.

Collaboration Offered

Discussed/agreed on case by case basis.

371 RDP Electronics Ltd

Grove Street
Heath Town
Wolverhampton WV10 0PY

Tel: +44 902 457512
Fax: +44 902 452000
E-mail:

Affiliation:
Contact: P J Smith, Managing Director

Organisation information
No of employees 26-100
Turnover (ECU*) 1m-5m
Advanced materials activities information
No of employees 26-100
Turnover (ECU*) 1m-5m
*1ECU=DM 1.89,FFr 6.30,£0.78,Lira 1909,Peseta 159.1

Organisation activities
Research or development on sensors, sensor materials or applications
Manufacture of finsihed sensors
Manufacture of sensor-based products

Application \ Signal Sensed	Radiation	Optical	Electrical	Magnetic	Thermal	Chemical	Mechanical
Automotive							•
Consumer products							
Process industries, mining, energy							•
Manufacturing, machinery							•
Medicine, health care							•
HVAC, fire control, security							•
Military, aerospace							•
Telecommunications,office equip.							
Meteorology, environment							•
Other *							

Main Activities

Design, development and manufacture of transducers and associated instrumentation. Parameters covered are: displacement, force, pressure,acceleration, velocity, torque and angular.

Specific Expertise

LVDT/inductive displacement measurement; strain gauge devices; capacitive sensors; digital instrumentation; analogue instrumentation.

372 Redman Controls & Electronics Limited

Malders Lane
Pinkneys Green
Maidenhead
Berks, SL9 6NQ

Tel: +44 1628 30514
Fax: +44 1628 25254
E-mail:

Affiliation:
Contact: D H Redman, MD
C Redman, Electronic Comp Design

Organisation information
No of employees 1-25
Turnover (ECU*) <1m
Sensor technology activities information
No of employees 1-10
Turnover (ECU*) <0.5m
*1ECU=DM 1.89,FFr 6.30,£0.78,Lira 1909,Peseta 159.1

Organisation activities
Testing, consulting, design.
Manufacture of sensor-based products

Application \ Signal Sensed	Radiation	Optical	Electrical	Magnetic	Thermal	Chemical	Mechanical
Automotive		•	•				
Consumer products							
Process industries,mining energy							
Manufacturing, machinery							
Medicine, health care							
HVAC, fire control, security							
Military, aerospace		•	•				
Telecommunications,office equip.							
Meteorology, environment							
Other *		•	•				

*Weld, seam following tube manuf. etc

Main Activities

We specialise in control and automatic welding systems, computer and video controlled automatic welding for the automative and aerospace industry. We developed a video controlled seam following system, and systems to customer specifications.

Specific Expertise

We use video cameras for character recognition to enable us to follow weld seams. We process the video signal to produce a server signal to drive a server system.

373 Remsdaq Limited

Parkway
Deeside Industrial Park
Deeside
Flintshire CH5 2NN

Tel: +44 1244 286495
Fax: +44 1255 286496
E-mail: sales@remsdaq.co.uk

Affiliation: Daqelectronics WC. New Jersey, USA.
Contact: Mr J Warner, Sales & Mkting
Mr B Griffiths, Security Systems
Mr M Parry, Europe & Mid. East

Organisation information
No of employees 101-1000
Turnover (ECU*) 5m-50m
Sensor technology activities information
No of employees 11-25
Turnover (ECU*)
*1ECU=DM 1.89,FFr 6.30,£0.78,Lira 1909,Peseta 159.1

Organisation activities
Manufacture of sensor-based products

Application \ Signal Sensed	Radiation	Optical	Electrical	Magnetic	Thermal	Chemical	Mechanical
Automotive							
Consumer products							
Process industries,mining energy							
Manufacturing, machinery							
Medicine, health care							
HVAC, fire control, security		•					
Military, aerospace							
Telecommunications,office equip.							
Meteorology environment							
Other (specify)							

Main Activities

Remsdaq design and manufacture fibre optic sensor perimeter security systems and a suite of access control, alarm management and integrated control equipment. Other areas of Remsdaq expertise lie in Scada and emergency service command and control systems.

Specific Expertise

The Sabre range of fibre optic sensor perimeter intruder detection systems can be utilised upon fencing, buried underground and within building fabric and operational under water.

Collaboration Offered

Please contact us directly for details.

Collaboration Sought

Please contact us directly for details.

374 Revere Transducers Europe

Midland Shires House
Stoutport Road
Kidderminster DY11 7QN

Tel: +44 1562 67060
Fax: +44 1562 8210540
E-mail:

Affiliation: Revere Transducers Europe - BV
Revere Transducers Inc - USA

Contact: A D Littleton, Regnl Mgr - N Eur
A Freire, Sales Director
A Boucharcat, Regnl Mgr - S Eur
K Heinz Janusen, Regnl Mgr - E Eur
J Kirsten, Product Mgr

Organisation information
No of employees 101-1000
Turnover (ECU*) 5m-50m

Sensor technology activities information
No of employees >100
Turnover (ECU*) >5m
*1ECU=DM 1.89,FFr 6.30,£0.78,Lira 1909,Peseta 159.1

Organisation activities
Research and development on sensors, sensor materials or applications.
Manufacture of finished sensors.

Application \ Signal Sensed	Radiation	Optical	Electrical	Magnetic	Thermal	Chemical	Mechanical
Automotive							
Consumer products			•				
Process industries,mining energy			•				
Manufacturing, machinery			•				
Medicine, health care							
HVAC, fire control, security							
Military, aerospace			•				
Telecommunications,office equip.							
Meteorology, environment			•				
Other *			•				

* Legality Protection

Main Activities

Wheatstone Bridge, Strain Gauge based load cells. Manufacturers.

Specific Expertise

Manufacture and development of high accuracy, high resolution, stainless steel products with legal approval for EC trade and hazardous area approval.

375 Rhopoint Limited

Sensor & Measurement Division
Holland Road
Hurst Green, Oxted
Surrey RH8 9BB

Tel: +44 883 722222
Fax: +44 883 712938
E-mail:

Affiliation:
Contact: D Bradley, Product Engineer

Organisation information
No of employees 26-100
Turnover (ECU*) 1m-5m

Advanced materials activities information
No of employees 11-25
Turnover (ECU*) 0.5m-1m
*1ECU=DM 1.89,FFr 6.30,£0.78,Lira 1909,Peseta 159.1

Organisation activities
Research or development on sensors, sensor materials or applications
Testing, consulting, design
Supply of raw materials or semi-finished products used to make sensors
Manufacture of systems used in the manufacture, testing or development of sensors
Manufacture of finished sensors
Manufacture of sensor-based products

Application \ Signal Sensed	Radiation	Optical	Electrical	Magnetic	Thermal	Chemical	Mechanical
Automotive					•		
Consumer products					•		
Process industries, mining, energy					•		
Manufacturing, machinery					•		
Medicine, health care					•		
HVAC, fire control, security					•		
Military, aerospace					•		
Telecommunications,office equip.					•		
Meteorology, environment					•		
Other *							

Main Activities

Rhopoint Sensor and Measurement currently specialises in: surface temperature sensors; heat flow sensors; temperature transmitters (scaleable and fixed) with intrinsic safety; process control hardware and software; process calibrators; temperature probes, thermowells and connection heads.

Specific Expertise

Type of signal sensed: surface temperature and heat flow vessels; aircraft, engines, components. Principle involved: surface temperature measurement using thermocouple technology and platinum resistance thermometer technology.

Collaboration Offered

Surface temperature sensors.
Sensor-to-computer interface modules.
Heat flow sensors.
Process control hardware/software.
Instruments for simulation, calibration and measurement.

376 Roditi International Corporation Limited

Carrington House
130 Regent Street
London W1R 6BR

Tel: +44 71 4394390
Fax: +44 71 4340896
E-mail:

Affiliation:
Contact: C Kenmore, Electro-optic Devices
J Brendel, Piezoeletric Crystals
J Heighway, Pyroelectrics & Consul tancy

Organisation information
No of employees 11-25
Turnover (ECU*) >5m

Advanced materials activities information
No of employees 1-10
Turnover (ECU*) 1m-5m
*1ECU=DM 1.89,FFr 6.30,£0.78,Lira 1909,Peseta 159.1

Organisation activities
Research or development on sensors, sensor materials or applications
Testing, consulting, design
Supply of raw materials or semi-finished products used to make sensors

Application \ Signal Sensed	Radiation	Optical	Electrical	Magnetic	Thermal	Chemical	Mechanical
Automotive							
Consumer products							
Process industries, mining, energy						•	
Manufacturing, machinery					•		
Medicine, health care							
HVAC, fire control, security							•
Military, aerospace							•
Telecommunications,office equip.							•
Meteorology, environment							
Other *							

Main Activities

Roditi is a development, engineering and sales activity, working on contracts for industrial clients as well as academic institutions. The company works in a range of areas including acoustics, optics and electronics with applications in sensors, signal processing and testing. The company supplies an extensive range of new materials.

Specific Expertise

Roditi specialises in the use of pyroelectric, piezoelectric and electro-optic crystals for use in the detection of heat, motion acceleration and pressure, gas/chemical detection and infrared detection. Infrared detectors/systems/applications. Acoustic devices, SAW and bulk fibre optic units.

Collaboration Offered

Electro-optic devices/crystals. Piezoelectric crystals. Pyroelectrics. Consultancy.

377 Rosemount Limited

Heath Place
Bognor Regis
West Sussex PO22 9SH

Tel: +44 243 863121
Fax: +44 243 867554
E-mail:

Affiliation:
Contact: J S Johnston, Director Research and Development
L Rolfe, Aircraft Instruments
R Migliorini, Process Control Instruments & Systems

Organisation information
No of employees >100
Turnover (ECU*) >5m
Advanced materials activities information
No of employees >100
Turnover (ECU*) >5m
*1ECU=DM 1.89,FFr 6.30,£0.78,Lira 1909,Peseta 159.1

Organisation activities
Research or development on sensors, sensor materials or applications
Manufacture of finished sensors
Manufacture of sensor-based products

Application \ Signal Sensed	Radiation	Optical	Electrical	Magnetic	Thermal	Chemical	Mechanical
Automotive						•	
Consumer products							
Process industries, mining, energy	•				•	•	•
Manufacturing, machinery							
Medicine, health care							
HVAC, fire control, security							
Military, aerospace	•				•		•
Telecommunications,office equip.							
Meteorology, environment						•	
Other *							

Main Activities

Development, manufacture and sale of aircraft and process control sensors, instruments and systems.

Specific Expertise

Aircraft: pitot and pitot static tubes, temperature sensors, ice-detectors etc. Process: sensors for temperature, pressure, flow, level, density, pH, oxygen, toxic and inflammable gases etc.

Collaboration Offered

Aircraft instruments. Process control instruments and systems.

378 Russel pH Limited

Station Road
Auchtermuchty
Fife KY14 7DP Scotland

Tel: +44 1337 828871
Fax: +44 1337 828972
E-mail:

Affiliation:
Contact: Keith Pledger, Plant Director
Richard Newbury, Regional Sales Manager
David Knight, Managing Director

Organisation information
No of employees 26-100
Turnover (ECU*) 1m-5m

Advanced materials activities information
No of employees 26-100
Turnover (ECU*) 1m-5m
*1ECU=DM 1.89,FFr 6.30,£0.78,Lira 1909,Peseta 159.1

Organisation activities
Research or development on sensors, sensor material or applications
Supply of raw materials or semi-finished products used to make sensors
Manufacture of sensor-based products

Application \ Signal Sensed	Radiation	Optical	Electrical	Magnetic	Thermal	Chemical	Mechanical
Automotive						●	
Consumer products						●	
Process industries, mining, energy							
Manufacturing, machinery						●	
Medicine, health care							
HVAC, fire control, security							
Military, aerospace							
Telecommunications,office equip.							
Meteorology, environment						●	
Other *						●	

* Education

Main Activities

Russel pH Ltd develops and manufactures chemical sensors for applications in the automotive, machinery and consumer goods industries.

Collaboration Offered

Product development. Sales Liaison. Co-operation and joint venture.

Collaboration Sought

Co-operation.

379 Sandhurst Scientific Instrument Company

68A High Street
Sandhurst, Berkshire
Surrey GU47 8ED

Tel: +44 252 879700
Fax: +44 252 875156
E-mail:

Affiliation:
Contact: Mr D Jack, Director

Organisation information
No of employees 1-10
Turnover (ECU*) <0.5m
Advanced materials activities information
No of employees 1-10
Turnover (ECU*) <0.5m
*1ECU=DM 1.89,FFr 6.30,£0.78,Lira 1909,Peseta 159.1

Organisation activities
Manufacture of finished sensors
Manufacture of sensor-based products

Application \ Signal Sensed	Radiation	Optical	Electrical	Magnetic	Thermal	Chemical	Mechanical
Automotive							•
Consumer products							
Process industries, mining, energy							•
Manufacturing, machinery							•
Medicine, health care							•
HVAC, fire control, security							
Military, aerospace							•
Telecommunications,office equip.							
Meteorology, environment							•
Other *							

Main Activities

Manufacture of transducers (sensors) for pressure, flow, force, load cells. Also signal conditioning instrumentation.

Specific Expertise

Strain gauge devices for pressure, load. Turbine meters for flow.

380 Scantron Industrial Products Limited

The Monarch Centre
Venture Way
Taunton
Somerset TA2 8DE

Tel: +44 823 333343
Fax: +44 823 333684
E-mail:

Affiliation:
Contact: Lars-H Lindstrand, Mng Director

Organisation information
No of employees 1-10
Turnover (ECU*) 0.5m-1m
Advanced materials activities information
No of employees 1-10
Turnover (ECU*) <0.5m
*1ECU=DM 1.89,FFr 6.30,£0.78,Lira 1909,Peseta 159.1

Organisation activities
Research or development on sensors, sensor materials or applications
Testing, consulting, design
Manufacture of sensor-based products

Application \ Signal Sensed	Radiation	Optical	Electrical	Magnetic	Thermal	Chemical	Mechanical
Automotive		•					
Consumer products							
Process industries, mining, energy							
Manufacturing, machinery		•					
Medicine, health care							
HVAC, fire control, security							
Military, aerospace							
Telecommunications,office equip.							
Meteorology, environment							
Other *		•					

* Machine tool

Main Activities

Systems for non-contact gauging, non-contact laser probes, on-line measurement, laser alignment.

Collaboration Offered

Zygo gauges. Limab lasers. Hamar alignment equipment. Feasibility studies. Alignment exercises.

381 Sensing Devices Limited

97 Tithebarn Road
Southport PR8 6AG

Tel: +44 1704 546161
Fax: +44 1704 546231
E-mail:

Affiliation:
Contact: L A Arnold, Sales Director

Organisation information
No of employees 26-100
Turnover (ECU*) 1m-5m
Sensor technology activities information
No of employees 26-100
Turnover (ECU*) 1m-5m
*1ECU=DM 1.89,FFr 6.30,£0.78,Lira 1909,Peseta 159.1

Organisation activities
Manufacture of finished sensors.

Application \ Signal Sensed	Radiation	Optical	Electrical	Magnetic	Thermal	Chemical	Mechanical
Automotive							
Consumer products							
Process industries,mining energy					•		
Manufacturing, machinery					•		
Medicine, health care					•		
HVAC, fire control, security							
Military, aerospace							
Telecommunications,office equip.							
Meteorology, environment							
Other *							

Main Activities

Manufacture of precision platinum resistance detectors as well as RTD thermometers of all types.

Specific Expertise

Specialists in manufacture of platinum resistance detectors. NAMAS accredited laboratory capable of uncertainties as low as ± 0.001K.

382 Sensor Technology Limited

PO Box 36
Banbury
Oxon OX15 6JB

Tel: +44 1295 730 746
Fax: +44 1295 730 746
E-mail: aji@sensors.demon.co.uk

Affiliation:
Contact: A J Ingham, Sales & Marketing
A Lonsdale, R&D Director
B Lonsdale, Technical Director

Organisation information

No of employees 1-25
Turnover (ECU*) <1m

Sensor technology activities information

No of employees 11-25
Turnover (ECU*) 0.5m-1m

*1ECU=DM 1.89,FFr 6.30,£0.78,Lira 1909,Peseta 159.1

Organisation activities

Research and development on sensors, sensor materials or applications.
Manufacture of finished sensors.
Manufacture of sensor-based products.

Application \ Signal Sensed	Radiation	Optical	Electrical	Magnetic	Thermal	Chemical	Mechanical
Automotive							●
Consumer products							
Process industries,mining energy							●
Manufacturing, machinery							●
Medicine, health care							
HVAC, fire control, security							
Military, aerospace							●
Telecommunications,office equip.							
Meteorology, environment							
Other *							

Main Activities

Sensor Technology Ltd has been involved in the design and manufacture of high performance torque transducers for 25 years. Recently the company originated a radical new technology for rotary torque measurement.

Developed under a DTI Link programme, TORQSENSE Transducers are the world's first low cost non contact rotary torque transducers designed specifically for OEM applications.

Specific Expertise

Traditionally rotary torque is difficult and expensive to measure, but, by using existing technology in a novel way, inexpensive transducers can be produced for situations where monitoring or control of drive mechanisms is required.

The transducers feature high resolution and accuracy, and excellent noise immunity.

Applications include automotive, such as electric power steering, and condition monitoring, where knowledge of torque is critical, as well as monitoring of torque during mixing processes where consistency of mix is required.

Collaboration Offered

A new type of rotary torque transducer.

Collaboration Sought

Applications for torque transducers.

383 Servomex (UK) Limited

Jarvis Brook
Crowborough
East Sussex TN6 3DU

Tel: +44 892 652181
Fax: +44 892 662253
E-mail:

Affiliation:
Contact: Dr R Kocache, Director, Technology

Organisation information
No of employees >100
Turnover (ECU*) >5m

Advanced materials activities information
No of employees 26-100
Turnover (ECU*) 1m-5m
*1ECU=DM 1.89,FFr 6.30,£0.78,Lira 1909,Peseta 159.1

Organisation activities
Research or development on sensors, sensor materials or applications
Testing, consulting, design
Manufacture of finished sensors
Manufacture of sensor-based products

Application \ Signal Sensed	Radiation	Optical	Electrical	Magnetic	Thermal	Chemical	Mechanical
Automotive							
Consumer products							
Process industries, mining, energy						●	
Manufacturing, machinery							
Medicine, health care						●	
HVAC, fire control, security						●	
Military, aerospace						●	
Telecommunications,office equip.							
Meteorology, environment							
Other *							

Main Activities

Servomex designs, develops, manufactures and markets worldwide, process analytical instruments, mainly for gas analysis.
The Sensors and Transducers Divsion concentrates on the basic measurement and develops sensors and transducers for use in Servomex Instruments and for certain targetted OEM applications.

Specific Expertise

Servomex has been known for decades for its paramagnetic oxygen sensors, upon which oxygen analyses are based. The company also has a range of zirconia oxygen sensors for different applications.
Infrared gas analysis is another area of expertise, and a range of infrared sensors as well as a range of catalytic, CO equivalent and flow sensors have been developed.
Thick film semiconductor gas sensors is another area of interest.

Collaboration Offered

Research and development for turning prototypes into manufacturable products.
Worldwide marketing organisation.

Collaboration Sought

Methods and devices for gas analysis that are in the feasibility stage.

384 Sheffield Hallam University

City Campus
Pond Street
Sheffield S1 1WB

Tel: +44 1142 533476
Fax: +44 1142 533306
E-mail: a.k.ray@shv.ac.uk

Affiliation:
Contact: Prof A K Ray, Head of Electronics

Organisation information
No of employees 1-25
Turnover (ECU*) <1m

Sensor technology activities information
No of employees 1-10
Turnover (ECU*) <0.5m
*1ECU=DM 1.89,FFr 6.30,£0.78,Lira 1909,Peseta 159.1

Organisation activities
Research and development on sensors, sensor materials or applications.
Testing, consulting, design.
Manufacture of systems used in the manufacture, testing or development of sensors.

Application \ Signal Sensed	Radiation	Optical	Electrical	Magnetic	Thermal	Chemical	Mechanical
Automotive		●	●				
Consumer products		●	●				
Process industries,mining energy							
Manufacturing, machinery							
Medicine, health care		●	●		●	●	
HVAC, fire control, security							
Military, aerospace							
Telecommunications,office equip.							
Meteorology, environment							
Other *							

Main Activities

An HEI establishment, engaged in research under the DTI and EPSRC support.

Specific Expertise

Sol-gel oxide materials; organic semiconductors such as phthalocyanine

Collaboration Offered

EC grant, pre competitive research.

Collaboration Sought

On site testing facilities.

385 Sheffield Hallam University

Environmental Research Centre
City Campus
Sheffield S1 1WB

Tel: +44 1142 533032
Fax: +44 1142 534262
E-mail:

Affiliation:
Contact: Prof M Cooke, Head of Chem.
Prof J Yarwood, Dir Materials Research Institute

Organisation information
No of employees >1000
Turnover (ECU*) >50m

Sensor technology activities information
No of employees 1-10
Turnover (ECU*) <0.5m
*1ECU=DM 1.89,FFr 6.30,£0.78,Lira 1909,Peseta 159.1

Organisation activities
Research and development on sensors, sensor materials or applications.

Application \ Signal Sensed	Radiation	Optical	Electrical	Magnetic	Thermal	Chemical	Mechanical
Automotive							
Consumer products							
Process industries,mining energy							
Manufacturing, machinery							
Medicine, health care							
HVAC, fire control, security							
Military, aerospace			•				
Telecommunications,office equip.							
Meteorology, environment							
Other *						•	

* Food

Main Activities
Higher education teaching at undergraduate and post graduate level. Research at postgraduate level. Consultancy across a wide range including materials research, chemical analysis, energy and waste management, environmental issues.

Specific Expertise
Fundamental science of structure v response, conducting polymer sensors, photochromic sensors, morphology, stability selectivity, sensory arrays based on polypyrole and poly aniline, neural networks.

Collaboration Offered
Full research facilities from fabrication to ß test and class 100 clean room facility.

Collaboration Sought
Proposals for collaborative research and development, commercial partnerships.

386 Silsoe Research Institute

Wrest Park
Silsoe
Bedford MK45 4HS

Tel: +44 15 25 86 00 00
Fax: +44 15 25 86 01 56
E-mail: david.smith@bbsrc.ac.uk

Affiliation: Biotechnology & Biological Serv ices Research Council
Contact: Dr D L Smith, Sen Business Mgr

Organisation information
No of employees 101-1000
Turnover (ECU*) 5m-50m

Sensor technology activities information
No of employees 1-10
Turnover (ECU*) <0.5m
*1ECU=DM 1.89,FFr 6.30,£0.78,Lira 1909,Peseta 159.1

Organisation activities
Research and development on sensors, sensor materials or applications

Application \ Signal Sensed	Radiation	Optical	Electrical	Magnetic	Thermal	Chemical	Mechanical
Automotive							
Consumer products							
Process industries,mining energy							
Manufacturing, machinery							
Medicine, health care							
HVAC, fire control, security							
Military, aerospace							
Telecommunications,office equip.							
Meteorology environment	•	•	•		•		•
Other							

Main Activities

Silsoe Research Institute aims to benefit biology-based industries by providing scientific understanding and engineering innovation through basic, strategic and applied research.

Specific Expertise

- Sensing for control of robots handling delicate biological materials
- Control and guidance of autonomous field vehicles
- Sensing state of animal health and development

Collaboration Offered

Engineering research (physics, mathematical modelling, simulation, experimental verification)

Collaboration Sought

Biological science (basic and applied), applications, commercial development.

387 Simrad Hydrospace Limited

Woolmer Way
Bordon
Hampshire GU35 9QE

Tel: +44 1420 983200
Fax: +44 1420 489073
E-mail:

Affiliation:
Contact: D Riches, General Manager

Organisation information
No of employees 11-25
Turnover (ECU*) 0.5m-1m

Advanced materials activities information
No of employees 1-10
Turnover (ECU*) <0.5m
*1ECU=DM 1.89,FFr 6.30,£0.78,Lira 1909,Peseta 159.1

Organisation activities
Manufacture of sensor-based products

Application \ Signal Sensed	Radiation	Optical	Electrical	Magnetic	Thermal	Chemical	Mechanical
Automotive							
Consumer products							
Process industries, mining, energy							
Manufacturing, machinery							
Medicine, health care							
HVAC, fire control, security							
Military, aerospace							
Telecommunications,office equip.							
Meteorology, environment							
Other *					•		•

* Underwater

Main Activities

Simrad Hydrospace specialises in the design, development and manufacture of underwater electronic modules controlled and monitored by through-water acoustics. These modules frequently contain sensors and are used in the remote monitoring of fishing nets, subsea well heads, ROVs amd acoustic fish fences.

Specific Expertise

Incorporation of sensors into underwater electronic packages which are then acoustically monitored and interrogated. The sensor units are battery powered and data is transferred to the surface via an underwater acoustic link.

Collaboration Offered

Fully proven methods of low to high data rate acoustic links for use in remote sensing of underwater parameters.
Applicable to sea or freshwater environments.

Collaboration Sought

Market and sensor information and expertise which could find application in an underwater environment, eg. remote acoustic monitoring of pollution sensors.

388 Sira Limited

Electro-Optics Division
South Hill
Chiselhurst
Kent BR7 5EH

Tel: +44 181 467 2636
Fax: +44 181 467 6515
E-mail: marketing@siraeo.co.uk

Affiliation:
Contact: John Hother, Marketing Manager

Organisation information
No of employees 101-1000
Turnover (ECU*) 5m-50m
Sensor technology activities information
No of employees 26-100
Turnover (ECU*) 1m-5m
*1ECU=DM 1.89,FFr 6.30,£0.78,Lira 1909,Peseta 159.1

Organisation activities
Research and development on sensors sensor materials or applications
Testing, consulting, design

Application \ Signal Sensed	Radiation	Optical	Electrical	Magnetic	Thermal	Chemical	Mechanical
Automotive		●					
Consumer products		●					
Process industries,mining energy		●					
Manufacturing, machinery		●					
Medicine, health care							
HVAC, fire control, security		●					
Military, aerospace		●					
Telecommunications,office equip.		●					
Meteorology environment		●					
Other							

Main Activities

Optical and electro-optic techniques for sensing a wide variety of parameters, e.g. position, temperature and pressure, force vibration, gas and liquid chemistry.

Collaboration Offered

We run and manage collaboratively funded research projects.

389 Spectrol Reliance Limited

Garrard Way
Swindon
Wiltshire SN3 3HY

Tel: +44 793 521351
Fax: +44 793 539255
E-mail:

Affiliation:
Contact: C Harris, Engineering Director
V Slawinski, Commercial Contracts
R Honeybun, Sensors

Organisation information

No of employees >100
Turnover (ECU*) >5m

Advanced materials activities information

No of employees 11-25
Turnover (ECU*) 1m-5m
*1ECU=DM 1.89,FFr 6.30,£0.78,Lira 1909,Peseta 159.1

Organisation activities

Research or development of sensors, sensor materials or applications
Testing, consulting, design
Manufacture of finished sensors
Manufacture of Sensor-based products

Application \ Signal Sensed	Radiation	Optical	Electrical	Magnetic	Thermal	Chemical	Mechanical
Automotive			•				•
Consumer products							
Process industries, mining, energy			•				•
Manufacturing, machinery			•				•
Medicine, health care							
HVAC, fire control, security							
Military, aerospace							
Telecommunications,office equip.							
Meteorology, environment			•				•
Other *							

Main Activities

Design and manufacture sensors using resistive and non-contacting technologies, utilising the most practical state-of-the-art technology. Spectrol is backed by the top quality service and reputation of U.T.C. The Spectrol Group has the total resources to meet the extensive sensor needs of the market: experienced engineers, dedicated production and quality personnel. Spectrol can put its full capabilities as a sensor system supplier to work for you now on a full partnership basis.

Specific Expertise

Potentiometric sensors using conductive plastic technology, applied in throttle position, exhaust gas recirculation (EGR), fuel tanks contents, life tested at 100°C exceeding 1 billion dither and 3 million full strokes. Rotational speed - two types of Eddy current sensors: a flat end with a detector coil functioning by measuring the modulation of RF in the coil; vane type which generates a high frequency magnetic field which induces Eddy currents in any conducting materials placed near the coils.
Current applications include ABS systems, automatic gear boxes.

Collaboration Offered

Design/development, endurance, reliability testing. Commercial. Potentiometric sensors, non-contacting sensors.

Collaboration Sought

Co-operation in joint ventures and development of systems.

390 Status Instruments Limited

Green Lane
Tewkesbury
Gloucester GL20 8HE

Tel: +44 1684 850985
Fax: +44 1684 293746
E-mail:

Affiliation:

Contact: N MacDonald, Marketing Dir
M White
J Rogers, Engineering Manager

Organisation information
No of employees
Turnover (ECU*)
Sensors technology activities information
No of employees
Turnover (ECU*)
*1ECU=DM 1.89,FFr 6.30,£0.78,Lira 1909,Peseta 159.1

Organisation activities
Manufacturer of sensor-based products
Manufacture of finished sensors

Application \ Signal Sensed	Radiation	Optical	Electrical	Magnetic	Thermal	Chemical	Mechanical
Automotive	•		•		•		
Consumer products			•		•		
Process industries, mining, energy	•		•		•		
Manufacturing, machinery	•		•		•		
Medicine, health care			•		•		
HVAC, fire control, security	•		•		•		
Military, aerospace							
Telecommunications,office equip.	•		•		•		
Meteorology, environment	•		•		•		
Other *							

Main Activities

The application of temperature, pressure and humidity measuring sensors to the necessary signal conditioning and data acquisition requirements of variable process measurement applications.

Specific Expertise

Temperature: thermocouples, platinum resistance thermometers. Humidity: capacitive sensor, wet and dry bulb sensor. Pressure: thin foil, strain gauge.

Collaboration Offered

Temperature, humidity, pressure sensing. Signal conditioning. Indication. Data acquisition (PC-based).

391 Strain Measurement Devices

Sharon Road
Blenheim Industrial Park
Bury St Edmunds
Suffolk IP33 3TZ

Tel: +44 1284 768337
Fax: +44 1284 763225
E-mail: smd@anglianet.co.uk

Affiliation:
Contact: W Mulholland, MD

Organisation information
No of employees 1-25
Turnover (ECU*) <1m
Sensor technology activities information
No of employees 1-10
Turnover (ECU*)
*1ECU=DM 1.89,FFr 6.30,£0.78,Lira 1909,Peseta 159.1

Organisation activities
Supply of raw materials or semi-finished products used to make sensors.

Application \ Signal Sensed	Radiation	Optical	Electrical	Magnetic	Thermal	Chemical	Mechanical
Automotive							
Consumer products							
Process industries,mining energy	•						
Manufacturing, machinery							
Medicine, health care	•						
HVAC, fire control, security							
Military, aerospace	•						
Telecommunications,office equip.							
Meteorology, environment							
Other *							

Main Activities

Applying thin film strain gauges to stainless steel beams and pressure diaphragms.

392 Strathclyde University

Dept of Electronic and Electrical Engineering
204 George Street
Glasgow GI 1XW

Tel: +44 14 15 52 44 00 x2543
Fax: +44 14 15 53 19 55
E-mail: b.cushaw@eee.strath.ac.uk

Affiliation:
Contact: Prof B Culshaw, Head Optoelec Div

Organisation information
No of employees >1000
Turnover (ECU*) >50m
Sensors technology activities information
No of employees 26-100
Turnover (ECU*) 1m-5m
*1ECU=DM 1.89,FFr 6.30,£0.78,Lira 1909,Peseta 159.1

Organisation activities
Research and Development
Testing, consulting, design

Application \ Signal Sensed	Radiation	Optical	Electrical	Magnetic	Thermal	Chemical	Mechanical
Automotive		●			●	●	●
Consumer products		●					
Process industries, mining, energy		●			●	●	●
Manufacturing, machinery		●			●		●
Medicine, health care		●				●	
HVAC, fire control, security		●			●	●	
Military, aerospace		●			●	●	●
Telecommunications,office equip.							
Meteorology, environment		●				●	●
Other *							

Main Activities

The Optoelectronics Division at Strathclyde specialise in the application of optical fibre technology in sensing and measurement. This also extends to implementation of structural interrogation systems and "smart structure" technologies, particularly those using optical fibre sensors. The available expertise ranges from optical component design and realisation through to system implementation, signal processing and extraction and design and test.

Specific Expertise

Principal areas of expertise include the design of guided wave and micro-optic sensor architectures relating to mechanical and chemical measurements, specialised spectral measuremnt systems designed to inerface to sensors. Signal processing data acquisition and system compensation algrethms to compensate for thermal and other environmental effects. Materials inerfaces and the impact on guided waver sensor design. Detection systems using optical fibres for static and dynamic strain, ultrasound, temperature fields, chemical fields, gas sensing. Sensor system design including multipoint architectures and distributed systems. Distributed (i.e. as a function of position along an optical fibre) measurement systems for strain, temperature and chemical detection (particular water and pH).

Collaboration Offered

Access to very substantial skill base in guided wave optical systems and the associated electronic and mechanical techniques required to implement a complete sensor system. A substantial research team internationally recognised for continued and significant contributions in guided wave optical measurement systems and techniques.

Collaboration Sought

Complementary expertise and in particular routes to exploitation via technology development programmes and appropriate licensing agreements.

393 Surrey Electro-Optics Ltd

Surrey House
London Road
Staines
Middlesex TW18 4HR

Tel: +44 784 461393
Fax: +44 784 469868
E-mail:

Affiliation: Surrey Group
Contact: Lt-Col J W Molyneux, Chairman
P K Howland, Marketing Services

Organisation information

No of employees	26-100
Turnover (ECU*)	1-5

Sensors technology activities information

No of employees	1-10
Turnover (ECU*)	<0.5m

*1ECU=DM 1.89,FFr 6.30,£0.78,Lira 1909,Peseta 159.1

Organisation activities
Research and Development
Manufacturer of sensor-based products
Testing, consulting, design
Manufacture of finished sensors

Application \ Signal Sensed	Radiation	Optical	Electrical	Magnetic	Thermal	Chemical	Mechanical
Automotive	●						
Consumer products	●						
Process industries, mining, energy	●						
Manufacturing, machinery	●						
Medicine, health care							
HVAC, fire control, security	●						
Military, aerospace	●						
Telecommunications,office equip.	●						
Meteorology, environment	●						
Other *							

Main Activities

Surrey Electro-Optics Ltd is one of 9 companies in the Surrey Group, which was formed 25 years ago. The main activity is optoelectronic components and electro-optic equipment manufacture.

Specific Expertise

Sensors and other photoelectric detectors are made in silicon, cadmium sulphite, cadmium selenide. Selenium technology with a wide range of encapsulation techniques provide environmental protection.

394 The Technology Partnership Plc

Melbourn Science Park
Cambridge Road
Melbourne, Royston
Hertfordshire SG8 6EE

Tel: +44 763 262626
Fax: +44 763 261582
E-mail:

Affiliation:
Contact: Dr J Teape
V Humberstone

Organisation information
No of employees 26-100
Turnover (ECU*) >5m
Sensors technology activities information
No of employees 11-25
Turnover (ECU*) 1-5m
*1ECU=DM 1.89,FFr 6.30,£0.78,Lira 1909,Peseta 159.1

Organisation activities
Research and Development
Manufacturer of sensor-based products
Testing, consulting, design
Manufacture of finished sensors

Application \ Signal Sensed	Radiation	Optical	Electrical	Magnetic	Thermal	Chemical	Mechanical
Automotive							
Consumer products							
Process industries, mining, energy							
Manufacturing, machinery							
Medicine, health care							
HVAC, fire control, security							
Military, aerospace							
Telecommunications,office equip.							
Meteorology, environment							
Other *							

Main Activities

The Technology Partnership is a major European provider of advanced technological services, working closely with many of the most successful companies worldwide.
Activities encompass all aspects of the development and manufacture of a wide range of sensor types including: optical, chemical, acoustic, electrical and magnetic, both as individual devices and as complete measurement and detection systems.
Technology Partnership personnel have been involved in the development of sensors for a variety of applications such as displacement measurement, mass flow, gas detection and pressure sensing. In addition to its hardware related activities, the Technology Partnership also provides expert advice to industry in devising and implementing their strategic plans.

Specific Expertise

Sensor Engineering: is concerned with the solution of practical problems associated with the design, manufacture and application of sensors and transducers of all types.
Sensor Development: in which advances in materials and processing technology are exploited to produce new types of sensors to meet application - specific requirements in the automotive, industrial and healthcare markets.

Collaboration Offered

Sensor engineering.Strategy and consultancy.Sensor Development.

Collaboration Sought

High resolution displacement sensors.Diffraction optics.Integrated sensors.Biosensors.
Diagnostics/healthcare.

395 Thermal Detection Limited

1-3 Mary Street
Stockton-on-Tees
Cleveland TS18 4AN

Tel: +44 642 602878
Fax: +44 642 618307
E-mail:

Affiliation: Cardfarth Control Engineering
Contact: N J Mitchell, MD

Organisation information
No of employees 1-10
Turnover (ECU*) <0.5m
Sensors technology activities information
No of employees 1-10
Turnover (ECU*) <0.5m
*1ECU=DM 1.89,FFr 6.30,£0.78,Lira 1909,Peseta 159.1

Organisation activities
Research and Development
Manufacturer of sensor-based products
Testing, consulting, design
Manufacture of finished sensors

Application \ Signal Sensed	Radiation	Optical	Electrical	Magnetic	Thermal	Chemical	Mechanical
Automotive					•		
Consumer products							
Process industries, mining, energy					•		
Manufacturing, machinery					•		
Medicine, health care					•		
HVAC, fire control, security					•		
Military, aerospace							
Telecommunications,office equip.							
Meteorology, environment					•		
Other *							

Main Activities

The development and design of temperature related instrumentation for industrial applications in general and the process industries in particular.

Specific Expertise

The design of a range of temperature sensor assemblies for use with pharmaceutical autoclaves that offer an enhanced reliability and operational life. Development of a multiple sensor assembly arrangement for the meat industry, to identify the temperature gradient in the flesh and muscle of recently slaughtered animals and to transmit this information to a data logger. The collaborative design and development of a temperature sensor assembly with integral short range radio transmission of the measured signal for application areas such as tunnel ovens, curing/cooking ovens and all types of rotating equipment.

Collaboration Offered

25 years of active experience in the process industry and a temperature calibration laboratory with direct traceability to the National Physical Laboratory.

Collaboration Sought

Open to discussion.

396 Thermo Electric International Limited

17E Upper Field Road
Dolphin Park
Eurolink, Sittingbourne
Kent ME10 3UP

Tel: +44 1795 479481
Fax: +44 1795 422174
E-mail:

Affiliation: Thermo Electric International (Inc) and in Holland, Belgium, France
Contact: A Flint, General Manager

Organisation information
No of employees 101-1000
Turnover (ECU*) 5m-50m
Sensor technology activities information
No of employees >100
Turnover (ECU*) >5m
*1ECU=DM 1.89,FFr 6.30,£0.78,Lira 1909,Peseta 159.1

Organisation activities
Research and development on sensors, sensor materials or applications.
Testing, consulting, design.
Supply of raw materials or semi-finished products used to make sensors.
Manufacture of systems used in the manufacture, testing or development of sensors.
Manufacture of finished sensors.
Manufacture of sensor-based products.

Application \ Signal Sensed	Radiation	Optical	Electrical	Magnetic	Thermal	Chemical	Mechanical
Automotive					•		
Consumer products					•		
Process industries,mining energy					•		
Manufacturing, machinery					•		
Medicine, health care					•		
HVAC, fire control, security					•		
Military, aerospace					•		
Telecommunications,office equip.							
Meteorology, environment					•		
Other *							

Main Activities

Manufacturers of temperature measurement sensors from 25mm dia to "multicouple" industrial designs. Additionally manufacturers of thermocouple wire and cable, instrumentation and connection systems. Fully NKO approved calibration laboratory. ISO 9000 approved.

Specific Expertise

Comprehensive range of products and application oriented expertise. Testing and calibration facilities.

Collaboration Offered

International infra structure. Wide customer base.

Collaboration Sought

Some areas of part supply and new sales area representation.

397 Thermoteknix Systems Limited

Mount Pleasant House
Mount Pleasant
Cambridge CB3 0RN

Tel: +44 1223 500 777
Fax: +44 1223 500 888
E-mail: info@thermoteknix.co.uk

Affiliation:
Contact: Dr Richard Salisbury, MD

Organisation information

No of employees 26-100
Turnover (ECU*) 1m-5m

Sensor technology activities information

No of employees 11-25
Turnover (ECU*) 0.5m-1m

*1ECU=DM 1.89,FFr 6.30,£0.78,Lira 1909,Peseta 159.1

Organisation activities

Manufacturer of systems
Manufacturer of sensor-based products

Application \ Signal Sensed	Radiation	Optical	Electrical	Magnetic	Thermal	Chemical	Mechanical
Automotive					●		
Consumer products							
Process industries,mining energy					●		
Manufacturing, machinery							
Medicine, health care							
HVAC, fire control, security							
Military, aerospace					●		
Telecommunications,office equip.							
Meteorology environment							
Other (specify)							

Main Activities

Manufacturer of infrared radiometers and associated electronics and software signal processing. Manufacturer of on-line process control systems using infrared linescanning and imaging technology.

Specific Expertise

Incorporation of current infrared detector technology into commercial products.

Collaboration Offered

Infrared sensor expertise together with advanced micro-electronics and software.

398 Tilt Measurement Limited

Horizon House
London Road
Baldock
Herts SG7 6NG

Tel: +44 1462 89456
Fax: +44 1462 895990
E-mail:

Affiliation:
Contact: A M B Shaw, Director

Organisation information
No of employees 1-25
Turnover (ECU*) <1m
Sensor technology activities information
No of employees 1-10
Turnover (ECU*) 0.5m-1m
*1ECU=DM 1.89,FFr 6.30,£0.78,Lira 1909,Peseta 159.1

Organisation activities
Research and development on sensors, sensors materials or applications.
Testing, consulting, design.
Manufacture of finished sensors.
Manufacture of sensor-based products.

Application \ Signal Sensed	Radiation	Optical	Electrical	Magnetic	Thermal	Chemical	Mechanical
Automotive			•	•	•		•
Consumer products							
Process industries,mining energy			•	•	•		
Manufacturing, machinery			•	•	•		•
Medicine, health care							
HVAC, fire control, security			•				•
Military, aerospace			•	•	•		•
Telecommunications,office equip.			•	•	•		•
Meteorology, environment							
Other *							

Main Activities

Development, design and manufacture of specialised sensors for mechanical electrical and other physical properties and the electronic systems for collecting, processing and displaying their data.

Specific Expertise

Very high accuracy tilt sensors. Very high accuracy capacitance sensors. Precision electronic measure systems. Signal processing hardware and software.

Collaboration Offered

Expert system analysis. Innovative design techniques. Development facilities for hardware and software.

Collaboration Sought

Market oriented problems to solve.

399 Titan Enterprises Ltd

Higher Kingsbury
Milborne Port
Sherborne
Dorset DT9 5EB

Tel: +44 963 251144
Fax: +44 963 251405
E-mail:

Affiliation:
Contact: J T Forster, MD

Organisation information
No of employees 1-10
Turnover (ECU*) <0.5m
Sensors technology activities information
No of employees 1-10
Turnover (ECU*) <0.5m
*1ECU=DM 1.89,FFr 6.30,£0.78,Lira 1909,Peseta 159.1

Organisation activities
Research and Development
Testing, consulting, design
Manufacture of finished sensors

Application \ Signal Sensed	Radiation	Optical	Electrical	Magnetic	Thermal	Chemical	Mechanical
Automotive							•
Consumer products							•
Process industries, mining, energy							•
Manufacturing, machinery							•
Medicine, health care							•
HVAC, fire control, security							
Military, aerospace							•
Telecommunications,office equip.							
Meteorology, environment							•
Other *							

Main Activities

Design and manufacture of chemically resistant flowmeters. Titan Enterprises also offers design and consultancy services for OEM applications.

Specific Expertise

Manufacturers of precision flowmeters which are suited to a wide range of applications. These include both translucent and opaque liquids, all acids and most alkalis and solvents. The company's flowmeters are used in both volume handling and laboratory situations and can be customised to specific needs. Flow rates from 0.001 L/min to 5000 L/min. New on the market is the high pressure range which operates up to 200 Bar and 125° C. While the quality of Titan Enterprises' products is paramount, prices are highly competitive.

400 Trolex Limited

Newby Road
Hazel Grove
Stockport
Cheshire SK7 5DY

Tel: +44 61 4831435
Fax: +44 61 4835556
E-mail:

Affiliation:
Contact: A Piera-Jones, Marketing Manager
P Dawson, OP Director

Organisation information
No of employees 26-100
Turnover (ECU*) >5m
Sensors technology activities information
No of employees 11-25
Turnover (ECU*) 1-5m
*1ECU=DM 1.89,FFr 6.30,£0.78,Lira 1909,Peseta 159.1

Organisation activities
Research and Development
Manufacturer of sensor-based products
Testing, consulting, design
Manufacture of finished sensors

Application \ Signal Sensed	Radiation	Optical	Electrical	Magnetic	Thermal	Chemical	Mechanical
Automotive	•				•	•	•
Consumer products							
Process industries, mining, energy	•				•	•	•
Manufacturing, machinery	•				•	•	•
Medicine, health care							
HVAC, fire control, security	•				•	•	•
Military, aerospace							
Telecommunications,office equip.							
Meteorology, environment	•				•	•	•
Other *							

Main Activities

Trolex designs, manufactures, and installs a wide range of sensors and control systems. Sensors are available for temperature, pressure, flow, level, vibration and gas in both industrial versions and those certified for use in hazardous areas. Special sensors and modifications to the company's standard sensors for specific applications are supplied.

Specific Expertise

Temperature sensors based on RTD, semiconductor thermocouple and giving 4-20mA or voltage outputs. Pressure sensors based on thin film technology having mV and 4-20mA outputs. Level sensors based on piezo accelerometers together with all-signal conditioning; specialist support systems analysis available. Vibration sensors based on catalytic, electro-chemical cell, and fuel cell technology having mV and mA output signals.

Collaboration Offered

Sensor design and applications. Expert vibration analysis systems. Process control and monitoring systems.

Collaboration Sought

All outputs and sensors particularly latest technology.

401 Ultrasonic Sciences Ltd

26 Albert Street
Fleet
Hants GU13 9RL

Tel: +44 1252 812010
Fax: +44 1252 812139
E-mail:

Affiliation:
Contact: C Gartside, Sales Director

Organisation information
No of employees 1-25
Turnover (ECU*) <1m
Sensor technology activities information
No of employees 1-10
Turnover (ECU*) <0.5m
*1ECU=DM 1.89,FFr 6.30,£0.78,Lira 1909,Peseta 159.1

Organisation activities
Research and development on sensors, sensor materials or applications.
Manufacture of finished sensors.
Manufacture of sensor-based products

Application \ Signal Sensed	Radiation	Optical	Electrical	Magnetic	Thermal	Chemical	Mechanical
Automotive			●				●
Consumer products							
Process industries,mining energy							
Manufacturing, machinery							
Medicine, health care			●				
HVAC, fire control, security							
Military, aerospace			●				●
Telecommunications,office equip.							
Meteorology, environment							
Other *							

Main Activities
Design, manufacture and application of ultrasonic nondestructive testing sensors, instruments and systems.

Specific Expertise
Manufacture and development of ultrasonic transducers,especially polymer (PVdF).

Collaboration Offered
Specific applications experience and development of instrumentation systems for ultrasonic sensors.

402 University of Aston

Speciality Materials Research Group
Aston Triange
Birmingham B4 7ET

Tel: +44 21 3593611
Fax: +44 21 3597358
E-mail:

Affiliation:
Contact: Prof B J Tighe

Organisation information
No of employees
Turnover (ECU*)
Advanced materials activities information
No of employees
Turnover (ECU*)
*1ECU=DM 1.89,FFr 6.30,£0.78,Lira 1909,Peseta 159.1

Organisation activities
Research or development on sensors, sensor materials or applications
Testing, consulting, design

Application \ Signal Sensed	Radiation	Optical	Electrical	Magnetic	Thermal	Chemical	Mechanical
Automotive							
Consumer products							
Process industries, mining, energy							
Manufacturing, machinery							
Medicine, health care					•	•	
HVAC, fire control, security							
Military, aerospace							
Telecommunications,office equip.							
Meteorology, environment					•	•	
Other *							

Specific Expertise

The research activity is concerned with the design and synthesis of permselective membranes which are capable of acting as reactive matrices in which (e.g.) colourimetric, fluorimetric, potentiometric signals are generated. It is an underlying aspect of the work that the polymers are designed to interface with biological systems with optimum compatibility. The greatest part of the activity is centred on fibre optic-based and FET-based sensors. The reactive membrane approach allows a great deal of versatility in the design of sensors to detect and quantify species in the (normally aqueous) environment. Thus sensors for organic and inorganic species, pH, temperature and tonicity (osmolality) may be constructed using this approach.

Collaboration Offered

Research and consultancy in the design and function of polymer membranes.

Collaboration Sought

Collaborative research activities in the development of sensors for specific species/applications.

403 University of Birmingham

Department of Metallurgy and Materials
PO Box 363
Birmingham B15 2TT

Tel: +44 121 4143344
Fax: +44 121 4143984
E-mail:

Affiliation:
Contact: Dr P S Dobson

Organisation information
No of employees
Turnover (ECU*)
Advanced materials activities information
No of employees
Turnover (ECU*)
*1ECU=DM 1.89,FFr 6.30,£0.78,Lira 1909,Peseta 159.1

Organisation activities
Research or development on sensors, sensor materials or applications
Testing, consulting, design

Application \ Signal Sensed	Radiation	Optical	Electrical	Magnetic	Thermal	Chemical	Mechanical
Automotive							•
Consumer products							
Process industries, mining, energy						•	
Manufacturing, machinery							
Medicine, health care						•	•
HVAC, fire control, security							
Military, aerospace						•	•
Telecommunications,office equip.							
Meteorology, environment							
Other *							

Main Activities

The activities are concerned with the fabrication of silicon sensors using micromachining techniques. The facilities available include mask making, photolithography, oxidation, diffusion, etching and sputter deposition. The major current interest is in resonant silicon devices for both physical and chemical sensors.

Specific Expertise

A variety of sensors for the measurement of both physical and chemical parameters are being investigated. These sensors are based on the very high strain or mass loading sensitivity of resonant structures. The sensors are fabricated by silicon micromachining techniques and both optical and electrical excitation/interrogation techniques are being developed.

Collaboration Offered

Sensor development. Micromachining techniques.

Collaboration Sought

Co-operation in sensor development.

404 University of Cranfield

Biotechnology Centre
Cranfield
Bedford MK43 0AL

Tel: +44 1234 754 131
Fax: +44 1234 752 401
E-mail: l.tigwell@cranfield.ac.uk

Affiliation:
Contact: Prof Anthony Turner, Head of School
Linda Tigwell, Mkting Manager

Organisation information
No of employees 26-100
Turnover (ECU*) 1m-5m
Sensor technology activities information
No of employees 11-25
Turnover (ECU*) 0.5m-1m
*1ECU=DM 1.89,FFr 6.30,£0.78,Lira 1909,Peseta 159.1

Organisation activities
Research and development on sensors sensor materials or applications
Testing, consulting, design
Manufacture of finished sensors

Application \ Signal Sensed	Radiation	Optical	Electrical	Magnetic	Thermal	Chemical	Mechanical
Automotive						•	
Consumer products						•	
Process industries,mining energy						•	
Manufacturing, machinery						•	
Medicine, health care		•	•			•	
HVAC, fire control, security						•	
Military, aerospace						•	
Telecommunications,office equip.						•	
Meteorology environment						•	
Other (specify)							

Main Activities

Cranfield Biotechnology Centre, backed by the full resources of Cranfield University, is an international centre of excellence in postgraduate education and industrially related research. Our work within the field of sensors and diagnostics encompasses: environmental monitoring, clinical diagnostics, food quality and control and process control and industrial fermentation monitoring. We offer a complete range of services including consultancy, feasibility studies, market research, contract research, manufacturing, postgraduate education and training courses.

Specific Expertise

Environmental monitoring: monitoring and detection of herbicides and pesticides in soil and water; gases and organic vapours; microbial contamination; pollutants.
Clinical diagnostics: chemical sensors for in-vivo monitoring; immunodiagnostics, biosensors for centralised and decentralised testing.
Process control and industrial fermentation monitoring: sensors for process analysis and protection; on-line monitoring of cell propagations; health and safety monitoring.

Collaboration Offered

Technological expertise: electroanalysis; electrochemistry in aqueous and gaseous media; optical and electrochemical instrumentation; integrated optics; enzyme manipulaiton; immobilisation and stabilisation; affinity and separation techniques; nucleic acid technology; catalysis and biomimetics; advanced printing techniques and microfabrication.

405 University of Durham

Centre for Molecular Electronics
Science Laboratories
South Road
Durham DH1 3LE

Tel: +44 91 3742389
Fax: +44 91 3747492
E-mail:

Affiliation:
Contact: Prof M C Petty, Co-Director
Prof M R Bryce

Organisation information
No of employees >100
Turnover (ECU*)
Advanced materials activities information
No of employees 11-25
Turnover (ECU*)
*1ECU=DM 1.89,FFr 6.30,£0.78,Lira 1909,Peseta 159.1

Organisation activities
Research or development on sensors, sensor material or applications

Application \ Signal Sensed	Radiation	Optical	Electrical	Magnetic	Thermal	Chemical	Mechanical
Automotive							
Consumer products							
Process industries, mining, energy	•		•		•	•	
Manufacturing, machinery							
Medicine, health care	•					•	
HVAC, fire control, security							
Military, aerospace					•		
Telecommunications,office equip.							
Meteorology, environment							
Other *							

Main Activities

Molecular electronics - development of new sensing materials/structures using organic thin films. Electrical and optical methods are used.

Specific Expertise

Langmuir-Blodgett films: synthesis of novel compounds, film deposition; conducting polymers: synthesis and thin film deposition; thin film characterisation: electron microscopy/diffraction, X-ray diffraction; UV/visible/infrared spectroscopy; electrical and optical techniques for sensing.

Collaboration Offered

As listed in Specific Expertise.

Collaboration Sought

Finance.

406 University of Edinburgh

Department of Chemistry
King's Buildings
West Mains Road
Edinburgh EH9 3JZ

Tel: +44 131 6501000
Fax: +44 131 6624054
E-mail:

Affiliation:
Contact: Dr H Girault

Organisation information
No of employees >100
Turnover (ECU*)
Advanced materials activities information
No of employees 1-10
Turnover (ECU*)
*1ECU=DM 1.89,FFr 6.30,£0.78,Lira 1909,Peseta 159.1

Organisation activities
Research or development on sensors, sensor materials or applications

Application \ Signal Sensed	Radiation	Optical	Electrical	Magnetic	Thermal	Chemical	Mechanical
Automotive							
Consumer products							
Process industries, mining, energy							•
Manufacturing, machinery							
Medicine, health care						•	
HVAC, fire control, security							
Military, aerospace							
Telecommunications,office equip.							
Meteorology, environment							
Other *							

Main Activities

Fundamental and applied research on the following topics:
CHEMFET (chemical field effect transistor) manufactured at the Micro Fabrication Unit of Edinburgh University; photopolymerisation of ion selective gate.
Ion selective electrodes: development of new ditropic ionophores.
Amperometric sensors: fabrication of ultra-microelectrode array by microelectronic fabrication techniques.
Electrochemistry: general study of ion transfer reactions across a liquid/membrane interface.

Specific Expertise

Expertise focused on electrochemical sensing.

Collaboration Offered

Research contracts.

Collaboration Sought

Electrochemical sensors.

407 University of Kent

Centre for Materials Research, Chemistry Department
Canterbury
Kent CT2 7NH

Tel: +44 1227 823519
Fax: +44 1227 827724
E-mail: j.d.wright@ukc.ac.uk

Affiliation:
Contact: Dr J D Wright,
Reader in Materials Chemistry
Dir of Chem. Sensors Research Grp
A Bentley, Assist Registrar

Organisation information
No of employees >1000
Turnover (ECU*) 5m-50m
Sensor technology activities information
No of employees 11-25
Turnover (ECU*) <0.5m
*1ECU=DM 1.89,FFr 6.30,£0.78,Lira 1909,Peseta 159.1

Organisation activities
Research and development on sensors, sensor materials or applications.

Application \ Signal Sensed	Radiation	Optical	Electrical	Magnetic	Thermal	Chemical	Mechanical
Automotive						•	
Consumer products						•	
Process industries,mining energy						•	
Manufacturing, machinery						•	
Medicine, health care						•	
HVAC, fire control, security						•	
Military, aerospace						•	
Telecommunications,office equip.							
Meteorology, environment						•	
Other *							

Main Activities

One of the UK's longest-standing sensor groups. It has received substantial long-term support from the UK Industry and government leading to patent applications for organic semiconductors for nitrogen dioxide and for organic nitro compounds and for metal oxide semiconductor sensors for flammable gases. The group has worked on optical chemical sensors for the past five years developing active collaborations with groups in Nijmegen and Moscow. The Department and University offer support facilities, notably analytical facilities including ICP-mass spectrometry, several chemical synthesis activities and an applied optics group in the Physics department.

Specific Expertise

Optical chemical sensors based on molecular receptors in porous sol-gel glass matrices, for detection of solvent vapours, metal ions and organic molecules (e.g. drugs, pesticides and herbicides). Surface plasmon resonance sensors, using monolayers of specific molecular receptors bound via Au-S interactions on a silicon-based Schottky-barrier grating-SPR device. Piezo-optic chemical sensors. Sol-gel-synthesised nanocrystalline tin oxide sensors. Organic semiconductor sensors.

Collaboration Offered

Research and development expertise for sensors and sensor materials. Both direct and indirect compensation/offset programmes are accepted

Collaboration Sought

Joint research projects with industry or government laboratories, or with sensor end-users.

408 University of Lancaster

Department of Engineering
Bailrigg
Lancaster LA1 4YR

Tel: +44 524 65201
Fax: +44 524 381707
E-mail:

Affiliation:
Contact: Prof R G Carter, Head of Dept

Organisation information
No of employees
Turnover (ECU*)
Advanced materials activities information
No of employees
Turnover (ECU*)
*1ECU=DM 1.89,FFr 6.30,£0.78,Lira 1909,Peseta 159.1

Organisation activities
Research or development on sensors, sensor material or applications
Testing, consulting, design

Application \ Signal Sensed	Radiation	Optical	Electrical	Magnetic	Thermal	Chemical	Mechanical
Automotive							•
Consumer products			•	•	•		•
Process industries, mining, energy							
Manufacturing, machinery			•	•	•		•
Medicine, health care			•				•
HVAC, fire control, security							
Military, aerospace							•
Telecommunications,office equip.							
Meteorology, environment							
Other *							

Main Activities

Research, development and design in the application of sensors including specialised integrated circuit design and development of systems using digital networking.

Collaboration Offered

Research and Development. Design consultancy.

409 University College of London

Electronic & Electrical Engineering Department
Torrington Place
London WC1E 7JE

Tel: +44 71 3877050
Fax: +44 71 3874350
E-mail:

Affiliation:
Contact: Dr Flanagan

Organisation information
No of employees
Turnover (ECU*)
Sensors technology activities information
No of employees
Turnover (ECU*)
*1ECU=DM 1.89,FFr 6.30,£0.78,Lira 1909,Peseta 159.1

Organisation activities
Research and Development
Testing, consulting, design

Application \ Signal Sensed	Radiation	Optical	Electrical	Magnetic	Thermal	Chemical	Mechanical
Automotive					•		•
Consumer products							
Process industries, mining, energy					•		•
Manufacturing, machinery					•		•
Medicine, health care							
HVAC, fire control, security							
Military, aerospace							
Telecommunications,office equip.							
Meteorology, environment							
Other *							

Specific Expertise

Optical fibre pressure sensors offering immunity to EMI. Pressure ranges from 0 to 10,000 psi over large temperature range; optical fibre/diaphragm pressure sensors for industrial applications with better than 0.5% linearity.

Collaboration Offered

Pressure sensors. Optical fibre sensor development. Optical fibre sensor research. Optical fibre component technology for sensor applications.

Collaboration Sought

Mechanical or thermal sensors.

410 University of Luton

Biosensors Research Group
Applied Research & Consultancy Centre
Putteridge Bury
Hitchin Road
Luton LU2 8LE

Tel: +44 582 482555
Fax: +44 582 482689
E-mail:

Affiliation:

Contact: Prof D M Rawson,
Director of Research

Organisation information
No of employees >100
Turnover (ECU*) >5m

Sensors technology activities information
No of employees 1-10
Turnover (ECU*) <0.5m
*1ECU=DM 1.89,FFr 6.30,£0.78,Lira 1909,Peseta 159.1

Organisation activities
Research and Development
Testing, consulting, design

Application \ Signal Sensed	Radiation	Optical	Electrical	Magnetic	Thermal	Chemical	Mechanical
Automotive							
Consumer products						•	
Process industries, mining, energy						•	
Manufacturing, machinery							
Medicine, health care						•	
HVAC, fire control, security							
Military, aerospace							
Telecommunications,office equip.							
Meteorology, environment						•	
Other *							

Main Activities

The Applied Research and Consultancy Centre is a commerical agency of Luton College of Higher Education HEC, located in a country estate at Putteridge Bury. The Centre co-ordinates a wide range of research and consultancy groups within the College. The Biosensor Research Group, housed in specialist accomodation, forms part of the Centre and is financed by contract research.

Specific Expertise

The main activity of the Biosensor Research Group is in the development of biosensors for use in pollutant and toxin detection; fermentation and process monitoring; food analysis and clinical application. Biosensors incorporating a variety of biocatalysts including microbial whole cells, biological membranes and immobilised enzymes are being developed to allow real-time monitoring of a wide range of chemical analytes.
Both broad spectrum and analyte specific devices are being developed. Amperometric electrodes including precious metals, carbons and conducting organic salts are employed in thin layers, wall-jet and diptype 2, 3 and 4 electrode configurations.

411 University of Manchester

Department of Medicine
Hope Hospital
Salford
M6 8HD

Tel: +44 161 787 4428
Fax: +44 161 787 4429
E-mail:

Affiliation:
Contact: P Vadgama, Prof of Clin. Biochem
Dr P Treloar, Tech Dir.
P Ternduth, Dev/Comm Mgr

Organisation information

No of employees	>1000
Turnover (ECU*)	<1m

Sensor technology activities information

No of employees	1-10
Turnover (ECU*)	<0.5m

*1ECU=DM 1.89,FFr 6.30,£0.78,Lira 1909,Peseta 159.1

Organisation activities
Research and development on sensors, sensor materials or applications

Application \ Signal Sensed	Radiation	Optical	Electrical	Magnetic	Thermal	Chemical	Mechanical
Automotive							
Consumer products							
Process industries,mining energy							
Manufacturing, machinery							
Medicine, health care						•	
HVAC, fire control, security							
Military, aerospace							
Telecommunications,office equip.							
Meteorology, environment						•	
Other *							

Main Activities

Development of polymeric membranes; in vivo monitoring;
enzyme electrodes; electrochemistry.

Specific Expertise

Membrane fabrication; membrane biocompatibioity evaluation.
Materials - PVC, polyurethane, cellulose, hydrofels.

Collaboration Offered

Packaging of biosensors so they work in real samples and 20 patented membrane systems.

Collaboration Sought

Scale up fabrication signal processing.

412 University of Manchester

Institut of Science and Technology - Umist
c/o DIAS
PO Box 88
Manchester M60 1QD

Tel: +44 161 200 4891
Fax: +44 161 200 4911
E-mail:

Affiliation: Dept in University
Contact: Dr R Narayanaswamy, Snr Lect

Organisation information
No of employees 101-1000
Turnover (ECU*) 1m-5m
Sensor technology activities information
No of employees 26-100
Turnover (ECU*) N/A
*1ECU=DM 1.89,FFr 6.30,£0.78,Lira 1909,Peseta 159.1

Organisation activities
Research and development on sensors, sensor materials or applications.
Testing, consulting, design.

Application \ Signal Sensed	Radiation	Optical	Electrical	Magnetic	Thermal	Chemical	Mechanical
Automotive							
Consumer products							
Process industries,mining energy							
Manufacturing, machinery							
Medicine, health care		●				●	
HVAC, fire control, security							
Military, aerospace							
Telecommunications,office equip.							
Meteorology, environment		●				●	
Other *		●					

* Chemical and Biosensors.

Main Activities

Optical chemical and biochemical sensors; electrochemical sensors; ultrasound and imaging; artificial nose systems.

Specific Expertise

Optical chemical sensors and biosensors; fibre optic systems; opto-electronic instrumentation; all research and development work.

Collaboration Offered

There is a special contract stating all conditions.

Collaboration Sought

Bilateral agreements.

413 University of Manchester

Dept of Electrical Engineering
PO Box 88
Manchester M60 1QD

Tel: +44 161 200 4702
Fax: +44 161 200 4781
E-mail: p.j.hicks@umist.ac.uk

Affiliation:
Contact: Prof P J Hicks
Dr JV Hatfield, Senior Lecturer
Dr R C Waterfall, Senior Lecturer
Dr T A York, Senior Lecturer

Organisation information

No of employees 26-100
Turnover (ECU*) 1m-5m

Sensor technology activities information

No of employees 11-25
Turnover (ECU*) 0.5m-1m

*1ECU=DM 1.89,FFr 6.30,£0.78,Lira 1909,Peseta 159.1

Organisation activities

Research and development on sensors sensor materials or applications
Testing, consulting, design

Application \ Signal Sensed	Radiation	Optical	Electrical	Magnetic	Thermal	Chemical	Mechanical
Automotive							
Consumer products						●	
Process industries,mining energy			●				
Manufacturing, machinery						●	
Medicine, health care							●
HVAC, fire control, security							
Military, aerospace							
Telecommunications,office equip.							
Meteorology environment							
Other	●						

Main Activities

UMIST is one of the UK's leading research-based universities speciaising in science and technology. The Department of Electrical Engineering and Electronics has 60 academic members of staff, covering all aspects of the subject from solid state devices to power systems engineering. The department was rated four in the last HEFCE Research Assessment Exercise.

Specific Expertise

Mixed analogue/digital integrated circuit design. Integrated sensors - examples include: position-sensitive electron detectors for scientific instruments; gas and odour sensors; integrated transmit/receive circuitry for phased-array ultrasound transducers.

Collaboration Offered

Mixed analogue/digital IC design expertise for sensor applications. Custom process technologies/layer processing/packaging techniques for novel sensor/transducer design.

Collaboration Sought

Industrial applications requiring novel sensor technologies, especially integrated sensors.

414 University of Oxford

Physical & Theoretical Chemistry Laboratory
South Parks Road
Oxford OX1 3QZ

Tel: +44 865 275413
Fax: +44 865 275410
E-mail:

Affiliation:
Contact: Prof R G crompton, Group Leader

Organisation information
No of employees
Turnover (ECU*)
Advanced materials activities information
No of employees
Turnover (ECU*)
*1ECU=DM 1.89,FFr 6.30,£0.78,Lira 1909,Peseta 159.1

Organisation activities
Research or development on sensors, sensor materials or applications

Application \ Signal Sensed	Radiation	Optical	Electrical	Magnetic	Thermal	Chemical	Mechanical
Automotive							
Consumer products							
Process industries, mining, energy							
Manufacturing, machinery							
Medicine, health care						●	
HVAC, fire control, security							
Military, aerospace							
Telecommunications,office equip.							
Meteorology, environment						●	
Other *							

Main Activities

University based research group involved in theory and practice of electroanalysis.

Specific Expertise

Fabrication, understanding and testing of both ion-selective electrodes for Li^+, Ca^{2+}, K^+, surfactants, and amperometric devices eg. for anaesthetic gases. Photoelectrochemical sensing: highly selective sensing of structurally similar compounds. The use of conducting polymers, such as polyperrole, in sensing methodology.

415 University of Reading

Department of Cybernetics
Whiteknights
Reading RG6 2AL

Tel: +44 734 318210
Fax: +44 734 318220
E-mail:

Affiliation:
Contact: Prof K Warwick, Head of Dept

Organisation information
No of employees >100
Turnover (ECU*) >5m
Advanced materials activities information
No of employees 11-25
Turnover (ECU*) 0.5m-1m
*1ECU=DM 1.89,FFr 6.30,£0.78,Lira 1909,Peseta 159.1

Organisation activities
Research or development on sensors, sensor materials or applications
Testing, consulting, design

Application \ Signal Sensed	Radiation	Optical	Electrical	Magnetic	Thermal	Chemical	Mechanical
Automotive			●				●
Consumer products			●				●
Process industries, mining, energy			●				●
Manufacturing, machinery			●				
Medicine, health care			●				
HVAC, fire control, security			●				
Military, aerospace			●			●	●
Telecommunications,office equip.			●		●		●
Meteorology, environment			●				
Other *							

Specific Expertise

Microcomputer based sensors; visual systems (neural net based) for high speed quality control (50 inspections per second); tactile sensors (optical, capacitive); multiparameter sensors; non-instrusive flow measurement; seismometers; optical filters.

Collaboration Offered

Consultancy. New product research and development. Research and development support services. Development of prototypes.

Collaboration Sought

Co-operative interest from industrial and academic concerns: intelligent sensors and related systems detailed in 'Specific Expertise'.

416 University of Southampton

Institute of Cryogenics
Cryogenics Advisory Unit
Southampton SO9 5NH

Tel: +44 703 593348
Fax: +44 703 593053
E-mail:

Affiliation:
Contact: P C McDonald, Tech & Projects Mgr

Organisation information

No of employees 26-100
Turnover (ECU*) 0.5m

Advanced materials activities information

No of employees 1-10
Turnover (ECU*) 0.5m

*1ECU=DM 1.89,FFr 6.30,£0.78,Lira 1909,Peseta 159.1

Organisation activities

Research or development on sensors, sensor materials or applications
Testing, consulting, design
Manufacture of finished sensors

Application \ Signal Sensed	Radiation	Optical	Electrical	Magnetic	Thermal	Chemical	Mechanical
Automotive							
Consumer products							
Process industries, mining, energy					•		•
Manufacturing, machinery							
Medicine, health care							
HVAC, fire control, security							
Military, aerospace					•		•
Telecommunications,office equip.							
Meteorology, environment							
Other *							

Main Activities

The Institute of Cryogenics is a teaching and research institute of the University of Southampton which specialises in engineering for low temperatures. The Cryogenics Advisory Unit is a subsidiary of the Institute which offers the facilities and expertise available within the Institute to industry in the form of R & D, test, design consultancy and short courses.

Specific Expertise

Work is being carried out within the Institute on the development of sensors eg. pressure and temperature sensors for use at low temperatures. A thermometer operating over the range -273°C to +150°C is now marketed through the Advisory Unit. Research is also being undertaken into the sensor applications of high temperature superconductors.

Collaboration Offered

Research and development facilities and expertise.

Collaboration Sought

Finance, hardware, manufacturing expertise.

417 University of Southampton

Institute of Transducer Technology (USITT)
University of Southampton
Highfield
Hants SO17 1BJ

Tel: +44 703 593545
Fax: +44 703 592865
E-mail:

Affiliation:
Contact: Dr S T Hughes, Manager
Dr M J Tudor, Lecturer

Organisation information
No of employees
Turnover (ECU*)
Advanced materials activities information
No of employees 1-10
Turnover (ECU*) <0.5m
*1ECU=DM 1.89,FFr 6.30,£0.78,Lira 1909,Peseta 159.1

Organisation activities
Research or development on sensors, sensor materials or applications
Testing, consulting, design
Supply of raw materials or semi-finished products used to make sensors

Application \ Signal Sensed	Radiation	Optical	Electrical	Magnetic	Thermal	Chemical	Mechanical
Automotive				•	•		•
Consumer products			•				•
Process industries, mining, energy	•			•	•	•	•
Manufacturing, machinery	•		•				•
Medicine, health care			•				•
HVAC, fire control, security	•		•		•	•	•
Military, aerospace					•	•	•
Telecommunications,office equip.			•	•		•	
Meteorology, environment						•	
Other *							

Main Activities

Following an initiative from industry, USITT was set up jointly by the Faculties of Science, Medicine and Engineering at Southampton University, to create a centre of excellence in transducer technology. The institute has therefore been established with the following aims:
Research: to undertake research into the development and production of more accurate, reliable and cost effective sensors and actuators.
Project Management: to organise and manage the diverse disciplines which are required for transducer development, so that the University can contribute effectively to industry.
Consultancy: to provide a pool of expertise and a data base of transducer information.
Teaching: through courses and regular colloquia, to provide instruction in the design and application of transducers.

Specific Expertise

Typical Projects - Development of NOx sensor for automotive use; superconductive thick films for IR detection; silicon micromachined sensor structures; design of automotive torque sensor; application of thick film techniques to the manufacture of pressure sensors and load cells; smart sensors/transmitters; optical sensors; finite element techniques applied to design of electromagnetic and mechanical sensors.

Collaboration Offered

Membership.
Course/colloquia enquiries.
Requests for USITT NEWS (Newsletter).
Research/consultancy

Collaboration Sought

Co-operation in all areas of sensor research and development.

418 University College of Swansea

Gas Analysis and Sensing Group
Electrical Engineering Group
Swansea SA2 8PP

Tel: +44 1792 295414
Fax: +44 1792 295686
E-mail: j.watson@swansea.ac.uk

Affiliation:
Contact: Dr J Watson, Chairman

Organisation information
No of employees 26-100
Turnover (ECU*) n/a
Sensor technology activities information
No of employees n/a
Turnover (ECU*) n/a
*1ECU=DM 1.89,FFr 6.30,£0.78,Lira 1909,Peseta 159.1

Organisation activities
Research and development on sensors sensor materials or applications
Testing, consulting, design

Application \ Signal Sensed	Radiation	Optical	Electrical	Magnetic	Thermal	Chemical	Mechanical
Automotive						●	
Consumer products						●	
Process industries,mining energy						●	
Manufacturing, machinery						●	
Medicine, health care						●	
HVAC, fire control, security						●	
Military, aerospace							
Telecommunications,office equip.						●	
Meteorology environment							
Other (specify)							

Main Activities

Holds three meetings per year and publishes proceedings and a newsletter.

Specific Expertise

All forms of gas sensors, gas alarms and gas monitors.

Collaboration Offered

Technical contacts, both industrial and academic.

Collaboration Sought

Joint R&D

419 University of Warwick

Department of Engineering
Coventry CV4 7AL

Tel: +44 1203 523523
Fax: +44 1203 418922
E-mail: j.w.garden@warwick.ac.uk

Affiliation:
Contact: Dr J W Gardner, Dir of Sensor Research Lab.

Organisation information
No of employees 26-100
Turnover (ECU*) 5m-50m

Sensor technology activities information
No of employees 11-25
Turnover (ECU*) 1m-5m
*1ECU=DM 1.89,FFr 6.30,£0.78,Lira 1909,Peseta 159.1

Organisation activities
Research and development on sensors, sensor materials or applications.

Application \ Signal Sensed	Radiation	Optical	Electrical	Magnetic	Thermal	Chemical	Mechanical
Automotive						●	●
Consumer products		●				●	
Process industries,mining energy						●	
Manufacturing, machinery							
Medicine, health care						●	
HVAC, fire control, security							
Military, aerospace		●				●	●
Telecommunications,office equip.							
Meteorology, environment						●	
Other *						●	

* Odours

Main Activities

Research and development of novel materials, devices and data processing techniques for microsensor technology. A wide range of activities take place from surface characterisation (microengineering) and precision metrology (nanotechnology), to odour analysis (electronic noses) and pattern recognition (neural networks).

Specific Expertise

Areas of expertise: micromachined sensors; low-power sensors; smart sensors; chemical microsensors; electronic noses; pressure sensors.
Materials: organic polymers; semiconducting oxides.

Collaboration Offered

R&D expertise and facilities.

Collaboration Sought

Complimentary facilities. Application, marketing, production engineering.

420 University College of Wales

Dept. Biological Science
Aberystwyth
Dyfed SY23 3DA

Tel: +44 1970 622334
Fax: +44 1970 622350
E-mail:

Affiliation:
Contact: Dr D Kell, Reader, Microbiology

Organisation information
No of employees 11-25
Turnover (ECU*) <0.5m
Advanced materials activities information
No of employees
Turnover (ECU*) <0.5m
*1ECU=DM 1.89,FFr 6.30,£0.78,Lira 1909,Peseta 159.1

Organisation activities
Research or development on sensors, sensor materials or applications
Testing, consulting, design
Supply of raw materials or semi-finished products used to make sensors
Manufacture of finished sensors
Manufacture of sensor-based products

Application \ Signal Sensed	Radiation	Optical	Electrical	Magnetic	Thermal	Chemical	Mechanical
Automotive							
Consumer products							
Process industries, mining, energy							
Manufacturing, machinery							
Medicine, health care			•				
HVAC, fire control, security							
Military, aerospace							
Telecommunications,office equip.							
Meteorology, environment			•				
Other *							

Main Activities

Linear and nonlinear electrical properties of biological systems.

421 University of West of England, Bristol

Molecular Electronics & Surface Science Group
Frenchay Campus
Bristol BS16 1QY

Tel: +44 1179 656261
Fax: +44 1179 9763871
E-mail:

Affiliation:
Contact: Prof C L Honeybourne

Organisation information
No of employees >100
Turnover (ECU*)
Sensors technology activities information
No of employees 1-10
Turnover (ECU*)
*1ECU=DM 1.89,FFr 6.30,£0.78,Lira 1909,Peseta 159.1

Organisation activities
Research and Development

Application \ Signal Sensed	Radiation	Optical	Electrical	Magnetic	Thermal	Chemical	Mechanical
Automotive							
Consumer products							
Process industries, mining, energy						•	
Manufacturing, machinery							
Medicine, health care							
HVAC, fire control, security						•	
Military, aerospace						•	
Telecommunications,office equip.							
Meteorology, environment						•	
Other *						•	

* Decay in foods

Main Activities
The Molecular Electronics and Surface Science Group at Bristol Polytechnic work on the theory and application of the semiconductive and non-linear optical properties of thin films of organic dyes. At present, the work focuses on: surface analysis of thin films; new materials for second-harmonic generation ; sensors for toxic gases and vapours.

Specific Expertise
The main effort with regard to sensors is directed towards the design, synthesis and thin film deposition of speciality organic materials with specific interactions with individual toxic, reactive gases or vapours or with individual classes of such gases or vapours (eg. powerful oxidising agents).
The chemical stimulus is at present detected in one or more of the following ways:
visible transmittance;
visible reflectance;
transmittance in the IR region (1500+/-500cm^{-1});
reflectance in the IR region (1500+/-500cm^{-1});
modifications to the propagation of surface plasmon;
polaritons across counted metal films.

Collaboration Offered
Research and development consultancies or collaborative proposals.

Collaboration Sought
All products aimed at gas and vapour detection.

422 Vector Instruments

115 Marsh Road
Rhyl
Clwyd LL18 2AB
Wales

Tel: +44 745 350700/330175
Fax: +44 745 344206
E-mail:

Affiliation:
Contact: R J Jones, Engineering Manager
J Penlington, Technical Sales Mgr

Organisation information
No of employees 11-25
Turnover (ECU*) 0.5m-1m
Advanced materials activities information
No of employees 11-25
Turnover (ECU*) 0.5m-1m
*1ECU=DM 1.89,FFr 6.30,£0.78,Lira 1909,Peseta 159.1

Organisation activities
Research or development on sensors, sensor materials or applications
Testing, consulting, design
Manufacture of finished sensors
Manufacture of sensor-based products

Application \ Signal Sensed	Radiation	Optical	Electrical	Magnetic	Thermal	Chemical	Mechanical
Automotive							
Consumer products							
Process industries, mining, energy					•	•	•
Manufacturing, machinery							
Medicine, health care							•
HVAC, fire control, security							•
Military, aerospace							
Telecommunications,office equip.					•	•	•
Meteorology, environment					•	•	•
Other *							

Main Activities

Design, development, manufacture of a range of windspeed, direction, air temperature and relative humidity sensors and manufacture of associated indication and control interface equipment. Supply of recorders and data-loggers for the above.

Specific Expertise

Development of rotary transducers; calibration of cup rotors for anemometers; sensors for true air temperature (avoiding radiation effects).

Collaboration Offered

Source of a novel range of windspeed and wind direction measuring instruments.
A market for other meteorological sensors.

Collaboration Sought

A market for wind measuring instrumentation in Europe.

423 Vishay Ltd

Pallion Industrial Estate
Sunderland
SR4 6SU

Tel: +44 191 514 4155
Fax: +44 191 567 8262
E-mail:

Affiliation:
Contact: Mr J Wheeler, MD
Mrs M Neil, Publicity Co-ord
F Welfare, Sales Dir.

Organisation information
No of employees 101-1000
Turnover (ECU*) 5m-50m
Sensor technology activities information
No of employees >100
Turnover (ECU*) 1m-5m
*1ECU=DM 1.89,FFr 6.30,£0.78,Lira 1909,Peseta 159.1

Organisation activities
Research and development on sensors, sensor materials or applications.
Testing, consulting, design.
Manufacture of systems used in the manufacture, testing or development of sensors.
Manufacture of finished sensors.

Application \ Signal Sensed	Radiation	Optical	Electrical	Magnetic	Thermal	Chemical	Mechanical
Automotive					•		
Consumer products					•		
Process industries,mining energy					•		
Manufacturing, machinery					•		
Medicine, health care					•		
HVAC, fire control, security					•		
Military, aerospace					•		
Telecommunications,office equip.					•		
Meteorology, environment							
Other *							

Main Activities

Manufacture of thermistors. Chip, leaded and assemblies.

Specific Expertise

Vishay manufacture curve tracking (NTC) and point matched thermistors, thermistor assemblies and probes. Curve tracking accurately down to 0.2°C over range eg. 20°C to 70°C.

Collaboration Offered

NTC thermistors, thermistor assemblies and thermistor probes.

424 WRC

Frankland Road
Swindon SNS8YF

Tel: +44 17 93 51 17 11
Fax: +44 17 93 51 17 12
E-mail: peters-d@wrcplc.co.uk

Affiliation:
Contact: Dr D C Peters, Mgr, Sensors & Instruments
P Whalley, Tech Specialist, Env Monitoring

Organisation information
No of employees 101-1000
Turnover (ECU*) 5m-50m
Sensor technology activities information
No of employees 11-25
Turnover (ECU*) 0.5m-1m
*1ECU=DM 1.89,FFr 6.30,£0.78,Lira 1909,Peseta 159.1

Organisation activities
Research and development on sensors, sensor materials or applications
Testing, consulting, design

Application \ Signal Sensed	Radiation	Optical	Electrical	Magnetic	Thermal	Chemical	Mechanical
Automotive							
Consumer products							
Process industries,mining energy		●	●	●		●	
Manufacturing, machinery							
Medicine, health care							
HVAC, fire control, security							
Military, aerospace							
Telecommunications,office equip.							
Meteorology environment		●	●	●		●	
Other (specify)							

Main Activities

WRC plc is a leading European research, technology and consultancy company. Our expertise lies in the fields of water, wastewater and environmental management in which we have over seventy years' experience. We seek to provide service as a problem solving organisation in these and related areas. WRC is the UK National Centre for Environmental Toxicology and the lead organisation for the European Topic Centre on Inland Waters in support of the European Environment Agency.

Specific Expertise

WRC's past and current activity related to sensors falls into the following headings:
- development of systems utilising sensors
- testing and evaluation of commercially available systems against user requirements
- preparation and drafting of standards at an industry, national and European level
- investigation of best practice in the application of sensor systems for specific applications
- provision of information services relating to sensors and their application.

Collaboration Offered

- deep understanding of requirements for sensors in the fields of environmental monitoring, water and waste water treatment and associated pipeline systems:
- system engineering capability to take advanced concepts to meet user requirements
- bespoke test capability
- experience of multi partner/multi national collaborative projects.

Collaboration Sought

- users able to provide a focus for development
- manufacturers able to provide technical skills and facilities and a route to market
- research specialists with leading edge technology looking for applications

425 Zellweger Analytics Limited

4 Stinsford Road
Poole
Dorset, BH17 0RZ

Tel: +44 202 676161
Fax: +44 202 678011
E-mail:

Affiliation:
Contact: Dr Prectort, Research Director

Organisation information
No of employees >100
Turnover (ECU*) >5m
Advanced materials activities information
No of employees >100
Turnover (ECU*) >5m
*1ECU=DM 1.89,FFr 6.30,£0.78,Lira 1909,Peseta 159.1

Organisation activities
Research or development of sensors, sensor materials or applications
Manufacture of sensor-based products

Application \ Signal Sensed	Radiation	Optical	Electrical	Magnetic	Thermal	Chemical	Mechanical
Automotive						•	
Consumer products							
Process industries, mining, energy						•	
Manufacturing, machinery						•	
Medicine, health care							
HVAC, fire control, security						•	
Military, aerospace						•	
Telecommunications,office equip.						•	
Meteorology, environment						•	
Other *							

Main Activities

Design, manufacture and supply of gas detection equipment for flammable and toxic gases.

Specific Expertise

Detection of combustible gases in the LEL or percentage v/v range, toxic gases in ppm range, together with associated control equipment and systems. Speciality is equipment certified for use in hazardous areas, such as encountered in oil and gas exploration and production, petrochemicals, mining, chemical storage and manufacture, solvent recovery or usage, biogas applications etc. Also liquid analysis in textile, petrochemical, food processing, water industry etc.

Collaboration Offered

As previously listed, gas detection and liquid analysis.

5. Cross References

Signal Sensed : Radiation
Application Area : Automotive

Signal Sensed : Radiation
Application Area : Consumer products

Signal Sensed : Radiation
Application Area : Process Industries, Mining, Energy

Signal Sensed : Radiation
Application Area : Manufacturing, Machinery

Signal Sensed : Radiation
Application Area : Manufacturing, Machinery - continued

Signal Sensed : Radiation
Application Area : Medicine, Health Care

Signal Sensed : Radiation
Application Area : HVAC, Fire Control, Security

Signal Sensed : Radiation
Application Area : Military, Aerospace

Signal Sensed : Radiation
Application Area : Telecommunications, Office Equipment

Signal Sensed : Radiation
Application Area : Meteorology, Environment

Signal Sensed : Optical
Application Area : Automotive

AG Electro-Optics Ltd	249
Baumer Electric Ltd	260
Brunel University, Brunel Centre for Manufacturing Metrology	267
Cambridge Consultants Ltd	268
Eltrotec Elektro GmbH	095
Fogale Nanotech	056
FRABA Management GmbH	096
Fraunhofer-Institut für Produktionstechnik & Automatisierung	101
City University, Dept. Electrical, Electronics & Information Engineering	276
Hadland Photonics Ltd	304
Heimann Optoelectronics GmbH	107
IMEC	005
Instituto de Bioingeniera - Fundacion CGH	212
Integrated Photomatrix Ltd	315
IRM SA	006
IROE - CNR	174
Jenasensoric e.V.	115
Lambda Photometrics Ltd	330
Laser Optronic SRL	175
Universität Erlangen-Nürnberg, Lehrstuhl Technische Thermodynamik (LTT)	152
Nitron Europe Ltd	167
Rank Taylor Hobson	370
Redman Controls & Electronics Ltd	372
Sheffield Hallam University, Electronics, Electronics	384
Sira Ltd	388
Strathclyde University, Department of Electronic & Electrical Engineering	392
Technische Universität Chemnitz Zwickau, Institut für Mikrosysteme	144
Université de Savoie, ESIGEC, Laboratoire CME	080
University of Transport, Rostov; Department of Automation	208
Universidad Complutense de Madrid	213
Weber Sensors GmbH	159

Signal Sensed : Optical
Application Area : Consumer Products

Ablewood Technology	242
Czech Technical University Prague, Faculty of Electrical Engineering	012
Eltrotec Elektro GmbH	095
FRABA Management GmbH	096
Fraunhofer-Institut für Siliziumtechnologie	102
City University, Dept. Electrical, Electronics & Information Engineering	276
Heimann Optoelectronics GmbH	107
Herga Electric Ltd	308
IMEC	005
Jenasensoric e.V.	115
KUBE Electronics Ltd	228
Lambda Photometrics Ltd	330
Porvair Technology Ltd	367
Rank Taylor Hobson	370
Sheffield Hallam University, Electronics	384
Sira Ltd	388
Strathclyde University, Department of Electronic & Electrical Engineering	392
Technische Universität Chemnitz-Zwickau, Institut für Mikrosysteme	144
Université de Savoie, ESIGEC, Laboratoire CME	080
Université de Troyes, IBT	082
Unitronic GmbH	150
University of Warwick, Department of Engineering	419
VTT - Electronics	030
Weinlich GmbH & Co	160

Signal Sensed : Optical
Application Area : Process Industries, Mining, Energy

Ablewood Technology	242
AG Electro-Optics Ltd	249
Baumer Electric Ltd	260
Bio-Orbit Oy	021
Brunel University, Brunel Centre for Manufacturing Metrology	267
Cambridge Consultants Ltd	268
City University, Dept. Electrical, Electronics & Information Engineering	276
CYLERGIE	044
Eltrotec Elektro GmbH	095
ETH Z, Centre for Chemical Sensors/Biosensors and Bioanalytical Chem.	224
Fogale Nanotech	056
Franco Corradi	173
Fraunhofer-Institut für Zerstörungsfreie Prüfverfahren	104
Gestra (UK) Ltd	302
H Warner & Son Ltd	311
Hadland Photonics Ltd	304
Heimann Optoelectronics GmbH	107
IMEC	005
INERIS - Verneuil-en-Halatte	063
INRA, Laboratoire de Génie des Procédés Alimentaires	065
Institut für Physikalische Hochtechnologie e.V.	112
Institute of Optical Research	215
Integrated Photomatrix Ltd	315
IRM SA	006
Jenasensoric e.V.	115
JMS Consultants	325
INPG - ENSERG, Laboratoire d'Electromagnétisme	064
Université Jean Monnet, Laboratoire Traitement du Signal et Instrumentation	081
Lambda Photometrics Ltd	330
Universität Erlangen-Nürnberg, Lehrstuhl Technische Thermodynamik (LTT)	152
Tampere University of Technology; Measurement & Information Technology	026
Neotronics Ltd	355
Nitron Europe Ltd	167
Parker Hannifin GmbH	129
Ploskon AT	209
Porvair Technology Ltd	367
Rank Taylor Hobson	370
Hungarian Academy of Sciences, ATKI-Institute for Materials Science	164
Research Centre for Ecological Technologies	200
Sira Ltd	388
Strathclyde University, Department of Electronic & Electrical Engineering	392
Teagasc	169
tec5 Sensorik und Systemtechnik GmbH	141
Universidad Complutense de Madrid	213
Université de Troyes, IBT	082
University of Cordoba, Analytical Chemistry Department	214
Valmet Automation Kavanni Ltd	028
Vision Systems Ltd	029
WRC	424

Signal Sensed : Optical
Application Area : Manufacturing, Machinery

AG Electro-Optics Ltd	249
Baumer Electric Ltd	260
Bio-Orbit Oy	021
Brunel University, Brunel Centre for Manufacturing Metrology	267
Cambridge Consultants Ltd	268
City University, Dept. Electrical, Electronics & Information Engineering	276
C.S.O. Mesure	043
Delft Sensor Systems	003
Eltrotec Elektro GmbH	095
FRABA Management GmbH	096
Fraunhofer-Institut für Siliziumtechnologie	102
Fraunhofer-Institut für Informations- & Datenverarbeitung	097
Fraunhofer-Institut für Zerstörungsfreie Prüfverfahren	104
Graticules Ltd	303
Herga Electric Ltd	308
IMEC	005
Integrated Photomatrix Ltd	315
Kollmorgen Motion Technologies Group	328
Université Jean Monnet, Laboratoire Traitement du Signal et Instrumentation	081
Lambda Photometrics Ltd	330
Laser Optronic SRL	175
Tampere University of Technology; Measurement & Information Technology	026
Ploskon AT	209
Rank Taylor Hobson	370
Russian Academy of Science, Institute of Automation & Electrometrics	206
Russian Academy of Science, Institute of Physics of Semiconductors	207
Sira Ltd	388
Strathclyde University, Department of Electronic & Electrical Engineering	392
tec5 Sensorik und Systemtechnik GmbH	141
Universidad Complutense de Madrid	213
Vision Systems Ltd	029
Zirox Sensoren & Elektronik GmbH	163

Signal Sensed : Optical
Application Area : Medicine, Health Care

AG Electro-Optics Ltd	249
AVL Medical Instruments	001
Bio-Orbit Oy	021
Cambridge Consultants Ltd	268
ETH Z, Centre for Chemical Sensors/Biosensors and Bioanalytical Chem.	224
University of Cranfield, Biotechnology Centre	404
Dantec/Invent Measurement Technology GmbH	092
Delta Lys & Optik	016
Università degli Studi di Firenze, Dip Sanità Pubblica Epid.	179
Ecole Nationale Supérieure des Mines de St-Etienne, SPIN	047
Fraunhofer-Institut für Integrierte Schaltungen	098
Graticules Ltd	303
City University, Dept. Electrical, Electronics & Information Engineering	276
Hadland Photonics Ltd	304
Heimann Optoelectronics GmbH	107
Herga Electric	308
Hidex Oy	022
Institut für Physikalische Hochtechnologie e.V.	112
Instituto de Bioingeniera - Fundacion CGH	212
Integrated Photomatrix Ltd	315
IROE - CNR	174
Jenasensoric e.V.	115
Kontron Instruments AG, Medical Sensors	227
Porvair Technology Ltd	367
Rank Taylor Hobson	370
Hungarian Academy of Sciences, ATKI-Institute for Materials Science	164
Research Centre for Ecological Technologies	200
Russian Academy of Science, Institute of Physics of Semiconductors	207
Sheffield Hallam University, Electronics	384
Strathclyde University, Department of Electronic & Electrical Engineering	392
University of Manchester, UMIST-Institute of Science and Technology	412
Université de Savoie, ESIGEC, Laboratoire CME	080
Université de Troyes, IBT	082

Signal Sensed : Optical
Application Area : HVAC, Fire Control, Security

Ablewood Technology	242
Baumer Electric Ltd	260
Dantec/Invent Measurement Technology GmbH	092
Heimann Optoelectronics GmbH	107
Herga Electric	308
Herga Electric	308
IMEC	005
INERIS - Verneuil-en-Halatte	063
Integrated Photomatrix Ltd	315
Jenasensoric e.V.	115
KUBE Electronics Ltd	228
Nitron Europe Ltd	167
Oldham	071
Porvair Technology Ltd	367
Remsdaq Ltd	373
Russian Academy of Science, Institute of Physics of Semiconductors	207
Sira Ltd	388
Strathclyde University, Department of Electronic & Electrical Engineering	392
VTT - Electronics	030
Weinlich GmbH & Co	160

Signal Sensed : Optical
Application Area : Military, Aerospace

AG Electro-Optics Ltd	249
Baumer Electric Ltd	260
Fogale Nanotech	056
FRABA Management GmbH	096

BitNet-Research Cente on Sensors & Systems Ltd	199
Brunel University, Brunel Centre for Manufacturing Metrology	267
Cambridge Consultants Ltd	268
City University, Dept. Electrical, Electronics & Information Engineering	276
C.S.O. Mesure	104
Delft Sensor Systems	003
Eltrotec Elektro GmbH	095
Franco Corradi	173
Fraunhofer-Institut für Siliziumtechnologie	102
Graticules Ltd	303
Hadland Photonics Ltd	304
IMEC	005
Institute of Optical Research	215
Integrated Photomatrix Ltd	315

Signal Sensed : Optical
Application Area : Military, Aerospace - continued

Jenasensoric e.V.	115
Kollmorgen Motion Technologies Group	328
Lambda Photometrics Ltd	330
Laser Optronic SRL	175
Minsk Research Institute of Radiomaterials	002
Ploskon AT	209
Porvair Technology Ltd	367
Rank Taylor Hobson	370
Redman Controls & Electronics Ltd	372
Russian Academy of Science, Institute of Physics of Semiconductors	207
Signaal USFA	187
Sira Ltd	388
Strathclyde University, Department of Electronic & Electrical Engineering	392
TNO - FEL, Physics & Electronic Laboratory	188
Universidad Complutense de Madrid	213
Université de Savoie, ESIGEC, Laboratoire CME	080
University of Warwick, Department of Engineering	419
VTT - Electronics	030
Weinlich GmbH & Co	160

Signal Sensed : Optical
Application Area : Telecommunications, Office Equipment

AG Electro-Optics Ltd	249
Baumer Electric Ltd	260
Cambridge Consultants Ltd	268
Department of Electronics	178
Fogale Nanotech	056
Franco Corradi	173
Fraunhofer-Institut für Siliziumtechnologie	102
City University, Dept. Electrical, Electronics & Information Engineering	276
Hadland Photonics Ltd	304
IMEC	005
Institut für Physikalische Hochtechnologie e.V.	112
Slovak Academy of Sciences, Institute of Electrical Engineering	210
Institute of Optical Research	013
Integrated Photomatrix Ltd	315
INPG - ENSERG, Laboratoire d'Electromagnétisme	064
Laser Optronic SRL	175
Minsk Research Institute of Radiomaterials	002
Ploskon AT	209
Hungarian Academy of Sciences, ATKI-Institute for Materials Science	164
Sira Ltd	388
Technische Universität Chemnitz-Zwickau, Institut für Mikrosysteme	144
Université de Savoie, ESIGEC, Laboratoire CME	222
Dibe University of Genoa	172
VTT - Electronics	030

Signal Sensed : Optical
Application Area : Meteorology, Environment

AG Electro-Optics Ltd	249
Baumer Electric Ltd	260
Bio-Orbit Oy	021
BitNet-Research Cente on Sensors & Systems Ltd	199
Cambridge Consultants Ltd	268
CEMAGREF, Groupement de Montpellier	038
CHROMDET/Analytical Instruments	201
City University, Dept. Electrical, Electronics & Information Engineering	276
CYLERGIE	044
Ecole Nationale Supérieure des Mines de St-Etienne, SPIN	047
Eltrotec Elektro GmbH	095
ETH Z, Centre for Chemical Sensors/Biosensors and Bioanalytical Chem.	224
F.U.L. - Fondation Universitaire Luxembourgeoise	004
Gestra (UK) Ltd	302
Hadland Photonics Ltd	304
Heimann Optoelectronics GmbH	107
Herga Electric	308
IMEC	005
Institut für Physikalische Hochtechnologie e.V.	112
Integrated Photomatrix Ltd	315
IROE - CNR	174
Jenasensoric e.V.	001
INPG - ENSERG, Laboratoire d'Electromagnétisme	064
Université Jean Monnet, Laboratoire Traitement du Signal et Instrumentation	081
Lambda Photometrics Ltd	330
Tampere University of Technology; Measurement & Information Technology	026
Nitron Europe Ltd	167
Parker Hannifin GmbH	129
Ploskon AT	209
Research Centre for Ecological Technologies	200
Revere Transducers Europe BV	185
Russian Academy of Science, Institute of Physics of Semiconductors	207
Silsoe Research Inst	386
Sira Ltd	388
Slovak Academy of Sciences, Institute of Electrical Engineering	210
Strathclyde University, Department of Electronic & Electrical Engineering	392
tec5 Sensorik und Systemtechnik GmbH	141
TNO - Nutrition & Food Research Institute	189
Technische Universität München, Lehrstuhl für Elektrische Messtechnik	148
Università degli Studi di Firenze, Dip Sanità Pubblica Epid.	179
Université de Savoie, ESIGEC, Laboratoire CME	080
Université de Troyes, IBT	082
University of Manchester, UMIST-Institute of Science and Technology	412
University of Cordoba, Analytical Chemistry Department	214
VTT - Electronics	030
WRC	424

Signal Sensed : Electrical
Application Area : Automotive

Signal Sensed : Electrical
Application Area : Consumer Products

Signal Sensed : Electrical
Application Area : Process Industries, Mining, Energy

Fraunhofer-Institut für Zerstörungsfreie Prüfverfahren	104
Gestra (UK) Ltd	302
H Warner & Son Ltd	311
Hawker Electronics Ltd	307
Heraeus Sensor	060
Rank Taylor Hobson	370
Research Centre for Ecological Technologies	200
Revere Transducers Europe	374
SFERE SA - Société Française d'Etudes et de Realisations Electroniques	077
Status Instruments Ltd	390

Signal Sensed : Electrical
Application Area : Process Industries, Mining, Energy - continued

tec5 Sensorik und Systemtechnik GmbH	141
Technische Universität Hamburg-Harburg, Optik & Messtechnik	146
Technology Partnership Plc	394
Technische Universität München, Lehrstuhl für Elektrische Meßtechnik	148
University of Manchester, UMIST-Institute of Science and Technology	413
Universität GH Siegen, Institut für Messtechnik	156
Universität Kassel, Fachbereich Physik	154
University of Lancaster, Department of Engineering	408
University of Reading, Department of Cybernetics	415
University of Southampton, Institute of Transducer Technology	417
Université de Troyes, IBT	082
Universiteit Twente, MESA Research Institute	190
Universidade Nova de Lisboa, Department of Applied Physics	198
University of Cordoba, Analytical Chemistry Department	214
Valmet Automation Kavanni Ltd	028
WRC	424
Zirox Sensoren & Elektronik GmbH	163

Signal Sensed : Electrical
Application Area : Manufacturing, Machinery

AEA Technology, TH2R2 Marketing & Sales	244
AFRC Silsoe Research Institute	247
Amot Controls	253
Applied Microengineering Ltd	256
Battelle Geneva Research Centre	218
Baumer Electric Ltd	260
Bedford Opto Ltd	261
Bio-Orbit Oy	021
Bolton Inst for Higher Education, Dept of Electronic Engineering	263
Bowers Internal Gauge Co Ltd	265
Bronkhorst High-Tech BV	182
Cambridge Consultants Ltd	268
Campbell Scientific Ltd	269
Castle Group Ltd	271
CSEM - Centre Suisse d'Electronique & de Microtechnique	221
DTH - Danmarks Tekniske Hojskole	018
Delft Sensor Systems	003
Università di Pavia, Department of Electronics	178
Digitron Instrumentation Ltd	285
Electromotif Electronic Ltd	290
Endress & Hauser (International) Holding AG	222
Entran	293
Groupe ESIEE	058
FGP Instrumentation	055
FR Sauter SA	225
Fraunhofer-Institut für Siliziumtechnologie	102
Fraunhofer-Institut für Zerstörungsfreie Prüfverfahren	104
Fraunhofer-Institut für Physikalische Messtechnik (IPM)	100
Geonor AS	195
Heraeus Sensor	060
Herga Electric Ltd	308
ICI Engineering	319
International Mining Consultants	314
IRD Condition Monitoring	317
Jackson Instruments & Controls Co Ltd	322
Jaltek Ltd	323
Kollmorgen Motion Technologies Group	328
KVARTA, Russia	202
Laaser & Co GmbH	119
Université Jean Monnet, Laboratoire Traitement du Signal et Instrumentation	081
Lucas Automotive Ltd	340
Tempere University of Technology; Measurement & Information Technology	026
Minsk Research Institute of Radiomaterials	002
Monitran Ltd	352
Murata Europe Management GmbH	126
PA Consulting Group	362
Panametrics Ltd	168
Parker Hannifin GmbH	129
PME - Meßsysteme H Zimmer GmbH	131
Precitec	133
Radio Frequency Investigation	369
Rank Taylor Hobson	370
Revere Transducers Europe	374
Russian Academy of Science, Institute of Automation & Electrometrics	206
Russian Academy of Science, Institute of Physics of Semiconductors	207
Sintef Delab	197
Spectrol Reliance Ltd	389
Status Instruments Ltd	390
tec5 Sensorik und Systemtechnik GmbH	141
Technische Universität Hamburg-Harburg, Optik & Messtechnik	146
Technology Partnership Plc	394
Tilt Measurement Ltd	398
Technische Universität München, Lehrstuhl für Elektrische Meßtechnik	148
University College of Galway, Deptartment of Physics	170
Universität GH Siegen, Institut für Messtechnik	156
Universität Kassel, Fachbereich Physik	154
University of Durham, Centre for Molecular Electronics	405
University of Edinburgh, Deptartment of Chemistry	406
University of Reading, Department of Cybernetics	415
Université de Troyes, IBT	082
Universidade Nova de Lisboa, Department of Applied Physics	198
Wenglor Sensoric GmbH	161
Zirox Sensoren & Elektronik GmbH	163

Signal Sensed : Electrical
Application Area : Medicine, Health Care

Bedford Opto Ltd	261
Bio-Orbit Oy	021
Bowers Internal Gauge Co Ltd	265
Bronkhorst High-Tech BV	182
CSEM - Centre Suisse d'Electronique & de Microtechnique	221
Franco Corradi	173
Fraunhofer-Institut für Physikalische Messtechnik (IPM)	100
H Tinsley & Co Ltd	310
Hawker Electronics Ltd	307
Heraeus Sensor	060

University of Cranfield, Biotechnology Centre 404
Dantec/Invent Measurement Technology GmbH 092
Digitron Instrumentation Ltd 285
Università degli Studi di Firenze, Dip Sanità Pubblica Epid. 179
Endress & Hauser (International) Holding AG 222
Entran 293
Groupe ESIEE 058
ETH Z, Centre for Chemical Sensors/Biosensors and Bioanalytical Chem. 224
Ferroperm A/S 019
FR Sauter SA 225
Herga Electric Ltd 308
Hidex Oy 022
Inserm U52, CNRS; Départ. Médecine Expérimentale 066
Instituto de Bioingeniera - Fundacion CGH 212
IEE - International Electronics & Engineering S. à r.l. 180
Jackson Instruments & Controls Co Ltd 322
Jaltek Ltd 323
Kontron Instruments AG, Medical Sensors 227
Laaser & Co GmbH 119
LEPMI-ENSEEG, Laboratory of Electrochemistry & Physical Chemistry 068

Signal Sensed : Electrical
Application Area : Medicine, Health Care - continued

Sheffield Hallam University, Electronics 384
Spectrol Reliance Ltd 389
Status Instruments Ltd 390
tec5 Sensorik und Systemtechnik GmbH 141
Technische Universität Hamburg-Harburg, Optik & Messtechnik 146
Technology Partnership Plc 394
TNO - Nutrition & Food Research Institute 189
Ultrasonic Sciences Ltd 401
Monitran Ltd 352
PA Consulting Group 362
Porvair Technology Ltd 367
Pulsar Developments Ltd 368
Radio Frequency Investigation 369
Rank Taylor Hobson 370
Research Centre for Ecological Technologies 200
Russian Academy of Science, Institute of Physics of Semiconductors 207
RWTH Aachen, WZL Laboratorium für Werkzeugmaschinen 135
Universität GH Siegen, Institut für Messtechnik 156
University of Lancaster, Department of Engineering 408
University of Reading, Department of Cybernetics 415
University of Southampton, Institute of Transducer Technology 417
Université de Troyes, IBT 082
Universidade Nova de Lisboa, Department of Applied Physics 198
University of Delft, Department of Electrical Engineering 191
Wenglor Sensoric GmbH 161

Signal Sensed : Electrical
Application Area : HVAC, Fire Control, Security

Amersham International Plc 254
Amot Controls 253
Baumer Electric Ltd 260
Bedford Opto Ltd 261
Bronkhorst High-Tech BV 182
Campbell Scientific Ltd 269
Castle Group Ltd 271
Dantec/Invent Measurement Technology GmbH 092
De Montfort University, Serc Centre 284
Digitron Instrumentation Ltd 285
Dublin City University, School of Chemical Sciences 165
Electromotif Electronic Ltd 290
Groupe ESIEE 058
Fraunhofer-Institut für Mikroelektronische Schaltungen & Systeme 099
H Warner & Son Ltd 311
Hawker Electronics Ltd 307
Heraeus Sensor 060
Jaltek Ltd 323
Kontron Instruments AG, Medical Sensors 227
KVARTA, Russia 202
Monitran Ltd 352
Murata Europe Management GmbH 126
Panametrics Ltd 168
Porvair Technology Ltd 367
Pulsar Developments Ltd 368
Radio Frequency Investigation 369
Russian Academy of Science, Institute of Physics of Semiconductors 207
Status Instruments Ltd 390
Technische Universität Hamburg-Harburg, Optik & Messtechnik 146
Technology Partnership Plc 394
Tilt Measurement Ltd 398
Technische Universität München, Lehrstuhl für Elektrische Meßtechnik 148
University College of Galway, Deptartment of Physics 170
University College of Wales, Deptartment of Biological Science 420
Universität Kassel, Fachbereich Physik 154
University of Edinburgh, Deptartment of Chemistry 406
University of Lancaster, Department of Engineering 408
University of Limerick, School of Engineering 171
University of Reading, Department of Cybernetics 415
Universiteit Twente, MESA Research Institute 190
Unitronic GmbH 150
Universidade Nova de Lisboa, Department of Applied Physics 198

Signal Sensed : Electrical
Application Area : Military, Aerospace

Ablewood Technology 242
Barnes & Associates 258
Baumer Electric Ltd 260
Bedford Opto Ltd 261
BitNet - Research Centre on Sensors & Systems Ltd 199
Bronkhorst High-Tech BV 182
Cambridge Consultants Ltd 268
Caradon Trend Ltd 270
Castle Group Ltd 271
EPFL - Ecole Polytechnique Fédérale de Lausanne. Dept. Electricité 223
Electromotif Electronic Ltd 290
Entran 293
PA Consulting Group 362
Panametrics Ltd 168
Porvair Technology Ltd 367
Pulsar Developments Ltd 368
Rank Taylor Hobson 370
Redman Controls & Electronics Ltd 372
Revere Transducers Europe 374
Russian Academy of Science, Institute of Physics of Semiconductors 207
Sheffield Hallam University, Environmental Research Centre 385
Signaal USFA 187
Sintef Delab 197
Status Instruments Ltd 390

Signal Sensed : Electrical
Application Area : Telecommunications, Office Equipment

Signal Sensed : Electrical
Application Area : Meteorology, Environment

Signal Sensed : Magnetic
Application Area : Automotive

3T BV	181
Ablewood Technology	242
Baumer Electric Ltd	260
Bronkhorst High-Tech BV	182
C4i, France	036
Cambridge Consultants Ltd	268
Czech Technical University Prague, Faculty of Electrical Engineering	012
Electromotif Electronic Ltd	290
Fraunhofer-Institut für Zerstörungsfreie Prüfverfahren	104
Gentech	301
IMEC	005
Institut für Physikalische Hochtechnologie e.V.	112
IEE - International Electronics & Engineering S. à r.l.	180
Lambda Photometrics Ltd	330
Monitran Ltd	352
Murata Europe Management GmbH	126
Panametrics Ltd	168
Ploskon AT	209
Radio Frequency Investigation	369
Siemens AG	139
Tilt Measurement Ltd	398
University of Transport, Rostov; Department of Automation	208
Unitronic GmbH	150
University of Modena	177
Zirox Sensoren & Elektronik GmbH	163

Signal Sensed : Magnetic
Application Area : Consumer Products

AB Elektronik GmbH	085
Ablewood Technology	242
C4i, France	036
Douglas Randall Ltd	286
Fraunhofer-Institut für Integrierte Schaltungen	098
Fraunhofer-Institut für Mikroelektronische Schaltungen & Systeme	099
Fraunhofer-Institut für Physikalische Messtechnik (IPM)	100
Gentech	301
IMEC	005
Lucas - Varity	341
Monitran Ltd	352
Murata Europe Management GmbH	126
Radio Frequency Investigation	369
Technology Partnership Plc	394
Universität GH Siegen, Institut für Messtechnik	156
Universität Kassel, Fachbereich Physik	154
University of Southampton, Institute of Transducer Technology	417
Universiteit Twente, MESA Research Institute	190
Unitronic GmbH	150
University of Delft, Department of Electrical Engineering	191
University of Modena	177

Signal Sensed : Magnetic
Application Area : Process Industries, Mining, Energy

Ablewood Technology	242
Alan Cobham Engineering	252
Amot Controls	253
Baumer Electric Ltd	260
Bronkhorst High-Tech BV	182
Cambridge Consultants Ltd	268
Czech Technical University of Prague	013
Douglas Randall Ltd	286
Electromotif Electronic Ltd	290
FRABA Management GmbH	096
Fraunhofer-Institut für Zerstörungsfreie Prüfverfahren	104
Fraunhofer-Institut für Mikroelektronische Schaltungen & Systeme	099
Gentech	301
H Warner & Son Ltd	311
IMEC	005
Institut für Physikalische Hochtechnologie e.V.	112
Slovak Academy of Sciences, Institute of Electrical Engineering	210
IRD Condition Monitoring	317
ITT Barton UK	321
Laaser & Co GmbH	119
Litre Meter	335
Littlemore Scientific Engineering Co	334
Tempere University of Technology; Measurement & Information Technology	026
Monitran Ltd	352
Murata Europe Management GmbH	126
PA Consulting Group	362
Panametrics Ltd	168
Ploskon AT	209
Radio Frequency Investigation	369
Hungarian Academy of Sciences, ATKI-Institute for Materials Science	164
Research Centre for Ecological Technologies	200
Technology Partnership Plc	394
Tilt Measurement Ltd	398
Universität GH Siegen, Institut für Messtechnik	156
Universität Kassel, Fachbereich Physik	154
University of Lancaster, Department of Engineering	408
University of Limerick, School of Engineering	171
Universiteit Twente, MESA Research Institute	190
WRC	424
Zirox Sensoren & Elektronik GmbH	163

Signal Sensed : Magnetic
Application Area : Manufacturing, Machinery

3T BV	181
ABB Kent Taylor Ltd	241
AEA Technology, Department TH2R2	244
Aerosonic Ltd	246
Alan Cobham Engineering	252
Bernstein Senso Plus GmbH	089
Bronkhorst Highn Tech BV	182
Cambridge Consultants Ltd	268
CSEM - Centre Suisse d'Electronique & de Microtechnique	221
DTH - Danmarks Tekniske Hojskole	018

Amot Controls 253
Bartington Instruments Ltd 259
Battelle Geneva Research Centre 218
Baumer Electric Ltd 260
Università di Pavia, Department of Electronics 178
Douglas Randall Ltd 286
Electromotif Electronic Ltd 290
Endress & Hauser (International) Holding AG 222

Signal Sensed : Magnetic
Application Area : Manufacturing, Machinery - continued

FR Sauter SA 225
Fraunhofer-Institut für Zerstörungsfreie Prüfverfahren 104
Fraunhofer-Institut für Physikalische Messtechnik (IPM) 100
Geonor AS 195
H Warner & Son Ltd 311
Hamos Elektronik 105
I/O Sensor Operations 316
ICI Engineering 319
IMEC 005
Institut für Physikalische Hochtechnologie e.V. 112
Slovak Academy of Sciences, Institute of Electrical Engineering 210
International Mining Consultants 314
IRD Condition Monitoring 317
Jackson Instruments & Controls Co Ltd 706 322
Laaser & Co GmbH 119
Litre Meter 335
Tempere University of Technology; Measurement & Information Technology 026
Monitran Ltd 352
PA Consulting Group 362
Panametrics Ltd 168
Ploskon AT 209
Radio Frequency Investigation 369
Siemens AG 139
SNT Sensortechnik AG 234
tec5 Sensorik und Systemtechnik GmbH 141
Technology Partnership Plc 394
Tilt Measurement Ltd 398
University College of Galway, Deptartment of Physics 170
Universität GH Siegen, Institut für Messtechnik 156
Universität Kassel, Fachbereich Physik 154
University of Southampton, Institute of Transducer Technology 417
University of Modena 177
Zirox Sensoren & Elektronik GmbH 163

Signal Sensed : Magnetic
Application Area : Medicine, Health Care

AB Elektronik GmbH 085
ABB Kent Taylor Ltd 241
Aerosonic Ltd 246
Bartington Instruments Ltd 259
Bernstein Senso Plus GmbH 089
Bronkhorst High-Tech BV 182
CSEM - Centre Suisse d'Electronique & de Microtechnique 221
Czech Technical University of Prague 013
Dewit Industrial Sensors BV 183
Douglas Randall Ltd 286
Endress & Hauser (International) Holding AG 222
FR Sauter SA 225
Fraunhofer-Institut für Mikroelektronische Schaltungen & Systeme 099
Fraunhofer-Institut für Physikalische Messtechnik (IPM) 100
Hamos Elektronik 105
Institut für Physikalische Hochtechnologie e.V. 112
Instituto de Bioingeniera - Fundacion CGH 212
Jackson Instruments & Controls Co Ltd 706 Littlemore Scientific Engineering 334
Monitran Ltd 352
PA Consulting Group 362
Ploskon AT 209
Radio Frequency Investigation 369
Research Centre for Ecological Technologies 200
RWTH Aachen, Institüt für Werkzeugmaschinen 135
SNT Sensortechnik AG 234
Technology Partnership Plc 394
Universität GH Siegen, Institut für Messtechnik 156
University of Lancaster, Department of Engineering 408
Université de Troyes, IBT 082
Universiteit Twente, MESA Research Institute 190

Signal Sensed : Magnetic
Application Area : HVAC, Fire Control, Security

Bartington Instruments Ltd 259
Baumer Electric Ltd 260
Bronkhorst High-Tech BV 182
De Montfort University, Serc Centre 284
Douglas Randall Ltd 286
Electromotif Electronic Ltd 290
H Warner & Son Ltd 311
Monitran Ltd 352
MR Sensors Ltd 353
Murata Europe Management GmbH 126
Radio Frequency Investigation 369
Technology Partnership Plc 394
University College of Galway, Deptartment of Physics 170
Universität Kassel, Fachbereich Physik 154

Signal Sensed : Magnetic
Application Area : Military, Aerospace

Ablewood Technology 242
Alan Cobham Engineering 252
Baumer Electric Ltd 260
Bronkhorst High-Tech BV 182
Cambridge Consultants Ltd 268
Czech Technical University Prague, Faculty of Electrical Engineering 012
Litre Meter 335
Monitran Ltd 352
National Defence Research Establishment, Dept of Sensor Technology 216
Oxford Instruments Ltd 361
PA Consulting Group 362
Panametrics Ltd 168

Signal Sensed : Magnetic
Application Area : Telecommunications, Office Equipment

Signal Sensed : Magnetic
Application Area : Meteorology, environment

Signal Sensed : Thermal
Application Area : Automotive

Ablewood Technology	242
AGEMA Infrared Systems	248
BICC Thermoheat	262
Bronkhorst High-Tech BV	182
Brunel University, Brunel Centre for Manufacturing Metrology	267
Burster Präzisionsmeßtechnik GmbH & Co kg	091
Campbell Scientific Ltd	269
Coreci SA	042
Cropico Ltd	278
Delta-T Devices Ltd	283
Fraunhofer-Institut für Siliziumtechnologie	102
Fraunhofer-Institut für Zerstörungsfreie Prüfverfahren	104
Hawco Ltd	386
Heimann Optoelectronics GmbH	107
IMEC	005
IMPAC Electronic GmbH	113
Jenasensoric e.V.	115
Labfacility Ltd	329
Lambda Photometrics Ltd	330
Universität Erlangen-Nürnberg, Lehrstuhl Technische Thermodynamik (LTT)	152
Tempere University of Technology; Measurement & Information Technology	026
Murata Europe Management GmbH	126
SA Thermibel	010
Solstat Operations	008
Strathclyde University, Department of Electronic & Electrical Engineering	392
Thermo Electric International Ltd	396
Thermoteknix Systems Ltd	397
Tilt Measurement Ltd	398
Technische Universität Berlin, Technologien der Mikropherik	143
Universidade Nova de Lisboa, Department of Applied Physics	198
Vishay Ltd	423
Weber Sensors GmbH	159
Zirox Sensoren & Elektronik GmbH	163

Signal Sensed : Magnetic
Application Area : Consumer Products

AB Elektronik GmbH	085
ABB Kent Taylor Ltd	241
Ablewood Technology	242
Battelle Geneva Research Centre	218
British Ceramic Research Ltd	266
Burster Präzisionsmeßtechnik GmbH & Co kg	091
Dantek Elektronik	015
Digitron Instrumentation Ltd	285
ERA Technology Ltd	295
ETI - Electronic Temperature Instruments	297
Fraunhofer-Institut für Siliziumtechnologie	102
Fraunhofer-Institut für Physikalische Messtechnik (IPM)	100
H Warner & Son Ltd	311
Hawco Ltd	306
Heimann Optoelectronics GmbH	107
Heraeus Sensor GmbH	109
IMEC	005
IMPAC Electronic GmbH	113
IEMN - Institut d'Electronique et de Microélectronique du Nord	061
Institut für Physikalische Hochtechnologie e.V.	112
Jenasensoric e.V.	115
Kayser Threde GmbH	116
Kobold Instruments Ltd	327
KUBE Electronics Ltd	228
Labfacility Ltd	329
Université Jean Monnet, Laboratoire Traitement du Signal et Instrumentation	081
Lambda Photometrics Ltd	330
Lucas Automotive Ltd	340
Lucas - Varity	341
Tempere University of Technology; Measurement & Information Technology	026
Microtherm GmbH	124
Murata Europe Management GmbH	126
Nulectrohms	357
Optilas Systems	072
Ped Ltd	363
PME - Meßsysteme H Zimmer GmbH	131
Rhopoint Ltd	375
SA Thermibel	010
Solstat Operations	008
Status Instruments Ltd	390
STW - Sensor Technik Wiedemann GmbH	140
Technische Uni Hamburg-Harburg, Inst Halbleitertechnologie	147
Technology Partnership Plc	394
Thermal Detection Ltd	395
Thermo Electric International Ltd	396
Trolex Ltd	400
Technische Universität Berlin, Technologien der Mikropherik	143
Ultrakust Electronic GmbH	149
Uni College of London, Electronic & Electrical Engineering Dept	409
Universität GH Siegen, Institut für Messtechnik	156
Universität Kassel, Fachbereich Physik	154
University of Limerick, School of Engineering	171
University of Southampton, Institute of Transducer Technology	417
Université de Troyes, IBT	082
Universidade Nova de Lisboa, Department of Applied Physics	198
University of Delft, Department of Electrical Engineering	191
University of Modena	177
Vaisala Oy	027
Vishay Ltd	423
VTT - Electronics	030
Weber Sensors GmbH	159
Xensor Integration BV	193

Signal Sensed : Thermal
Application Area : Process Industries, Mining, Energy

Ablewood Technology	242
AGEMA Infrared Systems	248
Amot Controls	253
Applied Microengineering Ltd	256
BICC Thermoheat	262
Bronkhorst High-Tech BV	182
Brunel University, Brunel Centre for Manufacturing Metrology	267
Burster Präzisionsmeßtechnik GmbH & Co kg	091
CYLERGIE	044
Czech Technical University of Prague	013
Delta T Devices Ltd	283
Digitron Instrumentation Ltd	285
ETI - Electronic Temperature Instruments	297
Franco Corradi	173
Fraunhofer-Institut für Zerstörungsfreie Prüfverfahren	104
Gestra (UK) Ltd	302

Campbell Scientific Ltd 269
Coreci SA 042
Cropico Ltd 278
H Warner & Son Ltd 311
Haenni & Co Ltd 226
Hawco Ltd 306

Signal Sensed : Thermal
Application Area : Process Industries, Mining, Energy - continued

Heimann Optoelectronics GmbH 107
Heraeus Sensor GmbH 109
Herga Electric Ltd 308
IMEC 005
IMPAC Electronic GmbH 113
IMT Frankfurt 114
INERIS, Verneuil-en-Halatte 063
INRA, Laboratoire de Génie des Procédés Alimentaires 065
IEMN - Institut d'Electronique et de Microélectronique du Nord 061
Institut für Physikalische Hochtechnologie e.V. 112
IRM SA 006
ITT Barton UK 321
Jenasensoric e.V. 115
KUBE Electronics Ltd 228
Laaser & Co GmbH 119
Labfacility Ltd 329
Lambda Photometrics Ltd 330
Laser Optronic SRL 175
Lee Dickens Ltd 332
Universität Erlangen-Nürnberg, Lehrstuhl Technische Thermodynamik (LTT) 152
Litre Meter 335
Tempere University of Technology; Measurement & Information Technology 026
Microtherm GmbH 124
Minsk Research Institute of Radiomaterials 002
Murata Europe Management GmbH 126
Novocontrol GmbH 127
Nulectrohms 357
Optilas Systems 072
PA Consulting Group 362
Parker Hannifin GmbH 129
Research Centre for Ecological Technologies 200
Rhopoint Ltd 375
SA Thermibel 010
Schmidt Feintechnik GmbH 137
Sensing Devices Ltd 381
SFERE SA - Société Française d'Etudes et de Realisations Electroniques 077
Status Instruments Ltd 390
Strathclyde University, Department of Electronic & Electrical Engineering 392
STW - Sensor Technik Wiedemann GmbH 140
TCS sprl 009
Teagasc 169
Technology Partnership Plc 394
Thermo Electric International Ltd 396
Thermoteknix Systems Ltd 397
Tilt Measurement Ltd 398
Universität GH Siegen, Institut für Messtechnik 156
Universität Kassel, Fachbereich Physik 154
University of Lancaster, Department of Engineering 408
Universidade Nova de Lisboa, Department of Applied Physics 198
Valmet Automation Kavanni Ltd 028
Vishay Ltd 423
Weber Sensors GmbH 159
WIKA Alexander Wiegand GmbH 162
Xensor Integration BV 193
Zirox Sensoren & Elektronik GmbH 163

Signal Sensed : Thermal
Application Area : Manufacturing, Machinery

ABB Kent Taylor Ltd 241
AEA Technology, Department TH2R2 244
AGEMA Infrared Systems 248
Amot Controls 253
Applied Microengineering Ltd 256
Automatic Systems Laboratories Ltd 257
Battelle Geneva Research Centre 218
BICC Thermoheat 262
British Ceramic Research Ltd 266
Bronkhorst High-Tech BV 182
Brunel University, Brunel Centre for Manufacturing Metrology 267
Burster Präzisionsmeßtechnik GmbH & Co kg 091
Campbell Scientific Ltd 269
CSEM - Centre Suisse d'Electronique & de Microtechnique 221
Coreci SA 042
Danfoss AS 014
Delft Sensor Systems 003
Digitron Instrumentation Ltd 285
Endress & Hauser (International) Holding AG 222
ETC Ltd 296
ETI - Electronic Temperature Instruments 297
FR Sauter SA 225
Fraunhofer-Institut für Integrierte Schaltungen 098
Fraunhofer-Institut für Zerstörungsfreie Prüfverfahren 104
Fraunhofer-Institut für Physikalische Messtechnik (IPM) 100
GEC Marconi Materials Technology Ltd 300
Haenni & Co Ltd 226
Hartmann & Braun AG 106
Hawco Ltd 306
Heraeus Sensor GmbH 109
Herga Electric Ltd 308
Kayser Threde GmbH 116
KUBE Electronics Ltd 228
KVARTA, Russia 202
Labfacility Ltd 329
Université Jean Monnet, Laboratoire Traitement du Signal et Instrumentation 081
Lambda Photometrics Ltd 330
Land Combustion 331
Laser Optronic SRL 175
Lee Dickens Ltd 332
Litre Meter 335
Lucas Automotive Ltd 340
Menvier Hybrids Ltd 348
Microsens SA 229
Mikro Mess-Vertriebs GmbH 125
Murata Europe Management GmbH 126
Nitron Europe Ltd 167
Novocontrol GmbH 127
Nulectrohms 357
Optilas Systems 072
PA Consulting Group 362
Parker Hannifin GmbH 129
Ploskon AT 209
PME - Meßsysteme H Zimmer GmbH 131
Rhopoint Ltd 375
Rosemount Ltd 377
SA Thermibel 010
Schiltknecht Messtechnik AG 232
Schmidt Feintechnik GmbH 137
Status Instruments Ltd 390
Strathclyde University, Department of Electronic & Electrical Engineering 392
STW - Sensor Technik Wiedemann GmbH 140

ICI Engineering	319
IMEC	005
IMPAC Electronic GmbH	113
IMT Frankfurt	114
IEMN - Institut d'Electronique et de Microélectronique du Nord	061
International Mining Consultants	314
ISA Controls Ltd	318
Jackson Instruments & Controls Co Ltd 706 Kobold Instruments Ltd	327
Technology Partnership Plc	394
Thermal Detection Ltd	395
Thermo Electric International Ltd	396
Tilt Measurement Ltd	398
Trafag AG	236
Trolex Ltd	400
Ultrakust Electronic GmbH	149
University College of Galway, Deptartment of Physics	170

Signal Sensed : Thermal
Application Area : Manufacturing, Machinery - continued

Uni College of London, Electronic & Electrical Engineering Dept	409
Universität GH Siegen, Institut für Messtechnik	156
Universität Kassel, Fachbereich Physik	154
University of Durham, Centre for Molecular Electronics	405
Uni of Southampton, Inst of Cryogenics	416
University of Southampton, Institute of Transducer Technology	417
Universidade Nova de Lisboa, Department of Applied Physics	198
University of Delft, Department of Electrical Engineering	191
Vaisala Oy	027
Vector Instruments	422
Vibro Meter SA	238
Vishay Ltd	423
WIKA Alexander Wiegand GmbH	162
Xensor Integration BV	193
Zirox Sensoren & Elektronik GmbH	163

Signal Sensed : Thermal
Application Area : Medicine, Health Care

ABB Kent Taylor Ltd	241
AGEMA Infrared Systems	248
Automatic Systems Laboratories Ltd	257
Battelle Geneva Research Centre	218
BICC Thermoheat	262
British Ceramic Research Ltd	266
Bronkhorst High TechBV	182
CSEM - Centre Suisse d'Electronique & de Microtechnique	221
Czech Technical University of Prague	013
Danfoss AS	014
Dantec/Invent Measurement Technology GmbH	092
Digitron Instrumentation Ltd	285
Endress & Hauser (International) Holding AG	222
ERA Technology Ltd	295
ETI - Electronic Temperature Instruments	297
FR Sauter SA	225
Franco Corradi	173
Fraunhofer-Institut für Integrierte Schaltungen	098
Fraunhofer-Institut für Siliziumtechnologie	102
Fraunhofer-Institut für Physikalische Messtechnik (IPM)	100
H Tinsley & Co Ltd	310
Haenni & Co Ltd	226
Heimann Optoelectronics GmbH	107
Heraeus Sensor GmbH	109
IMEC	005
IMPAC Electronic GmbH	113
IMT Frankfurt	114
Instituto de Bioingeniera - Fundacion CGH	212
Jackson Instruments & Controls Co Ltd 706 Kobold Instruments Ltd	327
Jenasensoric e.V.	115
Kayser Threde GmbH	116
KUBE Electronics Ltd	228
Laser Optronic SRL	175
Litre Meter	335
Menvier Hybrids Ltd	348
Microtherm GmbH	124
Murata Europe Management GmbH	126
Nitron Europe Ltd	167
IEMN - Institut d'Electronique et de Microélectronique du Nord	061
Nulectrohms	357
Optilas Systems	072
PA Consulting Group	362
Ped Ltd	363
Research Centre for Ecological Technologies	200
Rhopoint Ltd	375
Roditi International Corp Ltd	376
Sheffield Hallam University, Electronics	384
Solstat Operations	008
Status Instruments Ltd	390
STW - Sensor Technik Wiedemann GmbH	140
Technology Partnership Plc	394
Thermal Detection Ltd	395
Thermo Electric International Ltd	396
Trafag AG	236
Trolex Ltd	400
Ultrakust Electronic GmbH	149
Universität GH Siegen, Institut für Messtechnik	156
University of Lancaster, Department of Engineering	408
Université de Troyes, IBT	082
Universidade Nova de Lisboa, Department of Applied Physics	198
Vaisala Oy	027
Vishay Ltd	423
WIKA Alexander Wiegand GmbH	162
Xensor Integration BV	193

Signal Sensed : Thermal
Application Area : HVAC, Fire Control, Security

AGEMA Infrared Systems	248
Amot Controls	253
Applied Microengineering Ltd	256
Automatic Systems Laboratories Ltd	257
BICC Thermoheat	262
Bronkhorst High-Tech BV	182
Campbell Scientific Ltd	269
Dantec/Invent Measurement Technology GmbH	092
IMT Frankfurt	114
Institut für Physikalische Hochtechnologie e.V.	112
IRM SA	006
Jenasensoric e.V.	115
Kontron Instruments AG, Medical Sensors	227
KUBE Electronics Ltd	228
KVARTA, Russia	202
Labfacility Ltd	329

Digitron Instrumentation Ltd	285
ERA Technology Ltd	295
ETI - Electronic Temperature Instruments	297
Fraunhofer-Institut für Mikroelektronische Schaltungen & Systeme	099
Fraunhofer-Institut für Physikalische Messtechnik (IPM)	100
Gentech	301
H Warner & Son Ltd	311
Haenni & Co Ltd	226
Hartmann & Braun AG	106
Hawco Ltd	306
Heimann Optoelectronics GmbH	107
Herga Electric Ltd	308
IMEC	005
Laser Optronic SRL	175
Lee Dickens Ltd	332
Microsens SA	229
Microtherm GmbH	124
Murata Europe Management GmbH	126
IEMN - Institut d'Electronique et de Microélectronique du Nord	061
Nulectrohms	357
Optilas Systems	072
PA Consulting Group	362
Rhopoint Ltd	375
Schmidt Feintechnik GmbH	137
Sintef Delab	197
Solstat Operations	008

Signal Sensed : Thermal
Application Area : HVAC, Fire Control, Security - continued

Status Instruments Ltd	390
Strathclyde University, Department of Electronic & Electrical Engineering	392
Technology Partnership Plc	394
Thermal Detection Ltd	395
Thermo Electric International Ltd	396
TNO - FEL, Physics & Electronic Laboratory	188
Technische Universität Berlin, Technologien der Mikropherik	143
Ultrakust Electronic GmbH	149
University College of Galway, Deptartment of Physics	170
Universität Kassel, Fachbereich Physik	154
University of Aston, Speciality Materials Research Group	402
Universiteit Twente, MESA Research Institute	190
Universidade Nova de Lisboa, Department of Applied Physics	198
Vishay Ltd	423
Weber Sensors GmbH	159
Zirox Sensoren & Elektronik GmbH	163

Signal Sensed : Thermal
Application Area : Military, Aerospace

AGEMA Infrared Systems	248
Airflow Developments Ltd	251
Automatic Systems Laboratories Ltd	257
Battelle Geneva Research Centre	218
BICC Thermoheat	262
BitNet - Research Centre on Sensors & Systems Ltd	199
British Ceramic Research Ltd	266
Bronkhorst High-Tech BV	182
Brunel University, Brunel Centre for Manufacturing Metrology	267
Burster Präzisionsmeßtechnik GmbH & Co kg	091
Caradon Trend Ltd	270
CSEM - Centre Suisse d'Electronique & de Microtechnique	221
Cropico Ltd	278
Danfoss AS	014
Dantek Elektronik	015
Delft Sensor Systems	003
Digitron Instrumentation Ltd	285
EPFL - Ecole Polytechnique Fédérale de Lausanne, Dept. Electricité	223
Endress & Hauser (International) Holding AG	222
ERA Technology Ltd	295
ETC Ltd	296
ETI - Electronic Temperature Instruments	297
Franco Corradi	173
Fraunhofer-Institut für Physikalische Messtechnik (IPM)	100
GEC Marconi Materials Technology Ltd	300
Hartmann & Braun AG	106
Heraeus Sensor GmbH	109
IMEC	005
IMPAC Electronic GmbH	113
Institut für Physikalische Hochtechnologie e.V.	112
ISA Controls Ltd	318
Jackson Instruments & Controls Co Ltd 706 Kobold Instruments Ltd	327
Jenasensoric e.V.	115
KUBE Electronics Ltd	228
Labfacility Ltd	329
Lambda Photometrics Ltd	330
Laser Optronic SRL	175
Lee Dickens Ltd	332
Microsens SA	229
Microtherm GmbH	124
Mikro Mess Vertriebs GmbH	125
Monode Ltd	351
Murata Europe Management GmbH	126
Nulectrohms	357
Optilas Systems	072
PA Consulting Group	362
Rhopoint Ltd	375
Schiltknecht Messtechnik AG	232
Scientific-Technical Consulting	233
Signaal USFA	187
Solstat Operations	008
Status Instruments Ltd	390
Strathclyde University, Department of Electronic & Electrical Engineering	392
Technology Partnership Plc	394
Thermal Detection Ltd	395
Thermo Electric International Ltd	396
Thermoteknix Systems Ltd	397
Tilt Measurement Ltd	398
TNO - FEL, Physics & Electronic Laboratory	188
Trafag AG	236
Trolex Ltd	400
Ultrakust Electronic GmbH	149
Universität Kassel, Fachbereich Physik	154
University of Southampton, Institute of Transducer Technology	417
Vaisala Oy	027
Vishay Ltd	423
VTT - Electronics	030
WIKA Alexander Wiegand GmbH	162
Xensor Integration BV	193

Signal Sensed : Thermal
Application Area : Telecommunications, Office Equipment

ABB Kent Taylor Ltd	241
Automatic Systems Laboratories Ltd	257
Dantek Elektronik	015
Digitron Instrumentation Ltd	285
ERA Technology Ltd	295
Ferranti Technologies Ltd	298
Franco Corradi	173
Fraunhofer-Institut für Siliziumtechnologie	102
GEC Marconi Materials Technology Ltd	300
H Tinsley & Co Ltd	310
Hawco Ltd	306
Heraeus Sensor GmbH	109
Hollming Ltd	023
IMEC	005
ISA Controls Ltd	318
Kayser Threde GmbH	116
Lee Dickens Ltd	332
Lucas Automotive Ltd	340
Minsk Research Institute of Radiomaterials	002
Murata Europe Management GmbH	126
Nulectrohms	357
PA Consulting Group	362
PME - Meßsysteme H Zimmer GmbH	131
Rhopoint Ltd	375
Rosemount Ltd	377
TCS sprl	009
Technology Partnership Plc	394
Tilt Measurement Ltd	398
TNO - FEL, Physics & Electronic Laboratory	188
Trafag AG	236
Uni College of London, Electronic & Electrical Engineering Dept	409
University of Durham, Centre for Molecular Electronics	405
University of Reading, Department of Cybernetics	415
Uni of Southampton, Inst of Cryogenics	416
University of Southampton, Institute of Transducer Technology	417
Università di Pavia, Department of Electronics	178

Signal Sensed : Thermal
Application Area : Telecommunications, Office Equipment - continued

Vaisala Oy	027
Vibro Meter SA	238
Vishay Ltd	423

Signal Sensed : Thermal
Application Area : Meteorology, Environment

AGEMA Infrared Systems	248
Amot Controls	253
BICC Thermoheat	262
BitNet - Research Centre on Sensors & Systems Ltd	199
Bronkhorst High-Tech BV	182
Campbell Scientific Ltd	269
Coreci SA	042
Cropico Ltd	278
Delta - T Devices Ltd	283
Digitron Instrumentation Ltd	285
Endress & Hauser (International) Holding AG	222
ETI - Electronic Temperature Instruments	297
F.U.L. - Fondation Universitaire Luxembourgeoise	004
Franco Corradi	173
H Tinsley & Co Ltd	310
Haenni & Co Ltd	226
Hawco Ltd	306
Herga Electric Ltd	308
IMEC	005
IMPAC Electronic GmbH	113
IMT Frankfurt	114
IEMN - Institut d'Electronique et de Microélectronique du Nord	061
Institut für Physikalische Hochtechnologie e.V.	112
IRM SA	006
Jenasensoric e.V.	115
Kobold Instruments Ltd	327
KUBE Electronics Ltd	228
KVARTA, Russia	202
Université Jean Monnet, Laboratoire Traitement du Signal et Instrumentation	081
Laser Optronic SRL	175
Lee Dickens Ltd	332
Litre Meter	335
Tempere University of Technology; Measurement & Information Technology	026
Microtherm GmbH	124
Murata Europe Management GmbH	126
Novocontrol GmbH	127
Nulectrohms	357
Parker Hannifin GmbH	129
Ploskon AT	209
Research Centre for Ecological Technologies	200
Rhopoint Ltd	375
SFERE SA - Société Française d'Etudes et de Realisations Electroniques	077
Silsoe Research Institute	386
Status Instruments Ltd	390
Strathclyde University, Department of Electronic & Electrical Engineering	392
Technology Partnership Plc	394
Thermo Electric International Ltd	396
Technische Universität Berlin, Technologien der Mikropherik	143
Université de Troyes, IBT	082
Vaisala Oy	027
Weber Sensors GmbH	159
WIKA Alexander Wiegand GmbH	162
Xensor Integration BV	193

Signal Sensed : Chemical
Application Area : Automotive

Ai Cambridge	250
C4i, France	036
University of Cranfield, Biotechnology Centre	404
Ecole Nationale Supérieure des Mines de St-Etienne, SPIN	047
Fraunhofer-Institut für Siliziumtechnologie	102
Gas Analysis & Sensing Group	418
IMEC	005
INERIS, Verneuil-en-Halatte	063
Institute for Physical & Theoretical Chemistry	157
Jenasensoric e.V.	115
Lasmea, Uni Blaise Pascal, France	078
Universität Erlangen-Nürnberg, Lehrstuhl Technische Thermodynamik (LTT)	152
Motorola Sensor Products Div	070
Strathclyde University, Department of Electronic & Electrical Engineering	392
Temic Microsystems Sensors	142
TNO - FEL, Physics & Electronic Laboratory	188
Technische Universität Berlin, Technologien der Mikropherik	143
Unitronic GmbH	150
University of Kent, Centre for Materials Research	407
University of Warwick, Department of Engineering	419

Signal Sensed : Chemical
Application Area : Consumer Products

ABB Kent Taylor Ltd	241
AEA Technology Plc, Product Development Services Group	245
Ai Cambridge	250
Analytical Development Co Ltd	255
Applied Microengineering Ltd	256
Asea Brown Boveri AG	084
Battelle Geneva Research Centre	218
British Ceramic Research Ltd	266
C4i, France	036
City Technology Ltd	275
University of Cranfield, Biotechnology Centre	404
De Montfort University, Serc Centre	284
Digitron Instrumentation Ltd	285
Fraunhofer-Institut für Physikalische Messtechnik (IPM)	100
Gas Analysis & Sensing Group	418
Groupe Electrochimie, Universite de Nantes	079
IMEC	005
International Instruments Ltd	313
Jenasensoric e.V.	115
Lucas Automotive Ltd	340
Microsens SA	229
Moisture Control & Measurement Ltd	350
Motorola Sensor Products Div	070
National Physical Laboratory	356
Oldham	071
Porvair Technology Ltd	367
Rosemount Ltd	377
Russel pH Ltd	378
Technology Partnership Plc	394
Trolex Ltd	400
Technische Universität Berlin, Technologien der Mikropherik	143
Ultrakust Electronic GmbH	149
University of Manchester, UMIST-Institute of Science and Technology	413
University der Bundeswehr, Institute für Mess- und Automatisierungstechnik	151
Universität GH Siegen, Institut für Messtechnik	156
Universität Kassel, Fachbereich Physik	154
University of Southampton, Institute of Transducer Technology	417
Université de Troyes, IBT	082
Unitronic GmbH	150
University of Kent, Centre for Materials Research	407
University of Warwick, Department of Engineering	419
Zellweger Analytics Ltd	425

Signal Sensed : Chemical
Application Area : Process Industries, Mining, Energy

AEA Technology Plc, Product Development Services Group	245
Ai Cambridge	250
Airflow Developments Ltd	251
Applied Microengineering Ltd	256
Bio-Orbit Oy	021
British Ceramic Research Ltd	266
ETH Z, Centre for Chemical Sensors/Biosensors and Bioanalytical Chem.	224
Coreci SA	042
University of Cranfield, Biotechnology Centre	404
Crowcon Detection Instruments Ltd	279
CYLERGIE	044
Ecole Nationale Supérieure des Mines de St-Etienne, SPIN	047
Entran	293
ETI - Electronic Temperature Instruments	297
Fraunhofer-Institut für Produktionstechnik & Automatisierung	101
Fraunhofer Institute Solid State Technology	103
Gas Analysis & Sensing Group	418
Herga Electric Ltd	308
IMEC	005
INERIS, Verneuil-en-Halatte	063
INRA, Laboratoire de Génie des Procédés Alimentaires	065
Inst of Nuclear Physics, Uni of Moscow Capteurs Instrumentation Technique	203
International Instruments Ltd	313
IRM SA	006
Jenasensoric e.V.	115
Lee Dickens Ltd	332
Universität Erlangen-Nürnberg, Lehrstuhl Technische Thermodynamik (LTT)	152
LTH Electronics Ltd	339
Microsens SA	229
Nitron Europe Ltd	167
nstitute for Physical & Theoretical Chemistry	157
Oldham	071
PA Consulting Group	362
Porvair Technology Ltd	367
Research Centre for Ecological Technologies	200
Russel pH Ltd	378
Strathclyde University, Department of Electronic & Electrical Engineering	392
Technische Universität Hamburg-Harburg, Optik & Messtechnik	146
Technology Partnership Plc	394
Temic Microsystems Sensors	142
Technische Universität München, Lehrstuhl für Elektrische Meßtechnik	148
University der Bundeswehr, Institute für Mess- und Automatisierungstechnik	151
Universität GH Siegen, Institut für Messtechnik	156
Universität Kassel, Fachbereich Physik	154
University of Luton, Biosensors Research Group	410
Université de Troyes, IBT	082
University of Delft, Department of Electrical Engineering	191
University of Kent, Centre for Materials Research	407
University of Warwick, Department of Engineering	419
Valmet Automation Kavanni Ltd	028

JMS Consultants 325
Laboratoire Chimie Physique Environnement 067
Université Jean Monnet, Laboratoire Traitement du Signal et Instrumentation 081
Lasmea, Uni Blaise Pascal, France 078
VITO Flemish Institute for Technological Research 011
WRC 424
Xensor Integration BV 193
Zirox Sensoren & Elektronik GmbH 163

Signal Sensed : Chemical
Application Area : Manufacturing, Machinery

ABB Kent Taylor Ltd 241
AEA Technology Plc, Product Development Services Group 245
AEA Technology, Department TH2R2 244
Ai Cambridge 250
Applied Microengineering Ltd 256
Asea Brown Boveri AG 084
Battelle Geneva Research Centre 218
Bio-Orbit Oy 021
British Ceramic Research Ltd 266
City Technology Ltd 275
University of Cranfield, Biotechnology Centre 404
Crowcon Detection Instruments Ltd 279
Danfoss AS 014
Digitron Instrumentation Ltd 285
Dublin City Uni, School of Physical Sciences 166
Edinburgh Sensors Ltd 289
Envin Scientific Products Ltd 294
ETC Ltd 296
ETI - Electronic Temperature Instruments 297
Fraunhofer-Institut für Physikalische Messtechnik (IPM) 100
Fraunhofer-Institut für Produktionstechnik & Automatisierung 101
Gas Analysis & Sensing Group 418
GEC Marconi Materials Technology Ltd 300
Hartmann & Braun AG 106
Herga Electric Ltd 308
HSE - Health & Safety Exec 309
ICI Engineering 319
International Mining Consultants 314
Jandel Engineering Ltd 324
Land Combustion 331
Lee Dickens Ltd 332
Lee Integer Ltd 333
Loughborough Uni, Dept of Chemistry 338
Measuretech Ltd 346
Microsens SA 229
Mikro Mess Vertriebs GmbH 125
Moisture Control & Measurement Ltd 350
Monode Ltd 351
National Physical Laboratory 356
Oldham 071
PA Consulting Group 362
Roditi International Corp Ltd 376
Rosemount Ltd 377
Rotronic AG 231
Scientific-Technical Consulting 233
Sentron BV 186
Servomex (UK) Ltd 383
Technische Uni Hamburg-Harburg, Inst Halbleitertechnologie 147
Technische Universität Hamburg-Harburg, Optik & Messtechnik 146
Technology Partnership Plc 394
Trolex Ltd 400
Ultrakust Electronic GmbH 149
University of Manchester, UMIST-Institute of Science and Technology 413
University College of Galway, Deptartment of Physics 170
University der Bundeswehr, Institute für Mess- und Automatisierungstechnik 151
Universität GH Siegen, Institut für Messtechnik 156
Universität Kassel, Fachbereich Physik 154
University of Birmingham, Metallurgy & Materials Dept 403
University of Durham, Centre for Molecular Electronics 405
University of Edinburgh, Deptartment of Chemistry 406
University of Luton, Biosensors Research Group 410
University of Southampton, Institute of Transducer Technology 417
Université de Troyes, IBT 082
Uni of West of England, Bristol, Molecular Electronics & Surface Science Gro 421
Universiteit Twente, MESA Research Institute 190
University of Kent, Centre for Materials Research 407
Vector Instruments 422
Zellweger Analytics Ltd 425

Signal Sensed : Chemical
Application Area : Medicine, Health Care

3T BV 181
AEA Technology Plc, Product Development Services Group 245
Ai Cambridge 250
Analytical Development Co Ltd 255
Applied Microengineering Ltd 256
ASTEC GmbH 087
AVL Medical Instruments 001
Bio-Orbit Oy 021
British Ceramic Research Ltd 266
Cambridge Consultants Ltd 268
Capteurs Instrumentation Techniques Associees 219
ETH Z, Centre for Chemical Sensors/Biosensors and Bioanalytical Chem. 224
University of Cranfield, Biotechnology Centre 404
Crowcon Detection Instruments Ltd 279
ctronic Temperature Instruments 297
Università degli Studi di Firenze, Dip Sanità Pubblica Epid. 179
Ecole Centale de Lyon 046
Endress & Hauser (International) Holding AG 222
ETI-Electronic Temperature Instruments Ltd 297
Fraunhofer-Institut für Physikalische Messtechnik (IPM) 100
Fraunhofer Institute Solid State Technology 103
Fraunhofer-Institut für Siliziumtechnologie 102
Gas Analysis & Sensing Group 418
ger Ltd 333
KUBE Electronics Ltd 228
Laboratoire d'Energétique & Réactivité aux Interfaces 053
Lee - Integer Ltd 333
Minsk Research Institute of Radiomaterials 002
Moisture Control & Measurement Ltd 350
National Defence Research Establishment, Dept of Sensor Technology 216
National Physical Laboratory 356
nstitute for Physical & Theoretical Chemistry 157
PA Consulting Group 362
Porvair Technology Ltd 367
Research Centre for Ecological Technologies 200
Rotronic AG 231
Russel pH Ltd 378
Sheffield Hallam University, Electronics 384
Strathclyde University, Department of Electronic & Electrical Engineering 392
Technische Uni Hamburg 146
Technology Partnership Plc 394
TNO - Nutrition & Food Research Institute 189
Trolex Ltd 400
Ultrakust Electronic GmbH 149
University of Manchester, UMIST-Institute of Science and Technology 412
Université de Savoie, ESIGEC, Laboratoire CME 080
Universität GH Siegen, Institut für Messtechnik 156
Dibe University of Genoa 172

Signal Sensed : Chemical
Application Area : HVAC, Fire Control, Security

Signal Sensed : Chemical
Application Area : Military, Aerospace

Signal Sensed : Chemical
Application Area : Telecommunications, Office Equipment

Signal Sensed : Chemical
Application Area : Meteorology, Environment

Signal Sensed : Mechanical
Application Area : Automotive

Signal Sensed : Mechanical
Application Area : Consumer Products

Hottinger Baldwin Messtechnik GmbH	110
IMEC	005
Kayser Threde GmbH	116
Temic Microsystems Sensors	142
Titan Enterprises Ltd	399
TKM Microelectronic GmbH	145

Signal Sensed : Mechanical
Application Area : Consumer Products - continued

Trolex Ltd	400
Technische Universität Berlin, Technologien der Mikropherik	143
Uni College of London, Electronic & Electrical Engineering Dept	409
University der Bundeswehr, Institute für Mess- und Automatisierungstechnik	151
Universität GH Siegen, Institut für Messtechnik	156
Universität Kaiserslautern, Lehrstuhl Feinwerktechnik	153
Universität Kassel, Fachbereich Physik	154
University of Birmingham, Metallurgy & Materials Dept	403
University of Lancaster, Department of Engineering	408
University of Reading, Department of Cybernetics	415
University of Southampton, Institute of Transducer Technology	417
Université de Troyes, IBT	082
Universiteit Twente, MESA Research Institute	190
Universidade Nova de Lisboa, Department of Applied Physics	198
University of Modena	177
Vaisala Oy	027
VTT - Electronics	030
Weber Sensors GmbH	015
Wyler AG	240

Signal Sensed : Mechanical
Application Area : Process Industries, Mining, Energy

Ablewood Technology	242
Acam Instrumentation Ltd	243
Alan Cobham Engineering	252
Amot Controls	253
Automatic Systems Laboratories Ltd	257
Barnes & Associates	258
Baumer Electric Ltd	260
Baumer Electric Ltd	219
Beck GmbH & Co KG	088
Bedford Opto Ltd	261
BLH SR-4 Sensoren GmbH	090
Bourdon Sedeme Ltd	264
Bowers Internal Gauge Co Ltd	265
Brunel University, Brunel Centre for Manufacturing Metrology	267
Burster Präzisionsmeßtechnik GmbH & Co kg	091
Campbell Scientific Ltd	269
Castle Group Ltd	271
CEL Instruments Ltd	272
Coreci SA	042
Czech Technical University of Prague	013
Delta Controls Ltd	282
Douglas Randall Ltd	286
EFFA	048
Electromotif Electronic Ltd	290
Entran	293
Ferroperm A/S	019
Fraunhofer-Institut für Zerstörungsfreie Prüfverfahren	104
Gentech	301
H Warner & Son Ltd	311
Haenni & Co Ltd	226
IMO Industries Ltd	312
IMT Frankfurt	114
International Instruments Ltd	313
IRM SA	006
ITT Barton UK	321
Kytöla Oy	024
Laaser & Co GmbH	119
Lambda Photometrics Ltd	330
Universität Erlangen-Nürnberg, Lehrstuhl Technische Thermodynamik (LTT)	152
Litre Meter	107
Littlemore Scientific Engineering Co	334
Maywood Instruments Ltd	344
Tempere University of Technology; Measurement & Information Technology	026
Measurements Group UK Ltd	345
Menvier Hybrids Ltd	348
Micro Epsilon Messtechnik GmbH & Co	122
MKS Instruments	349
Monitran Ltd	352
Motorola Sensor Products Div	070
Murata Europe Management GmbH	126
Ometron Ltd	359
PA Consulting Group	362
Parsytec GmbH	130
Ploskon AT	209
Pressure Sensors (Marketing) Ltd	365
Rank Taylor Hobson	370
REAKONT, Russia	205
Hungarian Academy of Sciences, ATKI-Institute for Materials Science	164
Research Centre for Ecological Technologies	200
Revere Transducers Europe B.V.	185
Schmidt Feintechnik GmbH	137
Sensy SA	007
Strathclyde University, Department of Electronic & Electrical Engineering	392
STS Sensor Technik Sirnach AG	235
STW - Sensor Technik Wiedemann GmbH	140
Technische Universität Hamburg-Harburg, Optik & Messtechnik	146
Technology Partnership Plc	394
Temic Microsystems Sensors	142
Titan Enterprises Ltd	399
Technische Universität München, Lehrstuhl für Elektrische Meßtechnik	148
University der Bundeswehr, Institute für Mess- und Automatisierungstechnik	151
Universität GH Siegen, Institut für Messtechnik	156
Universität Kassel, Fachbereich Physik	154
University of Lancaster, Department of Engineering	408
University of Reading, Department of Cybernetics	415
University of Southampton, Institute of Transducer Technology	417
Université de Troyes, IBT	082
Universiteit Twente, MESA Research Institute	190
Universidad Complutense de Madrid	213
Universidade Nova de Lisboa, Department of Applied Physics	198
University of Delft, Department of Electrical Engineering	191
Vaisala Oy	027
Valmet Automation Kavanni Ltd	028
Weber Sensors GmbH	159
Whessoe Varec SA	083
WIKA Alexander Wiegand GmbH	162
Wyler AG	240

Signal Sensed : Mechanical

Application Area : Manufacturing, Machinery

Signal Sensed : Mechanical

Application Area : Manufacturing, Machinery - continued

Signal Sensed : Mechanical
Application Area : Medicine, Health Care

Signal Sensed : Mechanical
Application Area : Medicine, Health Care - continued

Signal Sensed : Mechanical
Application Area : HVAC, Fire Control, Security

Amot Controls	065
Automatic Systems Laboratories Ltd	257
Baumer Electric Ltd	260
Beck GmbH & Co KG	088
Bedford Opto Ltd	261
Bernstein Senso Plus GmbH	089
Campbell Scientific Ltd	269
CSEM - Centre Suisse d'Electronique & de Microtechnique	221
Contelec AG	220
Delta Controls Ltd	282
Douglas Randall Ltd	286
Druck Ltd	287
Dynamic Logic Ltd	288
EFFA	048
Electromotif Electronic Ltd	290
EME (UK) Ltd	292
ERA Technology Ltd	295
Ferraris Development & Engineering Co Ltd	299
Fraunhofer-Institut für Mikroelektronische Schaltungen & Systeme	099
Fraunhofer-Institut für Physikalische Messtechnik (IPM)	100
Gentech	301
H Warner & Son Ltd	311
Haenni & Co Ltd	226
Hengstler GmbH	108
Hollming Ltd	023
IMT Frankfurt	114
Kontron Instruments AG, Medical Sensors	227
Litre Meter	107
Marchbanks Measurement Systems Ltd	343
Maywood Instruments Ltd	344
Measurements Group UK Ltd	345
Mechtric Engineering Ltd	347
Microsens SA	229
Minsk Research Institute of Radiomaterials	002
MKS Instruments	349
Monitran Ltd	352

Signal Sensed : Mechanical
Application Area : HVAC, Fire Control, Security - continued

Motorola Sensor Products Div	070
National Defence Research Establishment, Dept of Sensor Technology	216
Ometron Ltd	359
Oxford Brookes Uni, Biomedical Instrumentation Group	360
PA Consulting Group	362
Pressure Sensors (Marketing) Ltd	365
Processor Scientific Instruments Ltd	366
RDP Electronics Ltd	371
Sandhurst Scientific Instrument Co Ltd	379
Schmidt Feintechnik GmbH	137
SensoNor AS	196
Sentron BV	186
STS Sensor Technik Sirnach AG	235
Technische Uni Hamburg -Harburg, Optik & Messtechnik	146
Technology Partnership Plc	394
Tilt Measurement Ltd	398
Titan Enterprises Ltd	399
TU Hamburg-Harburg, Inst Halbleitertechnologie	147
University College of Galway, Deptartment of Physics	170
Universität Kassel, Fachbereich Physik	154
University of Birmingham, Metallurgy & Materials Dept	403
University of Lancaster, Department of Engineering	408
University of Southampton, Institute of Transducer Technology	417
Universiteit Twente, MESA Research Institute	190
Universidade Nova de Lisboa, Department of Applied Physics	198
Universitätsspital Zürich	237
University of Delft, Department of Electrical Engineering	191
Weber Sensors GmbH	159
WIKA Alexander Wiegand GmbH	162

Signal Sensed : Mechanical
Application Area : Military, Aerospace

Ablewood Technology	242
Airflow Developments Ltd	251
Alan Cobham Engineering	252
Barnes & Associates	258
Battelle Geneva Research Centre	218
Baumer Electric Ltd	219
Baumer Electric Ltd	260
BLH SR-4 Sensoren GmbH	090
Bourdon Sedeme Ltd	264
Brunel University, Brunel Centre for Manufacturing Metrology	267
Burster Präzisionsmeßtechnik GmbH & Co kg	091
Caradon Trend Ltd	270
Castle Group Ltd	271
CEL Instruments Ltd	272
Contelec AG	220
Czech Technical University Prague, Faculty of Electrical Engineering	012
Danfoss AS	014
Delta Controls Ltd	282
Università di Pavia, Department of Electronics	178
Digitron Instrumentation Ltd	285
Douglas Randall Ltd	286
Druck Ltd	287
Dynamic Logic Ltd	288
EPFL - Ecole Polytechnique Fédérale de Lausanne, Dept. Electricité	223
EFFA	048
Electromotif Electronic Ltd	290
Entran	293
ETC Ltd	296
Ferroperm A/S	019
FGP Instrumentation	055
Microsens SA	229
Mikro Mess Vertriebs GmbH	125
Minsk Research Institute of Radiomaterials	002
MKS Instruments	349
Monitran Ltd	352
Motorola Sensor Products Div	070
Murata Europe Management GmbH	126
National Defence Research Establishment, Dept of Sensor Technology	216
Ometron Ltd	359
Penny & Giles Position Sensors Ltd	364
Ploskon AT	209
Polytec GmbH	132
Pressure Sensors (Marketing) Ltd	365
Processor Scientific Instruments Ltd	366
Rank Taylor Hobson	370
RDP Electronics Ltd	371
REAKONT, Russia	205
Reson System AS	020
Revere Transducers Europe B.V.	185
Roditi International Corp Ltd	376
Schiltknecht Messtechnik AG	232
Sensy SA	007
Strathclyde University, Department of Electronic & Electrical Engineering	392
Technische Universität Chemnitz-Zwickau, Institut für Mikrosysteme	144
Technische Universität Hamburg-Harburg, Optik & Messtechnik	146
Technology Partnership Plc	394
Temic Microsystems Sensors	142
Tilt Measurement Ltd	398
TNO - FEL, Physics & Electronic Laboratory	188
Trafag AG	236

Signal Sensed : Mechanical
Application Area : Telecommunications, Office Equipment

Signal Sensed : Mechanical
Application Area : Meteorology, Environment

6. List of Organisations

Austria

Belarus

Belgium

Czech Republic

Denmark

Finland

France

Additional Short Form Entries - France

CEA, Laboratoire des Systèmes Moléculaires Organisés
Centre d'Etudes du Bouchet, DGA/CEB
CNRS, LAAS - Laboratoire d'Analyse et d'Architecture des Systèmes
CNRS, Laboratoire d'Electrochimie et Chimie Analytique
CNRS, Laboratoire des Matériaux Organiques à Propriétés Spécifiques
CNRS, Laboratoire Physique et Métrologie des Oscillateurs
CNRS, Service Central d'Analyses, Vernaison
Ecole Nationale Supérieure Cachan, PPSM
Gaz de France, Direction de la Recherche
INSA Lyon, Laboratoire de Physique de la Matière, Microcapteurs
LIPHA SA
NEREIDES
PHILIPS, Laboratoire d'Electronique
PSA Peugeot-Citroën, DET MXT-CED
Radiometer Analytical
Rhône-Poulenc Industrialisation
SAT
SAGEM, Senseurs Optiques & Physique Appliquée
SAUR
Schlumberger, Gas Engineering Montrouge
SERES
Thompson - GSF, Laboratoire Central de Recherches
Université de Rennes 1, Laboratoire Radiocommunications
Université de Rouen, Laboratoire des Polymères et Biopolymères
Université du Maine, ENSIM

Germany

Additional Short Form Entries - Germany

ADUS Druckmesstechnik GmbH
AKTIV-ELECTRONIC GmbH
AMS THERMOTECH GmbH
Automation W+R GmbH
BARTEC Components und Systeme GmbH
Baumer Electric GmbH
Ernst Befort Wetzlar GmbH
BESTEC GmbH
betacontrol gmbh mess- und regeltechnik
Franz Binder GmbH & Co
BMC Dr. Schetter GmbH
Bonn Sensorik

Additional Short Form Entries - Germany (cont.)

Bourdon Sedeme GmbH
Brose Systeme GmbH
CIS Centrum für intelligente Sensorik Erfurt eV
CompuMess Elektronik GmbH
Corrsys Korrelatorsysteme GmbH
Datron Messtechnik GmbH
Delta OHM
Delta Regeltechnik GmbH
Disynet GmbH
DISYS Mess- und Testsysteme GmbH
DMT Druck-Messtechnik GmbH
Döhrer Elektrooptik GmbH i.Gr.
Dosch Messapparate GmbH
Drägerwerk AG
Driesen + Kern GmbH
Druck & Temperatur Leitenberger GmbH
ELKA-Elektronik GmbH
ELKUTEC Electronic GmbH
EUROPAVIA-Deutschland GmbH
FOS Messtechnik GmbH
Foundry Control
FPM Space Sensor GmbH
GELMA Industrieelektronik GmbH
Haenni-Mess-Systeme GmbH
Erwin Halstrup Multur GmbH
Hamamatsu Photonics Deutschland GmbH
Hamlin Electronics GmbH
Heckendorf Engineering GmbH
HEITRONICS infrarot Messtechnik GmbH
Hentschel System GmbH
Höntzsch GmbH
Honeywell AG
IBR Prozessautomation GmbH & Co Messtechnik KG
ifm electronic gmbh
IMM Messtechnologie GmbH
IMOS Gubela GmbH
Infra Tec GmbH
INOR Transmitter GmbH
Kaiser sensoren + systeme
Klipp & Zonen Deutschland GmbH
Kistler Instrumente GmbH
KOBOLD Messring GmbH
Krick Messtechnik & Partner KG
Dr. Kubelik Industrievertretungen
Kugler GmbH
Kulite Semi-Conducter GmbH
LAP GmbH Laser Applikationen

Additional Short Form Entries - Germany (cont.)

Lenord, Bauer & Co GmbH
Leuze electronic GmbH & Co
Lorenz Messtechnik GmbH
LUST-Oriel GmbH
MESA Systemtechnik GmbH
Messko A. Hauser GmbH & Co KG
MICRO-SENSYS GmbH
NATEC Schultheiss GmbH
National Semiconductor GmbH
Newport electronics GmbH
OEC GmbH
OPTOLAB GmbH
Penny & Giles GmbH
Pfaudler-Werke GmbH
Philips Thermocoax
Physik Instrumente (PI) GmbH & Co
PI Ceramic GmbH
Pier-Electronic GmbH
piezosystem jena
PIL Sensoren GmbH
PRÜFTECHNIK AG
Pulsotronic Merten GmbH & Co KG
Rheintacho GmbH & Co
RJM Rheinmetall Jena Image Technology GmbH
Rössel Messtechnik GmbH & Co
ROLEC
Klaus Schäfer Gesellschaft für Verfahrenstechnik mbH
Schäfter & Kirchhoff
Schönbach Electronic Hanesch GmbH & Co KG
Dr. Seitner System-Sensortechnik GmbH
Sensortechnics GmbH
Sentronic GmbH
SIE - Sensorik Industrie-Elektronik GmbH
Siemens AG, EC BW M/V
Steinbichler Optoelektronik GmbH
Entwicklungsbüro Dipl.-Phys. Gerhard Stresing
Telemeter Electronic
TESTEM Gesellschaft für Mess- und Datentechnik mbH
TSI GmbH
TWK-Elektronik GmbH
Tylan General GmbH
Vaisala GmbH
WACHENDORFF PROZESSTECHNIK GmbH & Co KG
Wallace & Tiernan GmbH

Hungary

Ireland

Italy

Luxemburg

Netherlands

Norway

Portugal

Romania

Russia

Slovakia

Spain

Sweden

Switzerland

United Kingdom

7. Participating Advertisers

CENTREDOC (Switzerland)

CMC2 - Club Microcapteurs Chimiques (France)

Istituto Superiore di Sanità, Annali (Italy)

OSCA - Optical Sensors Collaborative Association (United Kingdom)

VCH Weinheim (Germany)

Metra Martech Experience with Sensors

Metra Martech is a well established consultancy specialising in industrial market research and business development. This data base is a spin off from our work in Europe.

We have wide experience in the main industrial and manufacturing sectors and our clients include many world leaders in the electrical and controls markets. We also have experience in such areas as instrumentation and control systems, SCADA, machine tools, information systems and telecoms, transformers and control gear, oil field monitoring, environmental monitoring, food and chemical processing, and materials processing equipment.

More than half of Metra Martech's assignments and clients are in the markets outside the UK. We have a team with a balance of qualifications, skills and experience, well able to lead sensors-related assignments.

Senior Consultants in the team are generally engineering or science graduates, whose business experience has been gained from wide ranging project work, senior sales and marketing positions within the industrial and manufacturing sectors.

Our work is very varied. We show below typical project experience:

Chemicals: petrochemicals, fine chemicals, additives

Electrical: generation, transmission, distribution, industry users and producers

Engineering: process controls, machinery, components

Environment: water treatment, air, waste, conservation

IT: computer hardware, software, telecoms

Instrumentation: laboratory, controls, monitoring, sensors

Life Sciences: medical equipment and funding, biotechnology

Materials: steel, aluminium, polymers, composites etc.

Oil and Gas: exploration and extraction/production

Packaging: materials, systems, food related processes, recycling

Transport and Distribution: freight, airline, manufacture, marine equipment

Typical Metra Martech Studies

Technical Assessments - reaction to new or improved products can be assessed in existing or new markets. Justification for R&D budgets or commercialisation plans are prepared.

Market Opportunity Studies - a combination of the above and analysis of specific areas of opportunity for your company. These are often related to new or emerging markets, distribution channels, or technology. Practical action plans can be developed so that full advantage is taken of the opportunities identified.

Market Studies - general description of market size, trends, market segments, major players, distribution patterns, pricing structure etc. These studies can be tailor-made to provide you with the exact type of information you require. They form the basis for planning, targets, new investment and the development of business in new markets.

Market Entry Strategies - based on an analysis of the market conditions and requirements, together with your own company's strengths, Metra Martech can formulate strategies for successful market entry.

Acquisition Studies - searching for companies to acquire. These can vary from simple lists of companies operating in particular markets, to in-depth analyses of target companies' markets, technical and business situations, by extensive research and interviewing.

Image Studies - Metra Martech's image studies assess how your organisation compares with the competition on the key choice criteria. They can successfully pin-point areas where improvements in customer perceptions could lead to increased sales. Customer care programmes are developed from this work.

Distribution - identification of agents or distributors in new territories.

CPSIA information can be obtained
at www.ICGtesting.com
Printed in the USA
LVOW02s0830250316
480738LV00013B/332/P